Displacement, Belonging, and Migrant Agency in the Face of Power

This book centers the voices and agency of migrants by refocusing attention on the diversity and complexity of human mobility when seen from the perspective of people on the move; in doing so, the volume disrupts the binary logics of migrant/refugee, push/pull, and places of origin/destination that have informed the bulk of migration research.

Drawn from a range of disciplines and methodologies, this anthology links disparate theories, approaches, and geographical foci to better understand the spectrum of the migratory experience from the viewpoint of migrants themselves. The book explores the causes and consequences of human displacement at different scales (both individual and community-level) and across different time points (from antiquity to the present) and geographies (not just the Global North but also the Global South). Transnational scholars across a range of knowledge cultures advance a broader global discourse on mobility and migration that centers on the direct experiences and narratives of migrants themselves.

Both interdisciplinary and accessible, this book will be useful for scholars and students in Migration Studies, Global Studies, Sociology, Geography, and Anthropology.

Tamar Mayer is the Robert R. Churchill Professor of Geosciences at Middlebury College in Vermont, USA.

Trinh Tran is an Assistant Professor of Anthropology and Education Studies at Middlebury College in Vermont, USA.

Displacement, Belonging, and Migrant Agency in the Face of Power

Edited by Tamar Mayer and Trinh Tran

Routledge
Taylor & Francis Group

LONDON AND NEW YORK

First published 2022
by Routledge
4 Park Square, Milton Park, Abingdon, Oxon OX14 4RN

and by Routledge
605 Third Avenue, New York, NY 10158

Routledge is an imprint of the Taylor & Francis Group, an informa business

British Library Cataloguing-in-Publication Data
A catalogue record for this book is available from the British Library

Library of Congress Cataloging-in-Publication Data
Names: Mayer, Tamar, 1952- editor. | Tran, Trinh, editor.
Title: Displacement, belonging, and migrant agency in the face of power / edited by Tamar Mayer and Trinh Tran.
Description: Abingdon, Oxon ; New York, NY : Routledge, 2022. | Includes bibliographical references and index.
Identifiers: LCCN 2021062075 (print) | LCCN 2021062076 (ebook) | ISBN 9780367772932 (hardback) | ISBN 9780367772949 (paperback) | ISBN 9781003170686 (ebook)
Subjects: LCSH: Immigrants--Developing countries--Social conditions. | Refugees--Developing countries--Social conditions. | Internal migrants--Developing countries--Social conditions. | Imperialism--Social aspects. | Developing countries--Emigration and immigration--Social aspects.
Classification: LCC JV6225 .D57 2022 (print) | LCC JV6225 (ebook) | DDC 304.809172/4--dc23/eng/20220322
LC record available at https://lccn.loc.gov/2021062075
LC ebook record available at https://lccn.loc.gov/2021062076

ISBN: 978-0-367-77293-2 (hbk)
ISBN: 978-0-367-77294-9 (pbk)
ISBN: 978-1-003-17068-6 (ebk)

DOI: 10.4324/9781003170686

Typeset in Times New Roman MT Std
by SPi Technologies India Pvt Ltd (Straive)

To Shoshana Mayer and to the memory of Artur Mayer, Tai Tran, and Hia Quach, who all made the journey as migrants

Contents

Illustrations

Figures

Tables

Editors

Tamar Mayer is the Robert R. Churchill Professor of Geosciences at Middlebury College in Vermont. She is a political geographer whose research interests lie in two main areas: the interplay among nationalism, homeland, and memory, with a special focus on stateless ethnic nations; and the impact of global crises on local communities. She is the editor or coeditor of six books that focus on different dimensions of international and global crises. The two most recent of these are *The Crisis in Youth Unemployment* (Routledge, 2019) and *Food Insecurity: A Matter of Justice, Sovereignty, and Survival* (Routledge, 2021).

Trinh Tran is an Assistant Professor of Anthropology and Education Studies at Middlebury College in Vermont. Her work examines how school choice and housing policies affect migrant children's relationship to their home communities. At Middlebury, she teaches courses on global migration, global education, and urban sociology. Her research has been supported by the National Science Foundation and the Greater Good Science Center at the University of California Berkeley.

Contributors

Alejandra Aguilar received her BA in Sociology at California State University Stanislaus; during that time, she developed a pilot study focusing on the stereotyped portrayal of women in the media. She holds an MA from California State University Sacramento. As a graduate student, she conducted a project exploring the sense of belonging and legal vulnerability among undocumented youth. She has also presented at conferences such as California Sociological Association (CSA). Her focus of research includes education, gender inequalities, immigration, ethnicity, race, and racism.

Olayinka Akanle is a Lecturer in the Department of Sociology, Faculty of Social Sciences, University of Ibadan, Nigeria, and a research associate in the Department of Sociology, Faculty of Humanities, University of Johannesburg, South Africa. He was a Postdoctoral Fellow at the South African Research Chair Initiative (SARChI) in Social Policy, College of Graduate Studies, University of South Africa (UNISA), South Africa. He has held other scholarly positions, such as World Social Science Fellow (WSSF) of the International Social Science Council (ISSC), Paris, France, and Laureate of the Council for the Development of Social Science Research in Africa (CODESRIA), and he was awarded the University of Ibadan Postgraduate School Prize for scholarly publication. He is a scholar and an expert on international migration, diaspora studies, social policy and practice, and sustainable development in Nigeria and Africa. Dr. Akanle has attended many local and international scholarly conferences and has published widely in local and international journals, books, technical reports, and encyclopedias. He is the author of *Kinship Networks and International Migration in Nigeria* (Cambridge Scholar Publishers) and has coedited books including *The Development of Africa: Issues, Diagnosis and Prognosis* (Springer Publishing).

Naif Bezwan is Senior Researcher at the Ludwig Boltzmann Institute of Fundamental and Human Rights, Department of Constitutional and Administrative Law, Faculty of Law, University of Vienna, and Honorary Senior Research Associate in the Department of Political Science, University College London (UCL), London. He has worked, conducted research, and taught at diverse universities in Germany, Turkey, Austria, and the UK. Bezwan has published on politics, the political system and the modern history of Turkey, Kurdish politics, and the self-determination conflict in multiple languages. His areas of interest include Turkish foreign and domestic politics, Kurdish politics and the self-determination conflict, the Middle East, comparative migration and diaspora, political theory, and comparative politics.

Sandibel Borges is an Assistant Professor of Women's and Gender Studies at Loyola Marymount University. She has a PhD in Feminist Studies from the University of California Santa Barbara and completed a postdoctoral fellowship in the Center for Mexican American Studies at the University of Texas at Austin. Her work is situated in the fields of women of color and transnational feminisms, feminist and queer migrations, queer of color critique, disability justice, and critical ethnic studies. Her current book project uses ethnography, oral histories, and archival research to examine the colonial and genocidal logics of the U.S. immigration system.

Sabyasachi Basu Ray Chaudhury is Professor of Political Science, Rabindra Bharati University, Kolkata, India. He is a member of Calcutta Research Group (CRG). His areas of interest include global politics, South Asian politics, human rights, and refugee and migration studies. His latest publication is *The Rohingya in South Asia: People without a State* (2018). Earlier publications include *Sustainability of Rights after Globalization* (2011), *Indian Autonomies: Key Words and Key Texts* (2005), *Internal Displacement in South Asia: The Relevance of UN's Guiding Principles* (2005), and *Living on the Edge: Essays on the Chittagong Hill Tracts* (1997). He is a political commentator for many national and international electronic and print media.

Guo Chen is an Associate Professor in the Department of Geography, Environment, and Spatial Sciences and the Global Urban Studies Program at Michigan State University. She was a Wilson Center Fellow 2017–2018. She is author and coauthor of over 40 publications with a research focus on urbanization and inequality, urban poverty, slums, migrants, housing the poor, urban governance, land use, institutional bias, and social and environmental justice in China, the Asia-Pacific, and emerging countries. She coedited *Locating Right to the City in the Global South* (Routledge, 2013) and has edited special issues. Her articles have appeared in *PLOS One*, *Scientific Reports*, *Environment and Planning A*, *Urban Geography*, *Urban Studies*, *Cities*, *Habitat International*, *Area*, *Acta Geographica Sinica*, and many other journals. Her work has been widely funded and her public scholarship includes a documentary on hidden slums, op-eds, interviews, and many invited talks and organized panels.

Nafay Choudhury is the Jeremy Haworth Research Fellow at the University of Cambridge and a Social Sciences and Humanities Research Council Postdoctoral Fellow at the University of Toronto Faculty of Law. He is also an Affiliate Scholar with the Institute for Global Law and Policy at Harvard Law School. His PhD from King's College London looks at the interaction of state and nonstate legal systems in the production of legal order. Nafay is also a Senior Research Fellow at the Afghan Institute for Strategic Studies and was previously a Postdoctoral Fellow at Stanford Law School for the Afghanistan Legal Education Project.

Nasreen Chowdhory received her PhD in Political Science from McGill University, Canada, and is now an Associate Professor in the Department of Political Science, University of Delhi, India. Some of her significant publications include *Refugees, Citizenship and Belonging in South Asia: Contested Terrains* (Springer, 2018) and three edited volumes (as co-editor) titled *Deterritorialised Identity and Transborder Movement in South Asia* with Nasir Uddin (Springer, 2019), *Citizenship, Nationalism and Refugeehood of Rohingyas in Southern Asia* with Biswajit Mohanty (Springer, 2020), and *Gender, Identity and Migration in India* (Palgrave, 2022) with Paula Banerjee. She is the Vice President of Mahanirban Calcutta Research Group, Kolkata, India.

Kate Coddington is an Assistant Professor in the Department of Geography and Planning at the University at Albany, State University of New York. Her work focuses on borders, migration, and postcolonial governance in the Asia-Pacific. Her related work has been published in journals such as *Annals of the American Association of Geographers* (2020), *Transactions of the Institute of British Geographers* (2018), and *The Professional Geographer* (2017), as well as in several edited books.

Jia Feng is a Teaching Assistant Professor in the Department of Geography at the University of Nevada, Reno. He has a research interest in migration, marginality, urban enclaves, and urban-rural connections, with his specific research focused on recycling enclaves and "informal" recycling businesses developed by rural-to-urban migrants in large Chinese cities. He also collaborates in various urban and community research projects, such as bikeshare programs and mapping racial covenants using GIS, spatial data analysis, and statistical models.

Luisa Feline Freier is an Associate Professor of Political Sciences at the Universidad del Pacífico (Lima, Peru). She is a leading scholar of Latin American migration and refugee policies, south–south migration, and the Venezuelan displacement crisis. She holds a PhD in Political Science from the London School of Economics and Political Science (LSE) and an MA in Latin American and Caribbean Studies from the University of Wisconsin, Madison. She is Migration Research and Publishing High-Level Adviser of the International Organization for Migration (IOM).

Ben Suzuki Graves teaches in the Department of English and American Literatures at Middlebury College. His current research involves a recurring link in Black and Asian British literature between the caring functions of the state (housing, welfare provision, the National Health Service, etc.) and the making of emotional or "affective" subjects fit for British citizenship.

Dinko Hanaan Dinko is a PhD student in the Department of Geography and the Environment, University of Denver. His research focuses on the socio-political dimensions of water insecurity in northern Ghana.

Jennifer Hyndman is Associate Vice-President of Research at York University in Toronto, where she is also a Professor of Geography. Her research focuses on the geopolitics of forced migration, the biopolitics of refugee camps, humanitarian responses to displacement, and refugee resettlement in North America. Her most recent book is *Refugees in Extended Exile: Living on the Edge*, with Wenona Giles (Routledge, 2017). Hyndman is the author of *Dual Disasters: Humanitarian Aid after the 2004 Tsunami* (2011), *Managing Displacement: Refugees and the Politics of Humanitarianism* (2000), and co-editor, with W. Giles, of *Sites of Violence: Gender and Conflict Zones* (2004).

Anne Irfan is Lecturer in Interdisciplinary Race, Gender and Postcolonial Studies at University College London (UCL). She has published widely on subjects related to Palestinian refugee history, the UN refugee regime, and the historical conditionalities of forced migration. She has also spoken on these issues at the UN Headquarters in New York and Geneva, and the UK Parliament at Westminster. Her book *Refuge and Resistance: Palestinian Refugees and the UN Regime, 1948–82* is forthcoming with Columbia University Press.

Elena Isayev is a historian focusing on migration, hospitality, and displacement, which she has written about for the Red Cross and in her monograph *Migration Mobility and Place in Ancient Italy* (Cambridge, 2017). She also works with *Campus in Camps* in Palestine, and leads the project *Imagining Futures through Un/Archived Pasts* (funded by Arts and Humanities Research Council, Global Challenges Research Fund). She is a Professor of Ancient History and Place at the University of Exeter.

Janroj Yilmaz Keles is the editor of the British Sociological Association's journal *Work, Employment and Society* and a Senior Research Fellow in Politics at Middlesex University Law School, researching on peace and conflict, gender, political violence, ethnicity and nationalism, statelessness, migration, diasporas, (digital) social movements, and political communication. Keles has extensive experience of international education. He studied in Turkey, Germany, and the United Kingdom and received his PhD in Sociology and Communications from Brunel University, London. Keles has a proven track record of securing external research funding. He is one of the co-investigators of GCRF HUB – Gender, Justice and Security and has published several coauthored and single-authored articles in peer-reviewed journals. His monograph *Media, Conflict and Diaspora* (I.B. Tauris, 2015), was well received by the academic community.

Andreas Kossert is a Research Fellow at the Federal Foundation Flight, Expulsion, Reconciliation in Berlin, a position he has held since 2010. Between 2001 and 2009, he was a Research Fellow at the German Historical Institute in Warsaw, Poland, and from 2004 onward, he served as the institute's deputy director. Specializing in Central European history, Kossert has published widely on nationalism, borderlands, ethnic and religious minorities, forced migration, displacement, and refugees. He is the author of *Kalte Heimat. Die Geschichte der deutschen Vertriebenen nach 1945* (2008); *Ostpreußen: Geschichte und Mythos* (2007); and *Flucht. Eine Menschheitsgeschichte* (2020), among other titles.

Andrea Kvietok is a Research Consultant at the Universidad del Pacífico (Lima, Peru). Her research interests include transnational migration, ethnography, mental health, and gender. She was awarded a 2018–2019 Fulbright US Student grant to Peru, where she researched the lived experiences of return and reintegration for Peruvian migrants. She holds a BA in Anthropology from Macalester College (Saint Paul, Minnesota).

Emily Mitchell-Eaton is an Assistant Professor of Geography at Colgate University. Her work investigates the relationship between empire and migration, focusing on Pacific Islander diasporas within the United States. Her work has been supported by the National Science Foundation, the Andrew W. Mellon Foundation, the American Association of Geographers, and the journal *Human Geography* and has been published in *Political Geography*; *Society + Space*; *International Migration Review*; *Gender, Place & Culture*; *H-Migration*; and *Shima*, an international journal of research into island cultures.

Hanson Nyantakyi-Frimpong is an Assistant Professor of Geography at the University of Denver. His research focuses on the political ecology of rural development, the human dimensions of global environmental change, and sustainable agriculture and food systems. His work has been published in *Global Environmental Change, Journal of Peasant Studies, Land Use Policy, Ecology & Society, Geoforum, The Professional Geographer, Applied Geography*, and other interdisciplinary journals.

J. Santiago Palacios is an Associate Professor of Medieval History at the Universidad Autónoma de Madrid whose work has focused on the ideological and material aspects of the Reconquest, the frontier in Spanish medieval history, the history of the Crusades, and the military orders in Spain. He currently leads the research project called *Religious Violence in the Peninsular Middle Ages: War, Apologetic Discourse and Historiographic Narrative, 10th-15th Centuries*. He is also the author of several articles and books, the most recent titled *Crusades and Military Orders in the Middle Ages* (Madrid, 2017).

Shamna Thacham Poyil is a PhD candidate in the Department of Political Science at the University of Delhi, India. Her research focuses on the narrative of statelessness of the Rohingyas and the politics of exclusion. Her M-Phil dissertation titled *Birds of Freedom: Depiction of LTTE Militant Women in Tamil Cinema* explored the representation of militant women challenging the binary of agency and victimhood. Her recent publications about forced migration and refugee management in Southern Asia appear in books edited by or in collaboration with Dr. Nasreen Chowdhory.

Farhana Rahman is a Fellow at the Harvard University Asia Center. She received her PhD from the University of Cambridge and explored Rohingya refugee women's everyday lived experiences after forced migration. She co-founded Silkpath Relief Organization (silkpathrelief.org), a non-profit operating in Afghanistan and with Rohingya refugees in Bangladesh and Malaysia. She helped establish the first gender studies program in Afghanistan, based at the American University of Afghanistan in Kabul, where she was also a lecturer. Farhana was the 2021 recipient of the Paula Kantor Award from the International Center for Research on Women (ICRW).

Heidy Sarabia is an Associate Professor in the Department of Sociology at California State University, Sacramento. Her research focuses on globalization processes such as stratification, borders and borderlands, border violence, transnational social change, and immigrant adaptation and incorporation in the US. She received her PhD from the University of California, Berkeley, and holds a BA from the University of California, Los Angeles. She was a postdoctoral fellow at the University of Pennsylvania from 2014 to 2016. Her work has been published in several academic journals—*Sociological Forum, Ethnic and Racial Studies, Citizenship Studies, Analyses of Social Issues and Public Policy, American Behavioral Scientist, Latin American Perspectives, Migration Letters,* and *Carta Economica Regional.* She teaches Statistics, Methods, Sociology of Globalization, and Social Change and Migration in Latin America at CSU Sacramento.

Shawna Shapiro is an Associate Professor at Middlebury College (Vermont) who teaches courses in academic writing, linguistics, and education. Her research focuses on college transitions for immigrant and refugee-background students. Her work has appeared in peer-reviewed journals such as *Equity & Excellence in Education; Journal of Language, Identity, and Education;* and *TESOL Quarterly.* Her most recent book is *Educating Refugee-Background Students: Critical Issues and Dynamic Contexts* (2018). Shapiro also works on educational equity initiatives in her local community.

Doğuş Şimşek is an Assistant Professor of Sociology and Criminology at Kingston University London. Her research focuses on the intersections between race, class, and migration studies. Her work has been published in *Ethnic and Racial Studies, Journal of Refugee Studies, International*

Migration, Social Inclusion, and other interdisciplinary journals, as well as several edited books..

Laura Zaragoza completed her BA and MA in Sociology at California State University, Sacramento. As part of the Sac State Pathways Fellowship's initial cohort, she conducted a research project that explores the experiences of undocumented students in higher education. During her time at Sacramento State University, she has been a research assistant and has presented at conferences such as California Sociological Association (CSA), Pacific Sociological Association (PSA), and Mujeres Activas en Letras y Cambio Socia (MALCS). Her research focuses on immigration, stratification, social inequalities, race, and education.

Acknowledgments

The ideas for this book emerged at a conference titled *Migration, Displacement, and Belonging: Challenging the Paradigm*. With our colleague Prof. Darién Davis, we convened a group of interdisciplinary scholars to talk about pressing issues in migration research today. The seventh annual international and interdisciplinary conference conducted by Middlebury's Rohatyn Center for Global Affairs, the conference sparked critical conversations about voice and agency in the field of migration studies. These discussions inspired us to solicit additional essays from accomplished scholars who are both from the Global South and conduct research in that area. We thank the many departments at Middlebury and the Rohatyn Center for Global Affairs for generously supporting the conference. Special thanks go to our staff at the Center, who made the conference a success. We also owe our gratitude to Irene Bloemraad, Kyle Brudvik, and our student research assistants Kaleb Patterson and Georgia Vasilopoulos for their support. Lastly, we thank Greg Woolston for his cartographic contributions, Mary Bagg for her copyediting, and our editor at Routledge, Faye Leerink, for her support.

1 Displacement, belonging, and migrant agency in the face of power

Challenging paradigms in migration studies

Tamar Mayer and Trinh Tran

Vivid and gruesome images of migrants in search of asylum have been abundant in the last decades and have called attention to the magnitude and severity of the problem: black and brown bodies washing ashore; hundreds of thousands of asylum seekers from Africa, Asia, and Latin America sailing or marching to EU and US shores and borders; and subhuman living conditions in detention centers and makeshift dwellings, where migrants in limbo seek housing are the markings of the biggest human rights disaster of our time.

The number of migrants from the Global South seeking refuge is staggering. In 2019 alone, over 80 million people were dislocated in search of safety.[1] They fled war, persecution, ethnic cleansing, intense economic, and political instability, as well as the devastating impact of severe climate calamities (UNHCR 2020). And like more than 200 million migrants before them (IOM 2020), they left their homes with no clear idea if, when, and where they would put down roots again. These migrants are resilient and courageous, and in asserting their rights to flee for a better and safer life, they each "demand to be accepted as a distinctive human being" (Hamilakis 2016:125). They flee despite all the known and unknown dangers they may encounter en route: securitized borders and unforgiving terrain and weather conditions, for example, handlers and cayotes who might take advantage of them, and xenophobia borne by right-wing nationalism. At places of arrival, many migrants will encounter social, political, and economic discrimination, which often lead many to a life of poverty, precarity, and marginalization.

The focus of migration research has changed over time. If in most of the twentieth century, it focused on internal or international migrations—emphasizing migration flows, patterns, and processes and stressing demographics, statistics, and governance—then the focus in the 1990s became more attuned to individual-level factors and centered on gender, family, health, and diversity (Pisarevskaya et al. 2020). Since the 2000s, however, migration research has become even more attuned to individual migrants. It has incorporated migrants' narratives about their own experiences and paid attention to issues of diversity, racism, discrimination, and social-psychological issues (Pisarevskaya et al. 2020). In so doing, such research has given migrants a voice and acknowledged their agency. Our volume, *Displacement, Belonging,*

DOI: 10.4324/9781003170686-1

and Migrant Agency in the Face of Power, continues this approach by probing how migrants experience their dislocation and manage to survive (at times even to thrive) despite their horrific experiences. Our anthology aims to enrich migration research in various ways: by prioritizing migrants' voices and centering their agency; by seeing them as a heterogenous group with different experiences, intersectional identities, and multiple migration paths; and by presenting the work of transnational authors from different academic disciplines. Drawn from a range of disciplines and methodologies, broad in its regional scope, this volume challenges and pushes against the binary logic that has informed the bulk of previous migration research (such as migrant/refugee, push/pull, forced/voluntary, and places of origin/destination). Instead, we ask what we can learn about people on the move and their agency when we examine the complexities of mobility at different scales (both individual and community levels) and across different time points (from antiquity to the present) and geographies (not just the Global North but also the Global South).

Migrants' representation

Much of the migration literature, at least until recently, relied on European migration experiences and tended to center on the Global North, presenting static models of migration that were unidirectional (south–north, for example) (Collinson 1993; Adepoju et al. 2010; Filindra and Kovács 2012; Cerna 2014; Grosfoguel et al. 2015). It also reproduced the Eurocentric perspective of migrants as passive, voiceless, helpless, and unlucky victims (Kaye 2001; Collins 2007; Awad 2012; Blinder and Allen 2015). Such representations, which have been used to admit some migrants and refuse others, were simplistic and did not consider that migrants of all genders, ages, ethnicities, and classes are knowing actors: resilient authors of their own lives.

Further, much of the twentieth-century literature on migration referred to migrants as a homogenous group without acknowledging that migration differs greatly by gender, age, sexuality, race, class, ethnicity, and religion. These themes would take center stage starting in the 1990s (Pisarevskaya et al. 2020). The attention to heterogeneity (seen in the works of Hajdukowski-Ahmed et al. 2008; Jansen 2013; Fiddian-Qasmiyeh 2014; Sigona 2014) established what the migrants already knew, that women's journeys are often far more dangerous than those of men; that they are more often preyed upon and face sexism by fellow migrants, state agents, border patrol, and handlers on their route (see Chapter 5 in this volume); that gay and transsexual migrants too, as well as children who travel alone (see Chapters 17 and 19) have migration experiences unique to their identities and positionality; and that migrants who were economically better off before fleeing have easier journeys than poorer migrants (see Chapter 14).

This attention to heterogeneity has been important, but studying migration experiences, whether by gender, class, race, ethnicity, religion, or age, risks essentializing those experiences into discrete and separate social categories. People's social identities overlap and intersect, causing them to experience events in their lives differently, depending on the extent to which they intersect. In the context of this volume, it is important to consider that migration experiences for women will depend on their ethnicity, age, sexuality, class, and religion. A middle-class woman, for example, of a specific race or ethnicity, will most likely experience migration differently from a poor urban lesbian of the same ethnic group, whether documented or not (see Chapter 17). Indeed, theories of intersectionality help us complicate our understanding of the migration experience and offer an angle through which to focus on individuals as they are positioned both locally and globally, not just on the larger social categories.

The call to focus on individuals and their complicated migration stories comes from transnational writers who explore migrations from the Global South, including those whose paths involve many steps, perhaps even a return to the initial sending country. Following these authors,[2] our volume gives voice to the migrants from the Global South who have left homes in haste and who struggle through inhospitable environmental, social, political, and economic landscapes in search of a safer place to live. Focusing on the Global South, with its unique migration settings and patterns—rural to urban, intra-regional, and south to north—allows us to illuminate drivers of displacement, such as the impact of colonial legacy and climate disasters, that until recently have been largely invisible in works on the Global North. It also allows us to explore the ideas of belonging and re-creating new homes in places that are not always hospitable to nonwhite migrants.

Challenging migration paradigms

Despite the changes evident in migration studies since the 1990s, scholars today still adhere to several binary constructions (paradigms) used in the past to analyze, explain, and characterize the reasons, patterns, and directions of migratory flows—even if assigning such constructions different levels of importance. These binaries include push/pull factors, forced/voluntary migrations, and places of origin/destination. In this section, we explore and challenge the value of each, aiming to move scholarship toward a closer examination of the multiple and complex drivers and consequences of displacement by considering the spectrum of the migratory experience. Although extant work has neatly labeled migrants and their experiences into discrete and separate social categories, the chapters in this volume destabilize this binary logic by centering the agency and voice of individual migrants.

Push | pull

This path-breaking theory, pioneered in 1966 by sociologist Evertt Lee, attempted to explain that migration occurs because of the interplay between "push" and "pull," as conditions in one place repel and a potential new place attracts (Lee 1966). In addition to the conditions and the distance between two places, the theory's proponents recognized that intervening obstacles and opportunities affect the volume of migration. Although their studies were useful for mathematical modeling and for explaining south–north migration (see, for example, Jenkins 1977; Zimmermann 1996; Datta 2004; Dago and Barussaud 2021) and its continuous flow (e.g., Dorigo and Tobler 1983), they did not consider two other factors: (1) migration involves multiple physical, social, and symbolic locations, not simply two physical places; and (2) places can never be classified as simply attractive or not. Further, these studies did not consider how social networks and social capital affect migration. Finally, given its opportunities and disadvantages, place is experienced differently depending on one's social and cultural intersected positionalities. The push/pull construction provides no space for such considerations. By using feminist theory, paying attention to the strong ties between places left and places arrived at, and by focusing on agency, our volume complicates this binary and adds to the critique of the push/pull paradigm.

Forced | voluntary

This binary construction presupposes that migrants are either forced to leave their homes or choose to do so voluntarily. It follows the United Nation's 1951 Convention Relating to the Status of Refugees and its 1967 Protocols, which have been used since then by NGOs and governments to determine who will be granted entry and who will not (Betts 2013; Ottonelli and Torresi 2013). Because of this 1951 Convention, forced migration is associated with a real threat to people's lives and livelihood as evident in extreme violations of human rights resulting from war, ethnic cleansing, and genocide, or by severe political and economic instability. In contrast, therefore, voluntary migration refers to those who move for purely economic reasons, usually in search of a (better) job; their migration is considered a matter of free choice. As many scholars have already noted, this distinction is problematic because *all* migrations involve choice, including forced migrations (Bartram 2015), and because migrants are agents, involved in their own migration projects, who decide how much risk they can tolerate before choosing whether to flee or remain in place.

Because the UN and state organizations continue to use this binary construction to characterize migrations (either voluntary or forced), it is important to stress that contrary to the 1951 Convention, economics is a determinant in *all* migrations, not only in "voluntary" ones. When people's economic situation is so dire, as in the cases of Venezuela, Eritrea, Puerto Rico, and many other places; when crops fail because of drought and there is no food or water; when calamities of climate change strike and eviscerate the infrastructure and homes are leveled; and when political and economic systems collapse and there is no hope, many people choose to flee (see the case of Venezuelans' exodus in Chapter 5 and the Ghanaian internal migration case in Chapter 9). Since the categories established in the 1951 Convention and its 1967 Protocols were fixed, and since at the time there was no consideration given to climate-induced displacement or to economic refugees, those who dislocated because of environmental or economic reasons do not count among the refugees and do not qualify for assistance.[3]

A dichotomous construction of forced versus voluntary reifies legalistic distinctions between refugees and migrants. Such binary categories privilege the role of the state and speak more to its administering and management of borders than to the real experiences of migrants themselves. Instead of speaking to the reality of the migratory journey, arbitrary distinctions between forced and voluntary erect an unnecessary hierarchy of misery, serving as justification for the denial and receipt of aid by people on the move (Hamlin 2021).

Places of origin | destination

This category focuses on the initial place of departure and the final place of arrival. By using this binary construction, migration scholars assume that the migration track is linear and involves just two spatial references—origin and destination (e.g., Lewis 1954; Todaro and Maruszko 1987; Borjas 1989). The reality is far more complicated. Recent scholarship indeed shows that most migration journeys involve multiple locations and multiple border crossings. In fact, as Lok Siu (2012) argues, migrations are serial. Because migrants often have several places of origin and several different places of arrival (i.e., destinations), each destination becomes yet another place of origin and each stay, for whatever duration, is just one more chapter in a migrant's life story (Ossman 2013). Only when the search for safety is finally complete, if it ever is, can the last chapter of a migrant's story begin to be written.

Whatever the exact reasons for the migration, or the number of borders a displaced person crosses, migration is a long nonlinear process that takes a lifetime—and for most people, it never really ends (Innes 2016; Aciman 2021). If we look only at tables, matrices, and migration patterns (the macro level) but never listen to migrants' descriptions of their journey (the micro level), we will never know the complicated process by which migration unfolds and how memories of home, uprootedness, and trauma continue to shape their lives into the future. In other words, they never really reach their emotional and psychological destination, and they live (via memories) in their places of departure (see the examples of Palestinians and Kurds, the world's two largest stateless ethnonational groups, in Chapters 12 and 13). While some scholars tend to divide the migration process into three stages—leaving, arrival, and settlement—in reality, these are not clearly marked; they interconnect and bleed into one another, through social networks and chain migrations. If earlier literature examined migration at the macro level, and since the 1990s it has focused on the experiences of an individual migrant (the micro level), we now need to expand our understanding of migration; we must pay attention also to the meso level and the importance of social networks and social media in mediating between places of departure (origin) and arrival (destination) (Chapter 7). In other words, places of origin and destinations are inextricably tied to one another and should not be positioned vis-à-vis one another.

Further, in the binary construction of origin and destination, the former represents the place of bad experiences, from which one must leave, and the latter is constructed as its opposite, a place of hope, though not yet fully known. But places of origin may not be all bad and places of destination are never all good. In fact, because there are multiple places of destination/arrival, we know that some such places are not as safe as the migrants expect, otherwise they would not embark on further travel. This binary, then, is not useful, and also simply wrong.

Challenging the three binaries/paradigms of migration allows us to illuminate issues, such as migrants' agency, resilience, and power, which have been missing from the more traditional migration research. We focus on these issues by understanding how power operates at three different scales—global regimes and compacts, the state and its politics, and institutional membership and individual belonging. To this end, we divide the volume into several sections: regimes of belonging (Chapters 2–5), drivers of displacement (Chapters 6–10), re-creating a home away from home (Chapters 11–15), and gender/sexuality/age and belonging (Chapters 16–19). The last two sections focus on issues of agency and resilience. Several of the chapters belong in more than one of these sections: while the first two sections offer a macro perspective, the third and fourth zoom in on individual practices and everyday experiences and provide a nuanced micro-level picture. The last chapter (Chapter 20) challenges the methods migration scholars use when conducting research with asylum seekers and suggests that when we study people with trauma, which migrants are, we must pay close attention to their refusal to answer questions and to researchers' refusal to adhere to conventional social science practices. We include this chapter in the hopes that it will yield a discussion of methods employed by social science researchers studying migration, illuminate the limits of current research, and generate new research questions concerning migration.

The contributors to this volume discuss power and agency in migration in different historical periods and geographical settings and from multidisciplinary perspectives—classics, history, geography, political science, sociology, and literature. In their aggregate, their chapters offer a deeper and more integrated understanding of the complex migration process, which would not be possible from a single disciplinary approach, scale, time period, or region. By seeing migrants as a group with different experiences along gender, class, race, sexuality, and religion, the volume illuminates how complicated and nuanced the migration journey is and contributes to migration theory.

Regimes of belonging

To understand how power and agency operate at different scales, we adopt a *regimes of belonging* perspective; this model considers the complicated and ever-shifting dynamics of migrants' sense of belonging—interactions with histories, policies, laws, and practices developed at local, regional, national, and global levels. We thus ask: How do individuals on the move inhabit and navigate regimes of belonging that are situated at the global, state, and local level? For each of these dimensions, how is belonging constructed? Who decides who gets to belong?

The extensive literature on citizenship and migration takes the regime of the nation-state as its starting point for understanding migrants' sense of belonging and incorporation into their host countries (see Bloemraad 2006; Adamson et al. 2011; Bean et al. 2015). In this body of work, the attainment of citizenship, conferred by a single state, serves as a marker of belonging and signals the degree to which migrants and their children are integrated into their host societies. Irene Bloemraad (2006), for example, finds that to help migrants achieve citizenship inclusionary citizenship regimes like Canada's provide institutional support (e.g., helping people learn English, hosting citizenship classes so that people can get naturalized, teaching programs for economic integration that help immigrants who have college degrees get their professional credentials). Not only does attaining citizenship ease parents' membership to their host societies, but it also smooths the path of their children's integration (Bean et al. 2015). Children of parents who initially came without legal status but are later granted it experience educational gains on par with those whose parents came legally. Thus, in these models, the nation-state, which confers legal status and citizenship, plays a critical role in determining whether migrants belong.

Increasingly, however, scholars of transnationalism have recognized that forces like globalization are quickly eroding the nation-state as the primary basis of belonging. During an era of intense global movement driven by economic disparities, climate change, and regional conflicts, belonging is an increasingly a-spatial phenomenon that requires de-territorialized ways of understanding. Increasing connectivity and movements of people means that individuals can inhabit multiple locations and allegiances, complicating traditional notions of belonging that center on place (Soysal 2015). These scholars point out that migrants can form attachments to multiple places, belonging not just in their new place of settlement but also to the places they left behind (Castles and Davidson 2000; Levitt 2001). In her seminal study of migrants who originally came from the village of Miraflores in the Dominican Republic before settling in the Jamaica Plains neighborhood of Boston, Peggy Levitt (2001) points to remittances sent home and the role of transnational political, religious, and community organizations in facilitating territory-spanning modes of belonging. Additionally, transborder regimes like the World Trade Organization and European Convention on Human Rights offer non-nation-specific frameworks for understanding what it means to be a citizen, transcending territorially bounded definitions of what it means to belong (Soysal 2015).

In recognizing that "belonging has become a term that can no longer be linked to a fixed place or location but to a range of different locales in different ways" (Anthias 2016:183), we build on this work of transnational scholars. Contra this literature, however, we argue that de-territorialized forms of belonging predate modern forms of communication and the nation-state. As Elena Isayev highlights in Chapter 2, antiquity offers many compelling examples of how migrants navigated their sense of belonging prior to the regime of the nation-state. Drawing on evidence from the fifth-century BCE Greek polis and the Roman Empire in Late Antiquity, she finds that before the advent of the nation-state, in a world where territory was differently constructed and lacked clear physical boundaries, membership to communities was fluid, bound not by claims to place but by relations. Isayev points to Greek tragedies that detail how authorities claimed displaced persons to elevate their political standing. In doing so, they positioned themselves as hospitable to strangers and open to asylum seekers. This discourse of hospitality allowed state authorities to present themselves as honorable, heroic, and charitable, in contrast to their barbaric and inhumane enemies. Likewise, bishops in Late Antiquity paid the ransom of captives to display their own power, patronage, and moral superiority. Thus, out-of-place individuals such as asylum seekers and ransomed captives found belonging through the relational claims of authorities.

In shifting away from territory and boundaries, we join others in denaturalizing the nation-state and its borders. Anna Amelina and colleagues argue that "'migration,' 'mobility' and 'integration' are not naturally-given entities, but are processes continuously generated by nation-states, border-controls, administrative decision-making on entry and settlement and by dominant notions of 'migrant' deservingness to entry and settle" (Amelina et al. 2021:2). Politicians and policymakers use what Douglas Massey and Karen Pren refer to as "bureaucratic entrepreneurship" to craft policies and enforcement operations to reify and fortify state borders (Massey and Pren 2012:2). Likewise, indigenous anthropologists like Audra Simpson (2014) show how states create and concretize boundaries through bordering policies and surveillance technologies. When the Mohawks of Simpson's ethnography traverse the boundaries of Canada and the United States, border agents demand that they declare their membership to one nation or the other. In doing so, they ignore indigenous understandings of belonging that predate the nation-state. Collectively, this body of work shows that there is nothing "natural" or "given" about borders and belonging. Rather, states draw artificial boundaries around modern ideas of sovereignty and subsequently define insiders and outsiders (i.e., who belongs and who does not) around those borders.

Isayev's de-territorialized framework sheds light on the value placed on bodies by emphasizing the relational qualities of belonging and by highlighting the unique role that people in states of liminality play in elucidating inter- and intracommunity dynamics. Her approach spans spatial and temporal contexts and can help explain how liminal migrants today—those halted and stalled in protracted exile in camps, held in place as conditional deportees, or criminalized in prison cells—must navigate complicated and shifting regimes of belonging (Hernández García 2019). For more on these liminal migrants, see the discussion of Palestinians in Chapter 12, the examination of the enclave people (*chitmahals*) of the transition camps of India and Bengal in Chapter 8, and the study of Venezuelans in Peru in Chapter 5.

Acknowledging the complexities and diversity of regimes of belonging also means a serious consideration of the unique dynamics of belonging in regimes other than those found in the "bubble" of the Global North. Although extant theorizing has pushed thinking about belonging in the liberal democracies of the Global North, it has largely proven unsatisfactory for understanding the dynamics of inclusion and exclusion in the Global South (Natter 2018; Adamson and Tsourapas 2020). Drawn from the study of countries in post-1945 Europe and North America, scholarship, until recently, has largely failed to consider how colonial legacies and postcolonial economic developments have shaped the migration management regimes of the Global South. State management of borders and belonging in the Global South, as we see in Chapters 4 and 8, harkens back to postcolonial state-building, which anchored formal membership to society via citizenship in artificially demarcated territorial communities. In the Global South, which inherited colonially demarcated territories, this drawing of postcolonial states often fragmented earlier economic and cultural networks within ethnic communities. Instead of binding individuals to these ethnic communities, the drawing of arbitrary territories, based largely on ideas of the nation-state borrowed from the Global North, has thrust indigenous peoples who could not meet the evidentiary requirements of citizenship into a "world of unfreedom and precarity" (p. 78 in this volume).

Take, for example, the decolonization of the Indian subcontinent, which led to partition in 1947 and the emergence of two separate states, India and Pakistan. In Chapter 4, Sabyasachi Basu Ray Chaudhury shows how citizenship within the Indian subcontinent remains a contentious issue, even after seven decades of decolonization in the region. The newly enacted citizenship laws (as in Burma, now Myanmar, or in Bhutan in the early 1980s) and existing acts that were amended (as in India in 2019) remain hotly contested. Indigenous peoples who had long inhabited those territories were suddenly excluded. Likewise, in Chapter 8 Nasreen Chowdhory and Shamna Thacham Poyil discuss how the postcolonial state formation of India and Bangladesh impacts the *chitmahals*, who live on territory that spans India and Bangladesh. Forced to choose between citizenship imposed by the two newly created sovereign states of India and Bangladesh, the *chitmahals* are forced to live as either full citizens of India or Bangladesh or as "half" or noncitizens in transition camps. Colonialism, as Chaudhury as well as Chowdhory and Poyil separately show, haunts state-sponsored forms of belonging, as codified in citizenship laws.

Even when colonialism does not displace indigenous peoples, it reorganizes the sense of belonging that individuals have with their own home countries. Hanson Nyantakyi-Frimpong and Dinko Hanaan Dinko (Chapter 9) highlight this in their discussion of northern Ghanaians' feelings of "unbelonging" within their own country. British colonial rule, which initiated a policy of forced-labor recruitment from the north to increase the workforce in mines and railway construction projects in the south, privileged southern Ghanaians over northern Ghanaians. This colonial mindset still persists today in the form of economic exploitation of migrant workers from the north. By looking beyond the geographic contour of nation-states, we see how belonging in the Global South means "inclusion of certain communities within the fold of a postcolonial state and the simultaneous exclusion of the remaining communities from the same political space" (p. 80 in this volume). Decoupling belonging from territory reveals how colonial regimes imposed their own ideas of membership by drawing arbitrary state lines that mapped poorly onto existing ethnic communities.

The conception of regimes of belonging in this volume also looks at how international law and the refugee regime operate to maintain, permit, or inhibit migrants' sense of belonging. But our contributors refocus these discussions of power exercised at the international level from the *effects* of power (that is, how it unequally distributes recognition, rights, and resources) to how power is both navigated and contested. As a result, in their journey involving multiple dislocations and settlements, including loss of homes, livelihoods, and members of families, displaced people's sense of belonging shifts over time and place (Yeoh 2017; Crawley and Skleparis 2018). In one location, they may be a refugee, in another a migrant, and in yet another, they may adopt or be designated some other status. Thus, instead of fixed, permanent, and tied to place, we explore the "temporariness, transitoriness, impermanence, ephemerality, mutability and volatility" (Yeoh 2017) of migrants' sense of belonging.

To reveal the many ways that migrants seek belonging and membership, we shift the main focus of analysis from states to individuals. Joining recent calls to de-fetishize migrant categories imposed by international law and the refugee regime, we move away from state labels and categories of migrants that render migrant agency and empowerment invisible (Stasiulis et al. 2020). For example, Jennifer Hyndman's critique (in Chapter 3) of the recently released 2018 Global Compacts highlights how international law and the refugee regime obfuscates many of the actions that migrants themselves take in their search for belonging. Beyond being poor predictors of migrants' ability to secure protection and welfare, the legal, administrative status of "refugees" disempowers migrants (Landau and Duponchel 2011). Hyndman finds that "self-authorized security" and protection operate at other scales smaller than state-centric perspectives would suggest. Many grassroots efforts by civil society and local organizations offer support for, material assistance to, and legal protection of people who seek safety in other countries, thus providing strategies for security or protection. As Hyndman notes, refugee-migrant subjects are the authors of (and decision-makers in) these journeys, albeit under conditions not of their own making. By viewing migrants as active agents in the struggle for membership and belonging, we thus correct an imbalance in scholarship, which has long made displaced persons invisible by regarding them as silent, hapless victims rather than knowing, purposive actors (Malkki 1995, 1996; Abu-Lughod 2002; Turton 2005; Hammami 2010; Fiddian-Qasmiyeh 2014; Cabot 2016; Pugh 2018).

We also highlight migrant voices by detailing how they navigate spaces of inclusion and exclusion through their everyday decisions and practices. This focus on the everyday draws our attention to the strategies and resources that migrants use in their daily interactions and efforts to belong. Luisa Feline Freier and Andrea Kvietok's study (in Chapter 5) on displaced Venezuelans in Peru demonstrates how migrants renegotiate their membership into a community despite encountering mounting layers of exclusion and vulnerability (e.g., stricter legal entry requirements, reduced integration efforts, job instability, rising negative perceptions of Venezuelan immigration, and limited access to the Peruvian healthcare system during a pandemic). Freier and Kvietok show how displaced Venezuelans construct networks of material, informational, and affective support throughout the migratory process to cope with experiences of exclusion and vulnerability and to renegotiate notions of belonging and community. Some of these everyday strategies include relying on migrant networks as sources of information, support, and solidarity and keeping in touch with family members back home through phone and video calls. Freier and Kvietok highlight how migrants encounter and resist regimes of inclusion and exclusion that are framed around cultural discourses of assets and deficits. Although migrants make decisions within a set of constrained choices, the chapters in this volume point out how scholarship has often overlooked the aspirational assets and sense of resilience that migrants demonstrate in their everyday attempts to be included and recognized (Abrego 2014).

Our perspective on regimes of belonging also centers on migrant agency by viewing power as a broader relational struggle concerning societal questions about who belongs and why (Anthias 2008; Simpson 2014; Pugh 2018; Anthias and Lazaridis 2020; Amelina et al. 2021). In analyzing belonging as a relational contest of power exercised in multiple spaces, we consider how migrants navigate regimes constructed around identities based on race, ethnicity, class, gender, sexuality, and religion, and we examine how those identities intersect. Adopting an intersectional approach, we recognize that migrants' struggles for inclusion happen not only within particular social, cultural, economic, and political contexts but also when subjectively and intimately experienced along multiple axes of oppression (Yuval-Davis 2006; Bastia 2014; Anthias 2016; Stasiulis et al. 2020). Bodies and what they signify contend with hierarchized regimes of inclusion and exclusion that operate on and through migrant bodies or what Floya Anthias terms the "collision and collusion between inequalities and identities" (Anthias 2016:173). Although official statistics collected on migrants and discourses on identity politics would suggest otherwise, the social locations of migrants are rarely constructed along one power axis of difference. Freier and Kvietok's discussion of the unique vulnerabilities faced by female migrants from Venezuela to Peru (see Chapter 5) shows that migrants contend with multiple forms of hierarchical boundary-making. The hypersexualization and sexual harassment encountered by these Venezuelan female migrants point to the multiple layers of exclusion that migrants face in negotiating the many facets of their identity.

When seen through the lens of relational struggle, regimes of belonging emphasize how communities (and individuals' positions within them) are unstable and precarious because they are situated within wider social, cultural, economic, and political milieus (Taylor 2015; Paret and Gleeson 2016; Silvey and Parreñas 2020; Ewers et al. 2021). Migrants contend with an ever-changing landscape of nativist perception and reception (Chapter 5). They also face vulnerabilities when confronted by global regimes that seek to manage and contain individuals on the move via classifications that hinge on ideas of worthiness by juxtaposing refugees against migrants (Chapter 3) and citizens against outsiders (Chapters 4 and 8). Marcel Paret and Shannon Gleeson point out how "the migrant existence is often precarious in multiple, and reinforcing ways, combining vulnerability to deportation and state violence, exclusion from public services and basic state protections, insecure employment and exploitation at work, insecure livelihood" (Paret and Gleeson 2016:281). States play an active role in reproducing and/or ignoring the multiple vulnerabilities and exclusions that migrants face through stigmatizing discourses proliferated through policy and the media (Massey and Pren 2012; Krzyżanowski et al. 2018). Paradoxically, today's migrants face a fortification of borders, both internally and externally, and the relaxing of global capital markets (Massey 1999; Davis 2004). Viewing regimes of belonging relationally requires attention to the instability, precarity, and contradictions of this terrain. This contested terrain means that migrants' sense of belonging and membership is "always producing itself through the combined processes of being and becoming, belonging and longing to belong" (Yuval-Davis 2006:202).

Drivers of displacement

While acknowledging the precarity of the migratory journey and the challenges that migrants face in their search for belonging, we also consider how migrants act as deliberative agents of their own lives. That is, we consider how both choice and constraint act together to drive displacement. Our contributors recast migrants from passive reactors to deliberative actors who make choices contingent upon the intersection of their biography (e.g., race, ethnicity, sex, class, and religion) with the wider environment (e.g., social, political, economic, cultural, and physical context). In charting the drivers of displacement, we start by disrupting the neat typologies of voluntary versus forced migrants. Instead of separately theorizing their causes, we ask how these two forms of migration might instead be linked. How might we think of migration as deliberative choices made by migrants in response to circumstances not of their choosing?

We nuance understandings of the migratory process as primarily self-selecting and shaped by macrostructural forces like the world economy and state policies. In doing so, we depart from neoclassical models that explain migration as cost–benefit calculations made by individuals, with workers moving from low-wage, labor-surplus countries to high-wage, labor-scarce countries. Arthur Lewis (1954), an early proponent of this theory, viewed international migration as a result of responses to economic development and geographic differences in the supply and demand for labor. Microeconomists refined this argument by providing models to explain how individuals evaluate the returns to and risks of moving—weighing the investments needed for migration (such as funds for traveling and relocating, time to look for work, and adapting to the new language and culture) against the possibility of increased wages (Todaro and Maruszko 1987; Borjas 1989). This portrayal of migration, as self-selecting responses to economic conditions, is also evident among scholars of the new economics of migration who consider larger units of analysis, such as the household or the global economy (Stark and Bloom 1985). This model argues that migration is driven by decisions made and not by individuals, as under the neoclassical tradition, but by households seeking to diversify their incomes (Massey 1999). This body of work highlights how households allocate family workers to different labor markets in different geographic areas to maximize resources and minimize risks (Taylor 1987; Stark and Taylor 1991). In response to failures in the local economy, a household may send a few workers abroad to accumulate savings and to send home remittances. By working as a collective to circumvent local conditions, individuals maximize the returns to migration (as seen in increased wages) while minimizing the risks associated with moving. In both the neoclassical and new economics of labor migration frameworks, displacement is conceptualized as a microeconomic process shaped by the conditions of sending and receiving nations, whether guided by wage differentials, credit market failures, labor demands of modern economies, or the diversification strategies of households.

A tension that runs through these theoretical formulations is the role of agency versus constraint. Neoclassical economics and the new economics of migration ascribe too much choice to individuals, viewing migration as a rational cost–benefit exercise carried out by migrants, whether as individuals or as members of a household. Despite the alarming numbers of forcibly displaced persons, this form of mobility is curiously absent from theoretical models focused on the economic drivers of displacement. Instead, the literature on labor migrants treats involuntary movement as a separate category, relegating its examination to the subfield of forced migration (FitzGerald and Arar 2018). We complicate understandings of labor migration by exploring its connection to forced migration. Instead of viewing labor migration as a purely self-selecting process made in response to the economy, we examine how migrants make decisions to migrate within a set of choices constrained not just by the economy but also by the larger social and political environment. Within this wider environment, we focus, in particular, on the meso level of networks to consider how migrants' connections to family, friends, and acquaintances shape their decisions to move. As others working within the social capital framework have found, migrants leverage their connections to those who have already moved to diversify risk and maximize gains (Massey et al. 1994; Faist 2000; Menjívar 2000; Deshingkar et al. 2019).

These social ties act as a form of "social capital," originally theorized by James Coleman (1988), because migrants draw on these connections to acquire knowledge about a foreign land, access opportunities to jobs, and receive other resources to aid in relocating. Social ties drive migratory streams at the community level because each act of migration connects individuals to each other and groups to other groups, raising the odds of subsequent migrations. In certain contexts, migration chains are so entrenched that they become embedded in local traditions, such that the impulse to move becomes engrained in one's understanding of how to achieve the good life. In this way, social ties drive a "culture of migration" by affecting people's behaviors and the value they place on migration (Brettell and Hollifield 2014).

Building on insights from social capital theory and research on the migration industry, Olayinka Akanle's analysis of the drivers of displacement (Chapter 7) focuses on the meso layer by looking at the role that social networks play in constructing and sustaining migration chains of modern slavery. He notes that underdevelopment in Africa, facilitated by an exploitative network of migrant brokers, drives many into conditions of modern slavery in Europe. Myriad reasons, ranging from real needs and threats to survival and hopes for a better life elsewhere drive African migrants into exploitative working conditions abroad. Like social capital scholars who find that social ties ease the costs of moving to a different country, Akanle looks at how brokers facilitate the move of aspiring migrants by helping them circumvent layers of migration controls imposed by the state.

Akanle complicates the literature on social capital by showing how social ties between would-be migrants and brokers rest on both trust and deception. Brokers, who sometimes come from the same community as the migrants themselves, exploit existing bonds of trust to deceive and coerce migrants into conditions of modern slavery. He details how brokers use advances in IT—such as social network platforms like Facebook, WhatsApp, Skype, Zoom, Twitter, Instagram, emails, and websites—to recruit, retain, monitor, and exploit modern slaves. Akanle further complicates our understandings of forced versus voluntary migration by noting that some African migrants, driven by aspirations for a better life, *knowingly* enter into conditions of modern slavery. While focused on the meso layer and web of connections that bind would-be migrants to those who have already migrated, Akanle also considers the motivations and aspirations of individual migrants. Instead of rendering migrants speechless and lumping their various motivations to move into a homogenous and indistinct mass (see critiques by Malkki 1995, 1996), Akanle sees migrants as agents of their own lives, even though this agency sometimes means knowingly entering into the corrupt networks that foster modern slavery.

Just as migrant social networks facilitate the migration of Africans to Europe, various ties—whether formed through blood, place of origin, or work—Jia Feng and Gou Chen describe in Chapter 10 how migrant recyclers shuttle individuals from rural hometowns to China's urban centers. This displacement stems in part from China's adoption of a socialist market economy that encouraged the freed rural sector to seek opportunities in urban places. But the ability of rural migrants to realize their aspirations is highly constrained by the legal barriers posed by China's internal passport policy known as *hukou*, which links access to institutional resources such as education to one's permanent residence. Many of these migratory decisions are made at a household level; to ensure their children's access to education, families choose a split-family strategy, with either one parent remaining behind or both parents leaving childcare to a relative. Like Akanle's African migrants who knowingly enter into conditions of modern slavery in the hopes of achieving a better life, these migrant families make deliberative choices to move under highly constrained conditions, namely rural precarity and China's *hukou* policy environment.

Like the Henan migrants of Feng and Chen's analysis, who choose to leave their rural hometowns for Beijing in search of better economic prospects, northern Ghanaian migrants are compelled to move when faced with economic insecurity put in place during British colonial rule and exacerbated today by climate change. While crop failure, water scarcity, and food shortages drive the internal displacement of northern Ghanaians, Nyantakyi-Frimpong and Dinko caution against environmental determinism and show how climatic circumstances exist as only one of several factors that work in concert with other factors, such as historical context and economic policy, to drive migration (Klinenberg et al. 2020; Sherbinin 2020). Their contribution in Chapter 9 shows how slow-onset climate stress like drought exacerbates conditions for displacement already set in motion by British colonization and subsequent neoliberal economic reforms of the 1980s. British colonial rule initiated a policy of forced-labor recruitment from the north to increase the workforce in mines and railway construction projects in the south. This, combined with the more recent Structural Adjustment Programs of the 1980s, removed subsidies on fertilizers, seeds, and pesticides for agriculture, establishing patterns of north–south migration still evident in Ghana today. Thus, in Nyantakyi-Frimpong and Dinko's model, climate stress is just one of several factors driving the internal displacement of northern Ghanaians. The categories "voluntary versus forced" do not fully fit the predicament of these migrants; their decisions to move, while economically motivated and prompted by the realities of climate change, cannot be properly understood without also considering the historical legacy of colonialism.

Indeed, as extensive critique of the term "climate refugees" shows, labeling the migratory movement that results from environmental changes as "voluntary versus forced" remains contentious (Sherbinin 2020). Jon Barnett and W. Neil Adger concur: "The concept [environmental refugees] has been critiqued for being alarmist, counterproductive to addressing either the problem of environmental migration or the challenges facing refugees, and 'erroneous as a matter of law'" (Barnett and Adger 2018:135). Take, for example, displacement that results from climate events that are fast-onset (floods, storms, hurricanes) versus slow-onset (droughts and degradation). A situation in which inhabitants fleeing a village because of a sudden hurricane directly suggests forcible migration; however, discerning choice from force becomes more difficult in the case of farmers relocating because of decreased agricultural productivity as a result of progressive desertification caused farming practices. As Christina Cattaneo and colleagues note, "In practice, there is often a continuum between 'fast and slow onsets' as well as between 'direct and indirect impacts' and 'voluntary and involuntary movement'" (Cattaneo et al. 2019:3).

Thus far, we have examined the economic, social, political, and environmental causes of displacement within the context of movement and mobility. Yet as we see in Chapter 8 by Chowdhory and Poyil on the *chitmahals* of India and Bengal, and in Chapter 6 by J. Santiago Palacios on the Muslim Andalusi population of Medieval Iberia, these same insights can be applied to the displacement of sedentary people. For example, colonial partitioning of India and Bangladesh and the parameters of citizenship imposed by the two newly created sovereign nations have conditioned the displacement of enclave dwellers (*chitmahals*) who live on territory that spans those regions. Paradoxically, sedentariness, not movement, marks their dislocation from place. The *chitmahals*—who must choose to live as either full citizens of India or Bengal or as "half" or noncitizens in transition camps—defy neat categorizations as either voluntary or forced migrants. Similarly, sedentary displacement marks the condition of the Muslim Andalusi, who chose to remain in place after the conquest of their lands by Christians. Palacios explains how the domination of medieval Iberia by Christians forced a dilemma upon the Muslim Andalusi population, who could either endure exile or remain in their homes. Although the loss of al-Andalus resulted in the disappearance and displacement of a large segment of the Muslim population, some of these wartime migrants remained in place as subjugated Mudejars. For both the *chitmahals* and Muslim Andalusi populations, displacement following conquest comes in the form of choosing to remain in place. Indeed, neither example fits neatly into existing typologies of voluntary versus forced migrants.

In his review of the sociology of international migration, David Scott FitzGerald notes that theories on why people migrate have been hampered by "US dominance [which] has left major casualties on the field of knowledge. The entire enterprise is shot through with unstated and often mistaken assumptions of both universality and US exceptionalism" (FitzGerald 2014:116). Moreover, the preponderance of this scholarship has resulted in theories that primarily concern permanent, transoceanic migration. We avoid theoretical myopia in this volume by drawing on studies from across the globe and interrogating different forms of displacement. Akanle discusses migration that happens across international borders, with African migrants becoming entrapped into modern slavery in Europe, while contributions by Nyantakyi-Frimpong and Dinko and Feng and Chen examine the causes of temporary, internal displacement of northern Ghanaians and Chinese ruralites, respectively. Finally, Chowdhory and Poyil, and Palacios, separately, explore the drivers of sedentary displacement. Collectively, these contributors understand the causes of migration as diverse, dynamic, and context-specific; they show

how people of different social locations (e.g., race, gender, class, and religion) navigate choice and constraint in response to various power structures located at different levels (i.e., local, national, and global).

Re-creating home away from home

How do migrants find home following displacement? Dislocation always involves uprootedness from home and homeland. Whether migration occurs gradually or in haste, whether in response to economic and political instabilities or to other existential threats, migration means leaving behind loved ones, attachments, and all that was once familiar and represented comfort (Ahmed 1999). Many migrants who flee for their lives and who may never be able to return to the place they came from are often nostalgic for what they left behind, longing for their *home* and *land* and for what is no longer accessible to them (Dossa and Golubovic 2019). In fact, in their imaginary—with time and distance and the sense of loss—home and homeland become far more important than they ever were before; if it is not the physical home that migrants yearn for, then is it the *feeling* of home. They hold on to memories of the unique textures, distinctive smells, and sights of their homes and lands, and to the special feelings of love, comfort, and warmth that inhabited their home when they lived there (Feldman 2006; Perez Murcia 2019; Dossa and Golubovic 2019). So deep, formative, and comforting are these feelings that many migrants try to re-create "home" as a way to ease their integration in the receiving country. But because home does not have an essential meaning, and re-creating it means involving both past and future through inhabiting the grounds of the present (Ahmed et al. 2020:8, 9), re-creating home is a complicated process for any migrant, let alone those whose experiences of the past were traumatic.

Not all formative experiences and memories of the home/land are positive, and not all migrants feel an attachment to the home left behind, precisely because of their traumatic experiences there. They nevertheless must negotiate all these feelings, including trauma, when they rebuild and re-create their homes in places of arrival. One such way is through the arts—literature, poetry, or visual arts. [Smith et al. 2011 and Miyamoto and Ruiz 2021, among others, discuss the importance of expressing loss and trauma through metaphors.] The artistic expressions are often created by second- and third-generation refugees who might have been born already in new places of arrival but whose lives are anchored in their parents' national expulsion. They still live the trauma and express it through words and visual arts (Ankori 2020). A growing body of work focuses on participatory art projects, with art as part of the healing process of asylum seekers, but the works of a second generation may be less about healing and more about remembering.

Sentimentalizing home and communities of belonging

Those who were uprooted because of war are particularly attached to their lost home/land; they dream about it and describe it in ways that evoke the different senses. These descriptions of the lost lands as Paradise are a common theme of poetry and have been drafted as an important tool for nation-building, but they also raise both national consciousness and attachment to the homeland. It does not matter that such descriptions have little anchor in reality and that the real lands were sometimes poor and not always productive. In the imaginary of the displaced, Paradise was rich in water and fruit, and its smells, sights, and sites were simply a delight. Palacios, mentioned earlier for his work in Chapter 6, describes the loss of al-Andalus (the Arab lands of Spain) to the Christian Kingdoms of Iberia in the early Middle Ages. He quotes poetry from the eleventh and twelfth centuries that laments the loss of al-Andalus, describing it as a place of beauty, where the rivers flow, the sunsets are scented, and the air is filled with the scent of daffodils. Andalus is also remembered as mythical Arabia. So traumatic was the loss of the lands of the Arabian Paradise for the Muslim world that al-Andalus has remained a symbol of abundance, attachment, and ultimate loss. These images of al-Andalus were evoked hundreds of years later when the Arabs of Palestine were expelled from their lands in 1948 (Benvenisti 2002).

The loss of the homeland and the 1948 expulsion, so central to the life of the Palestinian nation, continue to define and shape both the nation and the Palestinian subject. To ease life in the camps located throughout the Middle East, Palestinian refugees have re-created communities of belonging to feel like home, decorating their temporary living units—tents and makeshift homes—in a way that reminded them of homes they left behind. Anne Irfan (Chapter 12) notes the extent to which Palestinians went in order to re-create Palestine in their camps. They named the camps, streets, and neighborhoods after villages and cities from which they fled, keeping Palestine alive. Further, they used decorative imagery in their makeshift homes, and women, even urban women, wore the traditional Palestinian peasant clothing as a way to denote attachment and belonging to the lost homeland. Although in 1948 these practices may not have been intended to foster nation-building, the camps not only provided a sense of home for these refugees, but they also became the site from which exiled Palestinian identity flourished and where Palestine, as an idea, was kept alive for second and third generations as well as for generations yet to be born.

Other refugees, like the Rohingya in Bangladesh, for example, have re-created home and achieved a sense of belonging in other ways. Rohingya women, for instance, rely on daily practices (Chapter 16). They, like the Palestinians, fled in haste and settled in makeshift refugee camps but have felt connected to others because of their shared experiences of displacement. Farhana Rahman and Nafay Choudhury show in this chapter how maintaining daily routines of communal praying, cooking, caring for children, and adorning themselves and their homes, together with other women neighbors, provide the Rohingya refugee women in Bangladesh a community, which is essential for their sense of belonging.

Community as home

Important to Rohingya women's ability to withstand life in tarp shelters, indeed to remaking them into homes, is the presence of community. Even prior to their exodus from Myanmar, home and belonging for the Rohingya meant the presence of a community made up of their extended families. In the camps, they sought to re-create such communities, but many family members, whether killed in Myanmar or on the journey, did not make it to Bangladesh. The community these women created now included extended family as well as others from their home villages who were alone with no living relatives. Regardless of who is now part of their community, its sheer existence provides Rohingya women a sense of home.

The importance of community as a way to re-create home and feelings of belonging is a theme running through Chapters 10, 13, 14, 15, and 17. Although the idea of community as home seems to be universal, the way community/home is constructed differs by geographical setting and circumstances of the migration. As the contributors in this volume show, many migrants re-create home not only by adorning their familial living quarters with mementoes from their past but by relying on social ties to create communities of support. This is the case for transboundary or intra-state migrants. Feng and Chen in Chapter 10 discuss the power of community for rural to urban migrants in Beijing based on their observations of garbage recyclers who came to the big city from Henan province and now reside in the same community in the city; the migrants experience the same challenges as urban villagers and find the community to be a source of strength. They have created a self-sufficient community, where neighbors are friends who share meals and who rely on one another for household furnishings and other needs. With this community, they have replicated a sense of home and eased life in the big city. Because this community is particularly transient and the migrants are reluctant to invest in improving their homes, the power of community and the sense of home it provides are magnified and essential for these rural migrants in Beijing. Equating the community with home is also the experience of Latina lesbians in Long Beach, California, discussed by Sandibel Borges in Chapter 17. These Latina migrants are harassed not just because of their ethnicity but also because they are lesbians. Karma Chávez calls on us to see migrants and LGBTQ issues as intertwined, arguing for the importance of understanding that as much as "queers and migrants are strange to the nation, they are strange to each other" (Chávez 2010:139); they are feared because they threaten not only the national borders but the moral borders as well. Therefore, as Borges shows, their own homes (as well as the homes of their friends) and the gay bar have become the arena where Latina lesbians build and find their community of comfort. In those places, they feel anchored and empowered to deal with daily harassment, rejection, isolation, prejudice, and displacement. But Borges cautions us not to romanticize the bar or the home as a necessarily happy place; in those spaces too, there are markers, such as linguistic markers, that divide the community and create a hierarchy of who belongs and who does not. Such criteria result in alienation, harassment, and rejection from within the Latina lesbian community, which migrants have to contend with as well.

Another institution that can help anchor migrants and provide them community, support, and a sense of home is the school system (Kia-Keating and Ellis 2007). But as Shawna Shapiro (Chapter 15) shows, schools in the northeast United States (she focuses on New England) often fail to provide students from refugee backgrounds the community/home that they need. Instead of getting assistance, these students experience systematic intellectual belittling and racially motivated prejudice from peers and teachers alike. Students describe a structural, social, and academic othering by their fellow students that often reinforces teachers' negative attitudes toward migrant students as being less prepared, or victims of trauma, and therefore deemed all-around inferior. With a more inclusive curriculum, Shapiro argues, schools can develop a culture of inclusion, thereby becoming institutions of support that will create community and provide a sense of home and belonging for these students.

Naif Bezwan and Janroj Keles show in Chapter 13 how Kurdish refugees in Turkey re-create home away from home by relying on the Kurdish diaspora to help them gain a sense of belonging. Kurds in Turkey use different media forms, such as radio and TV, social media, satellite broadcasting from London, desktop publishing, all in the Kurdish language, which help them adjust and (re)settle in the new country by providing a sense of home. At the same time, these media and various Kurdish institutions contribute to nation-building, to raising Kurdish national consciousness, and above all, to keeping the Kurdish nation alive in the minds of the Kurdish diaspora.

The examples discussed above focused on the different ways migrants re-create a sense of home in order to feel more settled and at ease in their places of arrival. Except for Chapter 17, which looks specifically at the sense of home for a subgroup of migrants, the other examples treat each group of migrants as a monolithic group and rely on a heteronormative discourse to belonging. The reality, however, is far more complicated. Experiences of the migration journey, as well as their efforts at resettlement and achieving a sense of belonging in the new places, differ by gender, sexuality, class, race, and religion. Doğuş Şimşek (Chapter 14) uses stories of four different recent Syrian migrants to Turkey as a way to discuss how class, race, and religion affect the migrant experience and the ability to adjust, arguing that a sense of belonging is articulated through its own hierarchy. Migrants who were of economic means in Syria and have had social ties in Istanbul prior to their exodus have been able to reside in better neighborhoods in Turkey and support themselves. These migrants felt welcomed and at home in Istanbul and did not long for their life in Syria. This has not been the case, however, for rural and poorer Syrian refugees who are harassed and feel out of place in Istanbul. Further, Şimşek shows that ethnicity is also an important determinant of the potential for migrants to develop a sense of belonging. She uses the examples of Armenians and Kurds, who feel the least at home of the groups she interviewed. Their history in Turkey was fraught with violence, especially as a result of the century-old systematic discrimination of these groups—ethnic cleansing and genocide in the case of the Armenians (from 1915 to 1917) and periodic massacres of Kurds (in the period between 1923 and 1991).[4]

Where is home? Here, there, or nowhere?

The need to re-create a home away from home is essential for most migrants because feeling at home in the country of arrival provides the kind of security that was missing during the migration journey. Home fosters a sense of belonging: some, as we see in this volume, re-create home away from home by engaging in everyday practices (Chapter 16); others achieve that sense through the use of mementoes and naming practices that evoke emotions and nostalgia (Chapter 12). But as important as keeping parts of the past alive in the present, and as crucial as the ethnic community is in the diaspora, both are double-edged swords. In one way, they keep migrants connected to their past, and on the other, they slow down, even inhibit, full engagement with the receiving society. The diaspora then helps keep migrants in a liminal space, and many feel that they belong neither here nor there. These feelings of not belonging are compounded by the fact that migrants are often unwelcomed, harassed, and experience economic, political, and social discrimination in the host society.

If citizenship is the marker of belonging, a coveted status for all refugees and migrants, then Chapters 11, 8, and 13, as examined respectively below, show that even when migrants are granted citizenship, that alone does not make them feel at home in their new countries of residence; in fact, many feel that they belong neither here nor there. Andreas Kossert and Tamar Mayer (Chapter 11) discuss the case of 14 million German refugees in the years following World War II, who were unwelcomed and discriminated against by none other than their co-ethnics in Germany. Over hundreds of years, ethnic Germans settled in Silesia, Prussia, Pomerania, and Brandenburg, which after 1945 were parceled into the countries that became part of the Communist bloc. Since the governments of the newly established states considered all ethnic Germans the enemy, they expelled them and forced their return to the German homeland. But the expellees had no history in Germany and had little in common with Germans in Germany other than a shared language and culture; they never lived in Germany and did not consider it their home. And since Germans living in Germany (i.e., the host society) were suspicious of the newcomers and reluctant to share their homes and communities with them despite government orders to the contrary, refugees came to a "cold home," where they endured hostility and resentment; it would take years and more than one generation before they would feel a real sense of belonging. For them and their offspring, home remained in the German eastern provinces, now the new state of Poland and Czechoslovakia. Though the postwar German governments congratulated themselves on the successful integration of their millions of refugees, the pain and resentment refugees had encountered did not abate, especially since those experiences were not acknowledged by German society until 2014. Citizenship, as Chapter 11 shows, was not enough for these refugees to feel at home and develop a sense of belonging.

While redrawing Europe's map forced ethnic Germans to flee their homes in the former German territories to the east, redrawing the map of South Asia at the end of the colonial era resulted in a different fate for the communities now divided by a border. As Chowdhory and Poyil show in Chapter 8, these communities were relegated to enclaves (*chit*) on both sides of the border. Bangladeshis were settled in enclaves in India and Indians settled in enclaves in Bangladesh. On both sides of the border enclave dwellers (*chitmahals*) lived in subpar conditions and experienced severe discrimination from the host communities and from the state apparatus. As a result, they belonged nowhere. Like the German refugees, enclave dwellers lived in a liminal space, considering no place as home. It would take years and political decisions to grant Indian citizenship to *chitmahals*, but even that would not make them feel a sense of belonging.

In Chapter 13, Bezwan and Keles also examine how structural constraints inhibit refugees' integration in their host country. They argue that being Kurdish is a state of mind and that statelessness has shaped the national consciousness and behavior of Kurdish migrants. Because registration in the arrival country is normally accorded only to those who come from an internationally recognized nation-state (Iraq or Syria, for example), and because Kurdistan does not physically exist, these Kurdish migrants do not officially exist. They do not appear in official statistics: they may count in the data as Iraqi, Syrian, and Turks, *but not as Kurds*. As a result, in some countries, these "invisible" Kurds cannot take advantage of recourses made available to other migrants in their respective languages; this prevents Kurdish migrants from enjoying multicultural policies and practices, or even knowing about them. Even though they are granted formal citizenship in their new places of residence, it does not provide them with recognition of communal existence and rights, nor does it lead to the enjoyment of citizens' rights and liberties. In many of these places, therefore, Kurdish refugees live in a liminal space—they do not feel that they belong in a place that does not count them as Kurds and more so, they have no homeland, except in their mind.

Home/land, as we have shown above, is often idealized and sentimentalized as a way to emphasize belonging. Its idealization is constructed in opposition to the migration journey and the experiences in the country of arrival. This binary construction of there (homeland) and here (country of arrival)—essentially of past and present—persists in migration literature even though the picture it paints is far from accurate. First, many migrants do not experience *home/land* and *there* as happy places. Despite its positive assets, home/land is a contested and ambivalent site (Blunt and Dowling 2006; Perez Murcia 2019). Home is often a place in flux, replete with interpersonal conflicts, which is never "exempt from the workings of gender, class, and power" (Dossa and Golubovic 2019:172). Many migrants also experienced the past as a place of economic hardship, political instability, and ethnic or religious tensions. Second, because migration is not simply a linear movement between two places—origin and destination—but rather involves several stops, migrants may have multiple sites of comfort and attachments where they feel at home (Ahmed 1999) and which they call home. In these situations, Sara Ahmed argues, the current place for migrants, the *here,* and not simply the past *there,* is where home is located. Third, as part of the migration process, home is not simply *there,* but is always *here* as well, because it entails feelings. After all, home and the past, together with the difficult experiences of place and time that propelled the move, continue to shape migrants' experiences in their new host countries. Dislocated exiled migrants never rid themselves of their past; it is forever located in their present. Further, for many migrants, home is located neither here nor there. As Luis Perez Murcia (2019) notes, for them the binary construction is between the former home and *nowhere,* and migrants who are nowhere are trapped in limbo, living with no place to call home.

Gender, sexuality, age, and belonging

The difficulties and dangers faced by migrants on their journey are well known, and many of the chapters in this volume explore these hardships in detail. Indeed, no matter the reasons for the migration, it is abundantly clear that the rupture migrants experience and the enormous dangers and difficulties they encounter on their journey differ greatly by gender, sexuality, age, race, or class. Yet, as we have already mentioned, the discourse of migration on the part of many scholars, policymakers, and NGOs remains mostly heteronormative and often deals with migrants as a homogenous group. Even though feminist scholars have addressed women migrants specifically by offering a different scale of analysis (Silvey 2006; Piper 2008; Afsar 2011), for the most part, their discourse has not been granulated enough to include the experiences of young, homosexual, or transgendered men and women. Chapters 5, 15, 16, 17, 18, and 19 attempt to begin and fill this lacuna with specific attention to gender, sexuality, and age. Though we divide the section here by the larger categories, we are cognizant that they overlap and intersect.

Gender

Migration is a gendered process; men and women experience it differently. In fact, "it can be argued that gender has the biggest impact on the migration experiences of men, women, boys, girls, and persons identifying as lesbian, gay, bisexual, transgender and intersex [LGBTI]" (IOM 2021). Although in the past women were considered "associated migrants" and "trailing wives" (Piper and Roces 2003, in Sahraoui 2020), this is hardly the case in most migrations of recent years. Women are active migrants who move for the same reasons as men and in similar numbers, although their numbers have shown a slight decline since the year 2000.⁵ Women migrants exercise their agency to escape their predicaments in hopes for safer and better lives despite their increased vulnerability at every leg of the journey; they are threatened by gangs, coyotes, and others who prey on them and endure gender-based discrimination by border regimes as well. Luisa Feline Freier and Andrea Kvietok highlight such hopes and efforts, as well as hardships, fears, and harsh treatments in Chapter 5, recounting the stories of Venezuelans who wanted to reach Peru and passed through multiple international crossings—— Colombia and Ecuador—where they were preyed upon, sometimes robbed of all their possessions, and abused by impervious border patrols who made their journey even more difficult. One of the interviewees, who was pregnant during part of the trek, explained that when she and her infant daughter arrived at the border at night, hoping to sleep in a tent provided by the IOM, border police forced her and her baby to sleep outside, where it often rained. Venezuelan female migrants also experienced discrimination once they reached Peru. As Freier and Kvietok note, all their interviewees experienced gender-based discrimination, hypersexualization, and sexual harassment in the workplace, which made seeking and keeping a job dangerous and degrading. Such harassments have not only economic ramifications but social and psychological ones as well.

In addition, women migrants suffer in other ways. Most migrations occur in or from the Global South, and many of the migrants come from traditional societies, where gender roles are clearly defined. During their migration, many encounter new and unsettling challenges. They experience profound loss—of traditional family structure (on which they had relied heavily) and of lines of authority. Along with the stresses associated with experiences during the journey itself, and the racism, sexism, and xenophobia at their host lands, the loss of tradition affects women's ability to adjust and feel a sense of belonging in their new places of residence. Ben Graves' analysis of Buchi Emecheta's semiautobiographical novel *Second Class Citizen* (in Chapter 18) illuminates the difficulties of belonging as experienced by a Nigerian woman who migrated in the 1960s to Britain, once Nigeria's colonizer, and who faces not only racism and xenophobia in housing but also in getting access to the healthcare system. Parts of the novel take place in the intimate setting of a maternity ward, where she experiences intervention by an army of Western medical practitioners—"primary care providers and OB/GYN specialists, emergency room doctors, pediatricians, nurses, counselors, and even ancillary workers, who reassert colonial hierarchies of race and stigmatize women who fall short of the prescribed standards of normative womanhood" (quote from Chapter 18). Her encounters with the British obstetrics practices, aimed to curtail women's reproduction, and her experiences in the maternity ward, which are so different from practices in unofficial communities of care among women in late-colonial Nigeria, crystalize for the protagonist the alienation an African migrant woman suffers in her new country of residence, especially within its institutions. But she develops solidarity with other black women who also experienced the system similarly. Through them and with them, she feels solidarity and a sense of belonging to a community with which she can confront racism and her treatment as a second-class citizen.

The importance of shared experiences, especially those born out of political and ethnic conflict, cannot be underestimated. These experiences catalyze the creation of tight-knit communities in the diaspora, in general, and in refugee camps in particular, and in all of them, women play a key role in keeping the community and the memories alive. Consequently, women have benefitted greatly from such communities as they are surrounded by those who understand their past and present and can provide the much-needed support. This very situation is described by Farhana Rahman and Nafay Choudhury, who in Chapter 16, write about the power of Rohingya women's communities in refugee camps in Bangladesh. Because their men were either killed or disappeared in Myanmar as part of the Burmese Muslim ethnic cleansing; Rohingya women escaped with their children and with other women from their home villages and from others in the region. They settled as a group and continued together many of their past practices. One of these practices, the women's praying circle, *taleem*, has given them a sense of community, anchor, and home and has empowered them to deal with their new circumstances.

Children

In 2019, the UNHCR estimated that among migrants worldwide, millions were children who accompanied their parents on their journey and endured the dangers and uncertainties of the migration process along with their families (Shubin and Lemke 2020). But there were also many children who were unaccompanied through the journey and were ever more vulnerable. Of the more than one million asylum seekers and refugees who arrived in Europe in 2015, one-third were children, and about 90,000 of them were unaccompanied (ISSOP 2017). These children have experienced not only the horrors of war, ethnic cleansing, violence, and exploitation but also have been separated from family and suffered from trauma and health problems, such as malnutrition and communicable diseases, perhaps even vaccine-preventable diseases (ISSOP 2017). Many got sick along the way, and their conditions worsened because they could not get medical help. Also, like most child migrants, they miss certain rites of passage, especially in education, which puts them at a disadvantage once they settle and are ready to enroll in school. As Shawna Shapiro discusses in Chapter 15, these deficits and disadvantages, along with the trauma, hinder students' ability to learn. Even though she uses the example of one school in a refugee resettlement community in the northeast region of the United States, her observations fit well in other schools throughout the country, especially in a state with predominantly white population. She shows that schools adhere to a "master narrative" that describes students from refugee backgrounds as trauma-ridden, and whose education and motivation are limited, not up to par with the abilities and drive of nonrefugee students. Instead of helping students overcome the deficit and associated stereotypes, many schools' policies and practices, whether in the classrooms, halls, or teachers' rooms, reinforce the marginalization of these students. Until schools adopt different policies and practices specific to students of refugee backgrounds, Shapiro argues, adjustment and sense of belonging in their new places of arrival will remain complicated, fluid, and flexible (Shubin and Lemke 2020).

Even though it is utterly important to understand the impact of migration on children of all ages because of the specific physical and emotional issues associated with their journey, many of the policies and most scholars treat child migrants as a monolithic group with little or no regard to age differences. But children of different ages experience the journey differently, and this affects the way they feel about their new home countries. For example, within the same families, migrants who left their birthplaces before or during their early teens experience the migration and the adjustment in the host countries differently than children who migrated, with or without their parents, at a much younger age. This group of young (teenaged) migrants are considered the 1.5 generation, the in-between generation, and exhibit different attachments, needs, and strategies for settlement than their younger siblings or their parents. Not surprisingly, the 1.5 generation migrants have a different sense of belonging (Bartley and Spoonley 2008; Lee and Kim 2020).

But the 1.5 generation itself is neither a monolithic nor homogenous group; young adults, whether male, female, or gender fluid, experience migration and adjustment to their new countries differently. Heidi Sarabia, Laura Zaragoza, and Alejandra Aguilar explore in Chapter 19 the experiences of 1.5-generation Mexican women migrants to the United States, who came as undocumented migrants and live there as illegal immigrants. They show how the circumstances of the migration—being a dependent "minor" or an independent "adult"—shape the migration journey, the experiences of these migrants in the host country, and therefore their sense of belonging. For instance, those who were younger and entered the school system had a greater sense of belonging than those who entered the labor market upon arrival. Further, the authors argue that regardless of the importance of family ties and the strength of the community, which usually help anchor migrants, the mere fact that they remain undocumented, and are often racialized, scars their ability to feel at home in their communities and in the United States.

Sexuality

As much as gender and age affect migrants' experiences, so does sexuality. Lesbians, gays, bisexuals, transexuals, and queer people (LGBTQ) are much more vulnerable and confront a set of obstacles that heterosexual migrants do not. In addition to all the threats and insecurities they do share with heterosexuals—that is, conditions driving them to escape their countries in search of a safer place to live—LGBTQ people have additional concerns related to their sexual orientation. Indeed, the source for the influx of migrants and refugees are troubled places where homosexuality is criminalized. LGBTQ people seeking refuge are often subject to rape, corrective rape (also called curative rape), torture, blackmailing, and murder, and many have been excluded from their communities and often from their families. When they leave home, their fears are not abated because the countries they pass through in Africa,[6] Asia, or the Middle East also outlaw homosexuality; once their sexuality is revealed, LGBTQ people face possible criminalization charges and are banned from entering the new country. If they reach the European asylum system,[7] as Johannes Gartner explains, "they are the most invisible migrants and are rarely offered asylum because they have to prove to immigration authorities that they are queer, that they fear persecution on the grounds of their sexuality, and that their fear is well-founded." This is hard to do because migration is perceived to be heteronormative and, as Gartner argues, "the outcome of their claims is largely dependent on the existence of usually non-existent evidence and their quests for refuge are easily made but impossible to prove" (Gartner 2015:n.p.).

LGBTQ migrants are not wanted in the United States either. They "challenge conventional belonging because they are both figured as strangers and threats to how the nation sees itself now, and, more importantly, how it hopes to see itself in the future" (Chávez 2010:151). Gay migrants are often on the margins before they flee, during their journey, and in their arrival places. They seek a homosocial community where they can belong, through which they can meet other gay men and assimilate into American cities, and where they can express their sexual desires and live an openly gay life (Carrillo 2018). At the same time, Héctor Carrillo shows that while many gay Mexicans seek to leave Mexico because of their sexual preference, many choose to remain in Mexico and live their lives there. Like all Mexicans who seek to migrate to the United States, gays face harassment and xenophobia at the border. And were they to reveal the reasons for their migration, they would certainly be subject also to homophobia.

The reasons for Latina lesbian migration to the United States are similar. They too want to escape homophobia, pressures for gender conformity, and the alienations, ridicule, harassment, and threats that they endure from their family and communities in response to their sexual orientation. As Sandibel Borges shows in Chapter 17, Latina lesbians' troubles do not abate once they cross the US-Mexico border. Even though anxieties of falling victim to sexual violence, being apprehended by border patrol, or getting lost or kidnapped seem to decrease slightly once they cross the border, the undocumented migrants now fear being discovered and deported not only because of their migration status or the color of their skin but because of their sexual orientation. These women live in two closets, the racial and the sexual, and seek their homosocial place where the two identities can dwell in harmony. They find the gay bar and the community to be places where the two identities dwell.

Researching people with trauma: a critique

The contributors in this volume focus on displacement and belonging among migrants, many of whom have experienced trauma before, during, and/or after settling in their host lands. These migrants have exhibited agency in face of power, without which their predicament would have undoubtedly been different, most likely worse. Many of the contributors have listened to migrants narrate their own experiences. Most used qualitative research methods to ensure that the narratives of displacement and belonging are as precise as possible. For historical and literary accounts, authors used literary texts and historical records (Chapter 2, 6, 11, and 18); for contemporary narratives, they used document analysis (Chapters 3 and 4), assuming the positions of participant-observers, analyzing conversations or comments and asking open-ended questions. As engaged readers, students, or scholars ourselves, we may gain important insights into the migrants' experiences from micro-level research, as well as meso-level analysis (of social media, Chapter 7), but our understanding of migrants' journey will be incomplete and partial unless we consider that not all migrants are willing or able to share their traumatic experiences. At least until very recently, researchers paid little attention to the way migrants communicated their trauma through art, whether storytelling, poetry, or through the visual arts (McKay and Bradley 2016; Lewis 2021). It is crucial to incorporate the arts for they "can highlight the particular and sensorial reality that is normally lost in our conceptualized understanding of migration" (Boon 2020:n.p.). In considering that not all migrants are able or willing to tell their stories, it is also very important to pay close attention to everyday artifacts used by migrants on their journey and left behind before crossing an international border, for they often tell of a more complicated journey than migrants are willing to disclose.

Some feminist scholars have explored the dilemma of conducting research with vulnerable populations when the researchers themselves "are implicated and participate in producing the politics and hierarchies of vulnerabilities" (Krystalli 2021:2, see also Cuomo and Massaro 2016; Smith 2016; Wolf 2021). Nevertheless, they conduct the research and seem to agree that in some cases, retelling one's experiences of trauma can be cathartic—especially when the researchers are sympathetic listeners and when they hope that the narrative can help bring change to policies and practices that made the journey so difficult and traumatic in the first place (Lakeman et al. 2013).

When research subjects refuse to participate in extractive research, their very refusal is important information that must be acknowledged in the research and reported in the results. Rarely, though, is this the case. Emily Mitchell-Eaton and Kate Coddington (Chapter 20) argue that taking refusal seriously affects the researcher or the research subjects as well as the research itself. Refusal, they argue, reframes the form and content of migration research and reveals not only the inherent power relations between researchers (mostly from the Global North) and research subjects (mostly from the Global South) but also larger "structures of domination." They explain that the refusal of trauma-ridden migrants to share their stories or to allow their narrative to become part of a coherent record has led them to see more clearly the relationship between migration, trauma, and empire. For the researcher, they argue, refusing to rely on existing migration theories, which are authored almost exclusively by white scholars from the north, opens new research avenues. Mitchell-Eaton and Coddington are now more open to theories devised by indigenous scholars, who see far more clearly the connections among trauma, empire, and migration; refusal, they argue, is "an ethical positioning," a "politics of questioning and resisting," and not only a research practice (p. 319 in this volume).

But refusal is neither the only politics nor the only research tool through which to understand migrants' trauma without explicitly asking them to recount their traumatic experience. There are other important ways to add to the story, perhaps more authentic ways, to which we must pay attention.

In exploring the "archeology of the contemporary" and paying attention to "the material histories that surround us and demand to be heard" (Hamilakis 2018:519), researchers should also incorporate two other forms that also express the hardships of the journey: the first is artistic and the second is materialistic. The artistic forms the experiential encounters into words, colors, movement, and images and relies on migrants' creative expressions; the materialistic relies on everyday objects left by refugees and asylum seekers on land and sea. Through these expressions, the story of trauma, loss, devastation, and tragedy associated with migrants' journey must be told.

Backpacks, shoes, articles of clothing, empty water bottles, and other objects left behind before crossing the borders tell us volumes about the migrants, the journey, the hostile terrain, and the haste to avoid the securitized and well-surveilled borders. Floating life jackets and detritus from shipwrecks washed ashore in the Mediterranean tell us about the grim fate of many who sought safety and about their fear of border regimes and detention centers. While migrants who drowned or died along the way remain unnamed and therefore unmourned, the material objects speak for them and their story. The other artifacts that have now begun to be curated memorialize events that some would rather forget. Nevertheless, when doing research on migration, silent stories must be incorporated, for they too speak volumes about migrants' journeys.

During a time of increasingly fortified borders in the face of global humanitarian crises, migration scholars play a unique role in bringing to light the vulnerabilities faced by migrants. Targeted and vilified globally, displaced people are silenced by nativist discourses and practices that push for their detention, deportation, surveillance, and social exclusion (Bloemraad and Menjívar 2021). By centering the agency of migrants, this volume brings their voices forward. In doing so, we present a more accurate understanding of the causes and consequences of their displacement. Yet centering the voices of migrants means attending to all the ways that migrants narrate their journey. Refusal as a research practice attunes migration scholars to the deliberative silences of their subjects. Artistic work by migrants as well as the materials left behind also speak to their experiences. Abiding by the ethics of migration research demands serious attention to this new terrain of practices and tools available to migration scholars.

Notes

1 According to migration data, global migration has seen a significant decline since early 2020 and the flows to OECD countries—measured by new permits issued—have been estimated to have fallen by 46% in the first semester of 2020, and 2020 is expected to be a historical low for migration to OECD countries (OECD 2020). Only five years earlier, in 2015, global migration involved 244 million, an increase of 30% since 2000 (United Nations 2016).

2 Zetter (1991); Malkki (1995, 1996); Soguk (1999); Abu-Lughod (2002); Turton (2005); Bohmer and Shuman (2007); Marfleet (2007); Fassin (2008, 2012); Jacquemet (2009); Fassin and Rechtman (2009); Betts (2010); Hammami (2010); Fiddian-Qasmiyeh (2014); and Cabot (2016).

3 Millions of people who have been severely affected by climate change would like to remain in place, but these aspirations are not fully addresses in migration research and climate adaptation policies (Mallick and Schanze 2020). Further, there is no legal category yet of climate refugees (Nabenyo 2020), though given their numbers, there perhaps should be.

4 The Armenian genocide, the systematic killing of Armenians, took place by the Ottomans during World War I, and predated the establishment of the Republic of Turkey in 1923.

5 According to the UN Population Division (UNDP 2020), female migrants during that year constitutes about 48.1% of the global international migrant stock. Their share among the international migrants has declined from 49.4% in 2000 to 48.1%.

6 According to the British Broadcasting Company (BBC 2021), there are 69 countries that have laws criminalizing homosexuality, and nearly half of these are in Africa. Even though some countries are in the process of decriminalizing same-sex relationships, the laws of these countries have not yet changed.

7 Asylum procedures within the European Union vary, in both substance and outcomes, from one member state to another.

References

Abrego, Leisy J. (2014). *Sacrificing Families: Navigating Laws, Labor, and Love across Borders*. Redwood City, CA: Stanford University Press.

Abu-Lughod, Lila (2002). "Do Muslim Women Really Need Saving? Anthropological Reflections on Cultural Relativism and Its Others." *American Anthropologist* 104(3): 783–90. doi: 10.1525/aa.2002.104.3.783.

Aciman, André (2021). *Homo Irrealis: Essays*. New York: Farrar, Straus and Giroux.

Adamson, Fiona B., Triadafilos Triadafilopoulos, and Aristide R. Zolberg (2011). "The Limits of the Liberal State: Migration, Identity and Belonging in Europe." *Journal of Ethnic and Migration Studies* 37(6): 843–59. doi: 10.1080/1369183X.2011.576188.

Adamson, Fiona B., and Gerasimos Tsourapas (2020). "The Migration State in the Global South: Nationalizing, Developmental, and Neoliberal Models of

Migration Management." *International Migration Review* 54(3): 853–82. doi: 10.1177/0197918319879057.

Adepoju, Aderanti, Femke Van Noorloos, and Annelies Zoomers (2010). "Europe's Migration Agreements with Migrant-sending Countries in the Global South: A Critical Review." *International Migration* 48(3): 42–75.

Afsar, Rita (2011). "Contextualizing Gender and Migration in South Asia: Critical Analysis." *Gender, Technology and Development* 15(3): 389–410.

Ahmed, Sara (1999). "Home and Away: Narratives of migration and estrangement." *International Journal of Cultural Studies* 2(3): 329–47.

Ahmed, Sara, Claudia Casteñeda, Anne-Marie Fortier, and Mimi Sheller (2020). "Introduction: Uprootings/Regroupings: Questions of Home and Migration," 1–19. In *Uprootings/Regroupings: Questions of Home and Migration*. Ahmed, Sara, Claudia Casteñeda, Anne-Marie Fortier, and Mimi Sheller (eds.). New York: Routledge.

Amelina, Anna, Jana Schäfer, and Miriam Friz Trzeciak (2021). "Classificatory Struggles Revisited: Theorizing Current Conflicts over Migration, Belonging and Membership." *Journal of Immigrant & Refugee Studies* 19(1): 1–8. doi: 10.1080/15562948.2020.1854918.

Ankori, Gannit (2020). "'Dis-Oientalisms': Displaced Bodies/Embodied Displacements in Contemporary Palestinian Art," 59–90. In *Uprootings/Regroupings: Questions of Home and Migration*. Ahmed, Sara, Claudia Casteñeda, Anne-Marie Fortier, and Mimi Sheller (eds.). New York: Routledge.

Anthias, Floya (2008). "Thinking through the Lens of Translocational Positionality: An Intersectionality Frame for Understanding Identity and Belonging." *Translocations: Migration and Social Change* 4(1): 5–20.

Anthias, Floya (2016). "Interconnecting Boundaries of Identity and Belonging and Hierarchy-making within Transnational Mobility Studies: Framing Inequalities." *Current Sociology* 64(2): 172–90. doi: 10.1177/0011392115614780.

Anthias, Floya, and Gabriella Lazaridis (2020). *Gender and Migration in Southern Europe: Women on the Move*. New York: Routledge.

Awad, Isabel (2012). "Desperately Constructing Ethnic Audiences: Anti-immigration Discourses and Minority Audience Research in the Netherlands." *European Journal of Communication* 28(2): 168–82.

Barnett, Jon, and W. Neil Adger (2018). "Mobile Worlds: Choice at the Intersection of Demographic and Environmental Change." *Annual Review of Environment and Resources* 43: 245. doi: 10.1146/annurev-environ-102016-060952.

Bartley, Allen, and Paul Spoonley (2008). "Intergenerational Transnationalism: 1.5 Generation Asian Migrants in New Zealand." *International Migration*. doi:10.1111/j.1468-2435.2008.00472.x.

Bartram, David (2015). "Forced Migration and 'Rejected Alternatives': A Conceptual Refinement." *Journal of Immigrant & Refugee Studies* 13(4): 439–56. doi: 10.1080/15562948.2015.1030489.

Bastia, Tanja (2014). "Intersectionality, Migration and Development." *Progress in Development Studies* 14(3): 237–48. http://dx.doi.org.ezproxy.middlebury. edu/10.1177/1464993414521330

BBC (2021). "Homosexuality: The Countries Where It Is illegal to Be Gay." May 12. https://www.bbc.com/news/world-43822234

Bean, Frank D., Susan K. Brown, and James D. Bachmeier (2015). *Parents without Papers: The Progress and Pitfalls of Mexican American Integration*. 1st edition. New York: Russell Sage Foundation.

Benvenisti, Meron (2002). *Sacred Landscape: The Buried History of the Holy Land Since 1948*. Berkeley: University of California Press.

Betts, Alexander (2010). "Forced Migration Studies: 'Who Are We and Where Are We Going?' Report on IASFM 12, Nicosia, Cyprus, June 28–July 2, 2009." *Journal of Refugee Studies* 23(2): 260–69.

Betts, Alexander (2013). *Survival Migration: Failed Governance and the Crisis of Displacement*. Ithaca, NY: Cornell University Press.

Blinder, Scott, and William Allen (2015). Constructing Immigrants: Portrayals of Migrant Groups in British National Newspapers, 2010–2012. *International Migration Review*, 49(3): 1–38.

Bloemraad, Irene (2006). "Becoming a Citizen in the United States and Canada: Structured Mobilization and Immigrant Political Incorporation." *Social Forces* 85(2): 667–95.

Bloemraad, Irene, and Cecilia Menjívar (2021). "Precarious Times, Professional Tensions: The Ethics of Migration Research and the Drive for Scientific Accountability." *International Migration Review*. doi: 10.1177/01979183211014455.

Blunt, Alison, and Robin Dowling (2006). *Home*. New York: Routledge.

Bohmer, Carol, and Amy Shuman (2007). *Rejecting Refugees*. New York: Routledge.

Boon, Errol (2020). *Art and Migration: On Cultural Internationalization in the Age of Displacement*. October 9. Dutch Culture. dutchculture.nl/en/news/art-and-migration.

Borjas, George J. (1989). "Economic Theory and International Migration." *International Migration Review* 23(3):457–85. doi: 10.1177/019791838902300304.

Brettell, Caroline, and James Frank Hollifield (2014). *Migration Theory: Talking across Disciplines*. Routledge: New York.

Cabot, Heath (2016). "'Refugee Voices': Tragedy, Ghosts, and the Anthropology of Not Knowing. *Journal of Contemporary Ethnography* 45(6): 645–72. doi: 10.1177/0891241615625567.

Carrillo, Héctor (2018). *Pathways of Desire: The Sexual Migration of Mexican Gay Men*. Chicago: University of Chicago Press.

Castles, Stephen, and Alastair Davidson (2000). *Citizenship and Migration: Globalization and the Politics of Belonging*. New York: Routledge.

Cattaneo, Cristina, Michel Beine, Christiane J. Fröhlich, Dominic Kniveton, Inmaculada Martinez-Zarzoso, Marina Mastrorillo, Katrin Millock, Etienne Piguet, and Benjamin Schraven (2019). "Human Migration in the Era of Climate Change." *Review of Environmental Economics & Policy* 13(2):189–206. doi: 10.1093/reep/rez008.

Cerna, Lucie (2014). "Attracting High-skilled Immigrants: Policies in Comparative Perspective." *International Migration* 52(3): 69–84.

Chávez, Karma (2010). "Border (In)security: Normative and Differential Belonging in LGBTQ and Immigrant Rights Discourse." *Communication and Critical/Cultural Studies* 7(2): 136–55.

Coleman, James S. (1988). "Social Capital in the Creation of Human Capital." *American Journal of Sociology* 94(s1): S95. doi: 10.1086/228943.

Collins, Jock (2007). "Immigrants as Victims of Crime and Criminal Justice Discourse in Australia." *International Review of Victimology* 14(1): 55–79.

Collinson, Sarah (1993). *Beyond Borders: West European Migration Policy towards the 21st Century*. London: Royal Institute of International Affairs.

Crawley, Heaven, and Dimitris Skleparis (2018). "Refugees, Migrants, Neither, Both: Categorical Fetishism and the Politics of Bounding in Europe's 'Migration Crisis'" *Journal of Ethnic and Migration Studies* 44(1):48–64. doi:10.1080/1369183X.2017.1348224.

Cuomo, Dana, and Vanessa A. Massaro (2016). "Boundary-making in Feminist Research: New Methodologies for 'Intimate Insiders'." *Gender, Place & Culture* 23(1): 94–106. doi: 10.1080/0966369X.2014.939157.

Dago, Franck, and Simon Barussaud (2021). "Push/Pull Factors, Networks and Student Migration from Côte d'Ivoire to France and Switzerland." *Social Inclusion* 9(1): 308–16.

Datta, Pranati (2004). "Push-pull Factors of Undocumented Migration from Bangladesh to West Bengal: A Perception Study." *The Qualitative Report* 9(2): 335–58.

Davis, Mike (2004). "Planet of Slums." *New Left Review: London* 26: 5–34.

Deshingkar, Priya, C. R. Abrar, Mirza Taslima Sultana, Kazi Nurmohammad Hossainul Haque, and Md Selim Reza (2019). "Producing Ideal Bangladeshi Migrants for Precarious Construction Work in Qatar." *Journal of Ethnic and Migration Studies* 45(14): 2723–38. doi: 10.1080/1369183X.2018.1528104.

Dorigo, Guido, and Waldo Tobler (1983). "Push-pull Migration Laws." *Annals of the Association of American Geographers* 73(1): 1–17.

Dossa, Parin, and Jelena Golubovic (2019). "Reimagining Home in the Wake of Displacement." *Studies in Social Justice* 13(1): 171–86.

Ewers, Michael C., Justin Gengler, and Bethany Shockley (2021). "Bargaining Power: A Framework for Understanding Varieties of Migration Experience." *International Migration Review* 01979183211007076. doi: 10.1177/01979183211007076.

Faist, Thomas (2000). "Transnationalization in International Migration: Implications for the Study of Citizenship and Culture." *Ethnic and Racial Studies* 23(2): 189–222. doi: 10.1080/014198700329024.

Fassin, Didier (2008). "The Humanitarian Politics of Testimony: Subjectification through Trauma in the Israeli–Palestinian Conflict." *Cultural Anthropology* 23(3): 531–58.

Fassin, Didier (2012). *Humanitarian Reason: A Moral History of the Present Times.* Berkeley: University of California Press.

Fassin, Didier, and Richard Rechtman (2009). *The Empire of Trauma: An Inquiry into the Condition of Victimhood.* Princeton, NJ: Princeton University Press.

Feldman, Ilana (2006). Home as a Refrain: Remembering and Living Displacement in Gaza. *History and Memory* 18(2): 10–47.

Fiddian-Qasmiyeh, Elena (2014). *The Ideal Refugees: Gender, Islam, and the Sahrawi Politics of Survival.* 1st edition. Syracuse, NY: Syracuse University Press.

Filindra, Alexandra, and Melinda Kovács (2012). "Analysing US State Legislative Resolutions on Immigrants and Immigration: The Role of Immigration Federalism." *International Migration* 50(4): 33–50.

FitzGerald, David (2014). "The Sociology of International Migration." In *Migration Theory: Talking Across Disciplines.* Caroline Brettell, James Frank Hollifield, and James F. Hollifield (eds.). New York: Routledge.

FitzGerald, David Scott, and Rawan Arar. 2018. "The Sociology of Refugee Migration." *Annual Review of Sociology* 44(1): 387–406. doi: 10.1146/annurev-soc-073117-041204.

Gartner, Johannes (2015). *(In)credibly Queer: Sexuality-based Asylum in the European Union.* Humanity in Action, Deutschland. https://www.humanityinaction.org/person/johannes-lukas-gartner/

Grosfoguel, Ramon, Laura Oso, and Anastasia Christou (2015). "'Racism,' Intersectionality and Migration Studies: Framing Some Theoretical Reflections." *Identities* 22(6): 635–52.

Hajdukowski-Ahmed, Maroussia, Nazilla Khanlou, and Helene Moussa (2008). *Not Born a Refugee Woman: Contesting Identities, Rethinking Practices*. New York: Berghahn Books.

Hamilakis, Yannis (2016). "Archaeologies of Forces and Undocumented Migration." *Journal of Contemporary Archaeology* 3(2): 121–39.

Hamilakis, Yannis (2018). "Decolonial Archaeology as Social Justice." *Antiquity* 92: 518–20.

Hamlin, Rebecca (2021). *Crossing: How We Label and React to People on the Move*. Redwood City, CA: Stanford University Press.

Hammami, Rema (2010). "Qalandiya: Jerusalem's Tora Bora and the Frontiers of Global Inequality." *Jerusalem Quarterly* 41(Spring).

Hernández García, César Cuauhtémoc (2019). *Migrating to Prison: America's Obsession with Locking Up Immigrants*. New York: The New Press.

Innes, Alexandria (2016). "In Search of Security: Migrants Agency, Narrative, and Performativity." *Geopolitics* 21(2): 263–83.

IOM (International Organization for Migration) (2020). *World Migration Report 2020*. New York: United Nations.

IOM (International Organization of Migration) (2021). *Gender and Migration*. Migration Data Portal. March 11, 2021.

ISSOP (2017). "Child: Care, Health and Development (Position Statement on Migrant Child Health)." ISSOP Migration Working Group, first published July 23, 2017. doi: 10.1111/cch.12485.

Jacquemet, Marco (2009). "Transcribing Refugees: The Entextualization of Asylum Seekers' Hearings in a Transidiomatic Environment." *Text & Talk—An Interdisciplinary Journal of Language, Discourse & Communication Studies* 29(5): 525–46.

Jansen, Sabine (2013). "Introduction: Fleeing Homophobia, Asylum Claims Related to Sexual Orientation and Gender Identity in Europe." 1–31. In *Fleeing Homophobia*. Thomas Spijkerboer (ed.). London: Routledge.

Jenkins, J. Craig (1977). "Push/Pull in Recent Mexican Migration to the U.S." *International Migration Review* 11(2): 149–77. doi: 10.1177/019791837701100202.

Kaye, Ronald (2001). "'Blaming the Victim'—An Analysis of Press Representation of Refugees and Asylum-seekers in the United Kingdom in the 1990s." 53–70. In *Media and Migration: Constructions of Mobility and Difference*. R. King and N. Wood (eds.). London: Routledge.

Kia-Keating, Maryam, and B. Heisi Ellis (2007). "Belonging and Connection to School Resettlement: Young Refugees, School Belonging, and Psychological Adjustment." *Clinial Child Psychology and Psychiatry* 21(1): 29–43.

Klinenberg, Eric, Malcolm Araos, and Liz Koslov (2020). "Sociology and the Climate Crisis." *Annual Review of Sociology* 46(1): 649–69. doi: 10.1146/annurev-soc-121919-054750.

Krystalli, Roxani (2021). "Narrating Victimhood: Dilemmas and (In)dignities." *International Feminist Journal of Politics* 23(1): 125–46. doi: 10.1080/14616742.2020.1861961.

Krzyżanowski, Michal, Anna Triandafyllidou, and Ruth Wodak (2018). "The Mediatization and the Politicization of the 'Refugee Crisis' in Europe." *Journal of Immigrant & Refugee Studies* 16(1–2):1–14. doi: 10.1080/15562948.2017.1353189.

Lakeman, Richard, Sue McAndrew, Liam MacGabhann, and Tony Warne (2013). "'That Was Helpful … No One Has Talked to Me About That Before': Research Participation As a Therapeutic Activity." *International Journal of Mental Health Nursing* 22(1): 76–84.

Landau, Loren B., and Marguerite Duponchel (2011). "Laws, Policies, or Social Position? Capabilities and the Determinants of Effective Protection in Four African Cities." *Journal of Refugee Studies* 24(1): 1–22.

Lee, Everett (1966). "A Theory of Migration." *Demography* 3(1): 47–57.

Lee, Jane Yeonjae, and Minjin Kim (eds.) (2020). *The 1.5 Generation Korean Diaspora: A Comparative Understanding of Identity, Culture, and Transnationalis*m. Lanham, MD: Lexington Books/Rowman & Littlefield.

Levitt, Peggy (2001). *The Transnational Villagers*. Berkeley: University of California Press.

Lewis, Rachael (2021). "Precarious Temporalities: Gender, Migration, and Refugee Art. Miyamoto." 133–52. In *Art and Migration: Revisioning the Borders of Community*. Bénédicte and Marie Ruiz (ed.). Manchester, UK: Manchester University Press.

Lewis, W. Arthur (1954). "Economic Development with Unlimited Supplies of Labour." *Manchester School* 22(2): 139–91.

Malkki, Liisa H. (1995). *Purity and Exile: Violence, Memory, and National Cosmology Among Hutu Refugees in Tanzania*. Chicago: University of Chicago Press.

Malkki, Liisa H. (1996). "Speechless Emissaries: Refugees, Humanitarianism, and Dehistoricization." *Cultural Anthropology* 11(3): 377.

Mallick, Bishawjit, and Jochen Schanze (2020). "Trapped or Voluntary? Non-migration Despite Climate Risks." *Sustainability* 12(11): 4718. doi: 10.3390/su12114718.

Marfleet, Philip (2007). "Refugees and History: Why We Must Address the Past." *Refugee Survey Quarterly* 26(3): 136–48.

Massey, Douglas S. (1999). "Why Does Immigration Occur? A Theoretical Synthesis." 34–52. In *The Handbook of International Migration: The American Experience*. C. Hirschman, P. Kasinitz and J. DeWind (eds.). New York: Russell Sage Foundation.

Massey, Douglas S., Luin Goldring, and Jorge Durand (1994). "Continuities in Transnational Migration: An Analysis of Nineteen Mexican Communities." *American Journal of Sociology* 99(6): 1492–533. doi: 10.1086/230452.

Massey, Douglas S., and Karen A. Pren (2012). "Unintended Consequences of US Immigration Policy: Explaining the Post-1965 Surge from Latin America." *Population and Development Review* 38(1): 1–29.

McKay, Samuel, and Jessica Bradley (2016). "How Does Arts Practice Engage with Narratives of Migration from Refugees? Lessons from 'Utopia'" *Journal of Arts and Communities* 8, (1–2): 31–46. doi: 10.1386/jaac.8.1-2.31_1.

Menjívar, Cecilia (2000). *Fragmented Ties: Salvadoran Immigrant Networks in America*. Berkeley: University of California Press.

Miyamoto, Bénédicte, and Marie Ruiz (2021). *Art and Migration: Revisioning the Borders of Community*. Manchester, UK: Manchester University Press.

Nabenyo, Ekai (2020). "Climate-induced Involuntary Migration: Nomadic-Pastoralists' Search for Elusive Pastures in Kenya." *Forced Migration Review* 64(43). https://www.fmreview.org/issue64/nabenyo

Natter, Katharina (2018). "Rethinking Immigration Policy Theory Beyond 'Western Liberal Democracies'." *Comparative Migration Studies* 6(1): 4. doi: 10.1186/s40878-018-0071-9.

OECD Library (2020). *International Migration Outlook*. https://www.oecd-ilibrary.org/social-issues-migration-health/international-migration-outlook-2020_ec98f531-en

Ossman, Susan (2013). *Moving Matters: Paths of Serial Migration*. Redwood City, CA: Stanford University Press.

Ottonelli, Valeria, and Tiziana Torresi (2013). "When Is Migration Voluntary?" *International Migration Review* 47(4): 783–813.

Paret, Marcel, and Shannon Gleeson (2016). "Precarity and Agency Through a Migration Lens." *Citizenship Studies* 20(3/4): 277–94. doi: 10.1080/13621025.2016.1158356.

Perez Murcia, Luis Eduaerdo (2019). "'The Sweet Memories of Home Have Gone:' Displaced People Searching for Home in a Liminal Space." *Journal of Ethnic and Migration Studies* 45(9): 1515–31.

Piper, Nicola (ed.) (2008). *New Perspectives on Gender and Migration: Livelihood, Rights, and Entitlements.* London: Routledge.

Piper, Nicola, and Mina Roces (2003). "'Introduction: Marriage and Migration in an Age of Globalization.'" 1–22. In *Wife or Worker? Asian Women and Migration.* Nicola Piper and Mina Roces (eds.). Lanham, MD: Rowman & Littlefield.

Pisarevskaya, Asya, Nathan Levy, Peter Scholten, and Joost Jansen (2020). "Mapping Migration Studies: An Empirical Analysis of the Coming of Age of a Research Field." *Migration Studies* 8(3): 455–81. https://doi.org/10.1093/migration/mnz031

Pugh, Jeffrey D. (2018). "Negotiating Identity and Belonging through the Invisibility Bargain: Colombian Forced Migrants in Ecuador." *International Migration Review* 52(4): 978–1010. doi: 10.1111/imre.12344.

Sahraoui, Nina (2020). "Gendering the Care/Control Nexus of the Humanitarian Border: Women's Bodies and Gendered Control of Mobility in the European Borderlands." *Environment and Planning D: Society and Space* 38(5): 905–22.

Sherbinin, Alex de (2020). "Climate Impacts as Drivers of Migration." *Migrationpolicy. Org.* https://www.migrationpolicy.org/article/climate-impacts-drivers-migration

Shubin, Sergei, and Melinda Lemke (2020). "Children Displaced across Borders: Charting New Directions for Research from Interdisciplinary Perspectives." *Children's Geographies* 18(5): 505–15. doi: 10.1080/14733285.2020.1781061.

Sigona, Nando (2014). "The Politics of Refugee Voices." In *The Oxford Handbook of Refugee and Forced Migration Studies.* Elena Fiddian-Qasmiyah, Gil Loescher, Katy Long, and Nando Sigona (eds.). New York: Oxford University Press.

Silvey, Rachel (2006). "Geographies of Gender and Migration: Spatializing Social Difference." *International Migration Review* 40(1): 64–81.

Silvey, Rachel, and Rhacel Parreñas (2020). "Precarity Chains: Cycles of Domestic Worker Migration from Southeast Asia to the Middle East." *Journal of Ethnic and Migration Studies* 46(16): 3457–71. doi: 10.1080/1369183X.2019.1592398.

Simpson, Audra (2014). *Mohawk Interruptus: Political Life across the Borders of Settler States.* Durham, NC: Duke University Press.

Siu, Lok (2012). "Serial Migration: Stories of Home and Belonging in Diaspora." 143–72. In *New Routes for Diaspora Studies* (21st Century Studies). Sukanya Banarjee, Aims McGuinness, and Steve McKay (eds.). Bloomington: Indiana University Press.

Smith, Laura, Brieahn DeMeo and Sunny Widmann (2011). "Identity, Migration, and the Arts: Three Case Studies of Translocal Communities." *The Journal of Arts Management, Law, and Society* 41(3): 186–97. doi: 10.1080/10632921.2011.598418.

Smith, Sara (2016). "Intimacy and Angst in the Field." *Gender, Place & Culture* 23(1): 134–46. doi: 10.1080/0966369X.2014.958067.

Soguk, Nevzat (1999). *States and Strangers: Refugees and Displacements of Statecraft.* Minneapolis: University of Minnesota Press.

Soysal, Yasemin Nuhoğlu (2015). *Transnational Trajectories in East Asia: Nation, Citizenship, and Region.* Abingdon, Oxon: Routledge.

Stark, Oded, and David E. Bloom (1985). "The New Economics of Labor Migration." *The American Economic Review* 75(2): 173–78.

Stark, Oded and J. Edward Taylor (1991). "Migration Incentives, Migration Types: The Role of Relative Deprivation." *The Economic Journal* 101(408): 1163–78. doi: 10.2307/2234433.

Stasiulis, Daiva, Zaheera Jinnah, and Blair Rutherford (2020). "Migration, Intersectionality and Social Justice—Guest Editors' Introduction." *Studies in Social Justice* 14(1). doi: 10.26522/ssj.v2020i14.2445.

Taylor, J. Edward. 1987. "Undocumented Mexico-U.S. Migration and the Returns to Households in Rural Mexico." *American Journal of Agricultural Economics* 69(3): 626–38. doi: 10.2307/1241697.

Taylor, Steve (2015). "'Home Is Never Fully Achieved … Even When We Are In It': Migration, Belonging and Social Exclusion within Punjabi Transnational Mobility." *Mobilities* 10(2): 193–210. doi: 10.1080/17450101.2013.848606.

Todaro, Michael P., and Lydia Maruszko (1987). "Illegal Migration and US Immigration Reform: A Conceptual Framework." *Population and Development Review* 13(1): 101–14. doi: 10.2307/1972122.

Turton, David (2005). *The Meaning of Place in a World of Movement: Lessons from Long-term Field Research in Southern Ethiopia.* Vol. 18. Oxford: Oxford University Press.

UNHCR (2020). *Forced Displacement in 2019.* https://www.unhcr.org/globaltrends 2019/

UNPD (United Nations Population Division) (2020). *International Migrant Stock.* https://www.un.org/development/desa/pd/content/international-migrant-stock

United Nations (2016). *International Migration Report, 2015.* New York: United Nations.

Wolf, Soja (2021). "Talking to Migrants: Invisibility, Vulnerability, and Protection." *Geopolitics*, 26(1): 193–214 doi: 10.1080/14650045.2020.1764540.

Yeoh, Brenda S. A. (2017). "Transient Migrations: Intersectionalities, Mobilities and Temporalities." *Transitions: Journal of Transient Migration* 1(1): 143–47. doi: 10.1386/tjtm.1.1.143_7.

Yuval-Davis, Nira (2006). "Belonging and the Politics of Belonging." *Patterns of Prejudice* 40(3): 197–214. doi: 10.1080/00313220600769331.

Zetter, Roger (1991). "Labelling Refugees: Forming and Transforming a Bureaucratic Identity." *Journal of Refugee Studies* 4(1): 39–62.

Zimmermann, Klaus (1996). "European Migration: Push and Pull." *International Regional Science Review* 19(1–2): 95–128. doi:10.1177/016001769601900211.

Part I
Regimes of belonging

2 Out of place in antiquity

Elena Isayev

Introduction

What compelled the Athenian tragedian Aeschylus, some 2,500 years ago, to stage a discourse on democracy—at the very moment of its instigation—through the predicament of a group of young women seeking asylum at the gates of mythical Argos? And what made captives, some thousand years later, a target of energies for ambitious bishops such as Caesarius of Arles, who sought them out for ransom? Underpinning these questions are historical settings, which expose the way that value was gained from people's condition of being out of place.[1] What it means to be out of place when it comes to rights, belonging, and protection depends on the temporal and sociopolitical context in which the condition exists, as does the role it takes on. Its meaning also depends on whether the out-of-placeness, or de-placement,[2] is understood from the perspective of the state and authorities, which are primarily locationally tethered, or from that of the people who experience it firsthand, often through a combination of forced uprooting and compromised mobility. Drawing on scenarios from two different points in antiquity—Classical Greece and the Late Antique period—these perspectives will frame an investigation into what being out of place meant for those not formally enslaved prior to the advent of the nation-state, which ushered in its own particular construction of space and migration.

In the surviving ancient writings of the Greco-Roman world, there is little interest in human mobility as a topic in itself, nor is it articulated as a distinct entity separate from the practices of the everyday. *Migration* as a general phenomenon does not appear as a matter of concern, either in terms of security or for the purposes of management and control (Isayev 2017a: ch. 1). Immigration has none of the connotations of its modern conception—as a move across a national border for the purpose of permanent residence—until it takes hold in the 1800s (Thompson 2003; Shumsky 2008: 195). Nothing survives within the ancient Mediterranean record that resembles immigration statistics, and little evidence exists for the presence of any state-boundary checkpoints where such data could have been collected. This is despite a sophisticated system of commercial treaties, taxation, and trade duties, which required monitoring and reporting (Moatti 2000; 2004; Moatti and Kaiser

DOI: 10.4324/9781003170686-3

2007; Moatti et al. 2009). Although rates of mobility are difficult to capture, what evidence there is suggests they were relatively high. The figures for ancient Italy for example, drawn from census and other data for the last two centuries BCE, indicate that the total number of individual migratory movements could have been 40 million.[3] Estimates for the number of male Romans over 45 born in a location different from their place of residence indicate that about 30% of them lived outside of their place of birth.[4] Additional forms of evidence, including in the material record, suggest that at any one time possibly a third of a community's population consisted of outsiders.[5] Even the seemingly timeless sedentary peasant has been exposed as a figure of myth-making.[6] All this is not to say that there was not a persistent fear of conquest, colonial enterprise, expulsion, displacement, or an interest in the outsider.

Human mobility was inevitably positioned differently in a world where state territory was not circumscribed and boundaries were differently conceived, lacking the kind of physical presence that could be traced on the Euclidean isotropic space of the modern map. There was a distinction made between accessing the land and accessing the membership privileges of the community occupying that land, hence the possibility of gaining the status of a *metic* (resident alien)—who had certain privileges and duties but without citizenship in fifth-century BCE Athens (Kasimis 2013, 2018). Centuries later, the Roman statesman Cicero, in his *De Officiis* (3.11.47), expressed the distinction more explicitly, by stating that while "it is right not to permit the rights of citizenship to one who is not a citizen ... to debar foreigners from using the city is clearly inhuman."[7] Both actions he refers to were the result of political infighting, not of any xenophobic or anti-migrant policies. Such a distinction would become difficult to articulate with the advent of the nation-state, once territory and membership overlapped. The resulting "trinity of state-people-territory," as Hannah Arendt (1968: 358) referred to it, created a particular form of statelessness, in which the dimension of physical placedness became part of the difficulty in accessing human rights (Gundogdu 2015: 2–5). It also created a sense of attachment to the land as elemental and migration as exceptional. Studies into ancient mobility challenge prevailing conceptions of a natural tie to the land and a demographically settled world, showing that much human mobility was ongoing and cyclical.[8] Furthermore, the places of departure and arrival, as well as access to them, were differently configured. For those who were forced to abandon their homes either through exile, expulsion, flight, or capture, the issue was more about the inability to return, rather than to find another place of settlement. It is what we today refer to as displacement. Yet ancient governing authorities were less concerned about how to keep civilian foreigners out than how to control their own subjects, how to keep them from moving for long enough to tax them, and recruit them into the army. This chapter, framed by the two questions posed at the very start, explores the ancient meaning of place and attitudes to those on the move, in order to expose how people, who are transformed into bodies out of place, become valuable in negotiations of power and identity formation.

Shaping place through asylum in the ancient Greek polis

The starting point here is the relation of people to place, whether in terms of ties of home and belonging, or a civic membership with its community obligations and reciprocal protection—if not precisely "rights" as understood today. Specifically, the focus is on the intrinsic value held by people who are denied access and ability to enact such relations to place as a result of being restricted from it by those in power. In different guises over the *longue durée*, the forms of that value are affected by the demarcation of place—or the entity one is restricted from accessing—which is itself unstable. This is the case even in contexts such as the Greek polis (city-state), where sociopolitical parameters seem clear and appear similar to the twenty-first-century system of sovereign states with their associated citizenships (which they are not). As we will see, the situation in Late Antiquity has no semblance of such a framework. The Greek polis in the fifth century BCE is presented as the defining unit of allegiance, citizenship, and politics for free-born men. Yet the wandering Cynic philosopher Diogenes, exiled from his home polis Sinope (Desmond 2020), could subvert this through appeals to cosmopolitanism and the denial of his allegiance to any single state. None deserved his exclusive devotion, not even Athens, where he took refuge and which was a crucible for his philosophical ideology that led to the setting up of the Cynic and Stoic Schools.[9] Taking advantage of his out-of-placeness, he made it into a value that allowed for a critical perspective on the world and his own repositioning within it.[10]

For Diogenes as for many ancient exiles, the focus of emotional energies in displacement was on the inability to return to one's place of home (Isayev 2021). It was less challenging to find somewhere that would give refuge, as distinct from acquiring a status elsewhere that would be equivalent to one initially held in the home polis—for example, that of a citizen or *metic*. We have no recorded circumstances, either in myth or historical writings, which indicate a refusal of asylum to civilians on the sole basis of them being foreign. Technically, entry into the land under the jurisdiction of another community was not sought after, but rather subsistence, patronage, and protection in the broadest sense. The possibilities for finding alternative places to exist, as Arendt observed, were more widespread in the ancient world than in the territorially constrained globe of abutting nation-states.[11] Still, there were circumstances that might make refuge difficult to find. For those who committed blood crimes, the threat of pollution for the hosts—which could bring divine vengeance—made them reluctant to give asylum. For people escaping from conflict or oppression, gaining refuge could become difficult if those responsible for causing their flight or exile threatened potential hosts with violence or coercion. From the recorded cases of such instances, as in the predicament of those from Plataea (an ancient Greek city in southeastern Boeotia), or those explored in the tragedies below, this seems to have been the key reason for rejecting asylum requests.

Where and by whom is the decision made, whether to grant or reject the pleas of asylum seekers? The difficulty of responding to this question of who is the host—or who within a given society holds the power to maintain, permit, or inhibit access to place, of rights and protection—exposes the ambiguity in demarcating ancient place and its threshold. In the face-to-face society of elite-warriors of Homer's epics, written in the eighth century BCE prior to the emergence of the polis, an individual could request asylum and hospitality at the threshold of the head of household—who alone made the decision whether to grant it. Several hundred years later, in the fifth-century world of city-states with democratic institutions, the responsibilities and obligations became less clear, at times resulting in tensions between the *demos* (people) and their leaders. As appeals were now made to the community as a whole, the position of the host became ambiguous, and the responsibility to provide hospitality more diffuse. The situation was further complicated by groups who made appeals not at domestic thresholds but at other liminal places, such as altars and sanctuaries, which offered temporary refuge under divine protection (Herodotus, *Histories* 1.159.3; Sinn 1993; Chaniotis 1996: 69; Isayev 2017b: 81–82). The possibility of severe punishment by the most supreme gods—as Zeus with attributes of either Xenios (protector of guests), or Hikesios (protector of suppliants)—put fear into those who might transgress the rules of hospitality.[12] Through the presence of such divine designations, we can gauge the persistence of people requesting hospitality and the obligation to accommodate them. Such encounters underpin the narratives of the earliest Greek epics. Homer's world of the *Odyssey* is constructed through its protagonist's experience as a guest and suppliant among the inhabitants dwelling on the shores of the Mediterranean. During his travels, his hosts are most keen to know whether those he met were kind or hostile to strangers (e.g., Homer, *Odyssey*, Book 7). It mattered, since the actions and decisions within host-guest encounters determined (and perhaps still determine) the positioning of a society on the spectrum of just, civilized, or barbarian. For Derrida ([1997] 2000), this was the essence of culture. In the ancient Mediterranean, hospitality was a sought-after badge that ancient *poleis* strove to gain and hold onto, irrespective of their actual policies and attitudes toward the outsider seeking refuge.

The tragedies of power and hospitality

The out-of-place condition allowing opportunities for hospitality provides the grounds for a discourse on the meaning of place, sovereignty, and its custodians. Ancient writers recognized the value of hospitality as a vehicle for articulating community and autonomy, especially during transformative moments such as the emergence of the Greek polis and its city-state citizenship in the fifth century BCE. The encounters and negotiations that people out of place engendered lay at the boundary of civic society and the international community, in the space between citizen rights and universal rights. It is precisely this space that was of interest to such Greek tragedians as Aeschylus, whose *Suppliants* was performed in the 460s BCE—the formative years of Classical Athenian democracy. The play reveals both the agency of those who are out of place and the value that others draw from their predicament when it comes to formulating power and identity.[13] Set in the mythical past of the Bronze Age (c. 3000–1000 BCE), the *Suppliants* tells the story of the 50 daughters of Danaeus (the Danaids), who fled Egypt to escape a forced marriage, seeking refuge in the land of the Argives. *Suppliants* opens with the women clinging to sacred altars on the shore. By taking sanctuary at public shrines, they have made themselves suppliants not of an individual (King Pelasgos) but of the Argive state—as the king's response to their supplication indicates (Aeschylus, *Suppliants*: lines 365–70).[14]

> You are not sitting at the hearth of my house.
> If the city as a whole is threatened with pollution,
> It must be the concern of the people as a whole to work out a cure.

But the Danaids, refusing to accept their predicament forcefully challenge this (Aeschylus, *Suppliants*: lines 370–75):

> You are the city, I tell you, you are the people!
> A head of state, not subject to judgement,
> You control the altar, the hearth of the city.

Through their challenge, we witness a clash between the outdated aristocratic frameworks of an oligarchic regime and the rise of the new democratic state that appeals to the will of the people (*demos*) (Zeitlin 1992; Cole 2004: 63; Bakewell 2013: 13, 30–32). There is a particular interest here in the tensions of balancing the obligation to provide hospitality and asylum, underwritten by the gods (as it is perhaps today by the UN), and the responsibilities of security and well-being owed to the citizens of the host community by their leaders. As such, it is also an expression of the ambivalent role of those in power as representatives of the common will.

The play proceeds with the king going between sanctuary and city, to consult his people, until finally they agree to take in the suppliants, despite the risk of ensuing war with the Egyptian suitors (Aeschylus, *Suppliants*: lines 930–52). This scenario is the product of the historical context of mid-fifth-century Athens. In this period, we can perceive an ideological move away from the Archaic oligarchic mindset of supra-state elite networks, toward a more exclusive, Classical democracy of Periclean Athens (from 461 BCE). It was a new setting that did not tolerate internal class divisions. With this change, one can witness a shift of the private ties of hospitality to the more public ones of asylum, which now required a *proxenos*—a sponsor or intermediary (Walbank 1978: 2–3; Bakewell 2013: 30–31; Garland 2014: 13). The new location of appeals from the distance of public shrines created different conditions to those of the private threshold and hearth, where Odysseus performed his supplications. Through the interchange between the Danaids and Pelasgos, Aeschylus showcases how the private, exclusive guest-friendship, which underpinned inter-elite ties and authority, no longer had a place in the new polis society.

People who are out of place, in a juxtaposition with those who are "placed," provide the privileged view of the outsider that affects "insider" communities' self-definition and articulation of boundaries.[15] They also hold the power to influence how such communities are perceived globally, as the treatment of people seeking refuge becomes a gauge for levels of "civilization" or humanity. In this sense, people under such circumstances have "contingent agency," a term referring to an unforeseen mode of agency, with latent qualities that are conditionally activated at the moment of displacement (Isayev 2017b: 88–92). A state's reputation can be created or destroyed depending on its response to appeals for asylum, which becomes a tool for glorification by friends or vilification by enemies. In their own myths, the Athenians prided themselves for not giving in to external pressure to give up their asylum seekers or deny them shelter: this attitude seems to echo the spirit of today's UN non-refoulement clause (Article 33 of the UN 1951 Geneva Convention on Refugees). The tragedy of Euripides' *Children of Heracles*, performed circa 430 BCE, is in part about the community tensions that result from having to make decisions about how to respond to such appeals. Within this play, despite the threat of war, there is resistance to giving up the suppliants sheltering in the sanctuary of Zeus at Marathon, and ultimately they are given protection. It is not a simple resolution for the city's king, Demophon, who has to balance competing obligations between the will of the people, the gods, and the suppliants (Euripide, *Children of Heracles*: lines 413–22):

> Now you will see crowded assemblies being held, with some maintaining that it was right to protect strangers who are suppliants, while others accuse me of folly. In fact, if I do as I am bidden, civil war will break out.
> Therefore, consider these facts and join with me in discovering how you yourselves may be saved and this land as well, and how I may not be discredited in the eyes of the citizens.

The historical context of the play's performance, in the second year of the Peloponnesian War, is relevant here. It may be a comment on Athens' reassertion of sovereignty in the face of aggression from its rival Sparta, where the ability to grant asylum becomes a statement of state power and autonomy. Part of the ambition for any polis in the Greek Classical period was how to secure a position at the "civilized" or virtuous end of the spectrum. It is here that people who are out of place and needing asylum become valuable. By providing the opportunity for a polis to honor the duties of hospitality, it showed itself to be respecting the will of the gods and hence in their favor. Conversely, enemies could be charged with disregarding their responsibilities toward suppliants, as Athens and Sparta had done reciprocally prior to their clash in the Peloponnesian War in the last decades of the fifth century (Thucydides 1.126–28). Athens' self-presentation was of a polis open to refugees and outsiders, and a protector of the weak and destitute.[16] Athenians sought to maintain this image despite their imperial ambitions, which brought autonomous states to submission,[17] and despite their guarded exclusive citizenship and the distinctive myth of autochthony (the belief that the city's primordial inhabitants sprung from the land) (Purcell 1990; Horden and Purcell 2000: 384; Wilson 2006: 32). Athens' hospitable character was showcased in contrast to that of their Spartan enemies.[18] These were political claims. In real terms, there may have been little difference in how the two adversaries behaved. We know that because generations later, on the takeover of their homeland by Thebans, the Plataeans came to request asylum from Athens for the second time in 373 BCE. Their appeals were denied. As Isocrates records (*Plataicus* 14: 1–2, 53), one of their points of appeal was to stress that such a denial would risk harming Athens' long-standing reputation of hospitality. In this case, apparently, its reputation mattered less than the diplomatic and military necessities of the situation, which required that Athens reject the Plataean plea.

In the ancient context, the literature reveals that communities were keen to position themselves as being hospitable to strangers and open to those who came to seek asylum. This was an important measure of society, and it allowed states to present themselves as "civilized." By casting their enemies as doing the opposite, states could show their enemies to be barbaric and inhuman. Ancient authors also reveal the challenges of making such decisions. They operationalize people who are out of place: to provide an outsider as a contrast to the insider, and through the discourse of hospitality, to reflect on societal relations and the place of power in the moral dilemmas of leaders and communities. Proposing that in certain ancient-world contexts, the treatment of outsiders was a core measure of society does not mean that decisions to welcome and protect guests and suppliants were necessarily favorable. Rather, it shows that these considerations shaped the strategies for appeal as well as the deliberations of appropriate action. It also indicates that the refusal of hospitality is framed as the exception needing explaining, rather than its opposite. Cicero's powerful statement, that to deny outsiders access to the city is inhuman, echoes this sentiment. Such an outlook fits a society in which mobility was perceived not as exceptional but as an everyday norm.

Late antique place, status, mobility, and the captive body

Venturing forward some thousand years into Late Antiquity, being out of place took on its own quality.[19] As new status categories came to the fore, their relationship to territory and place was increasingly tenuous. Following the enfranchisement of all free members of the Roman Empire in 212 CE,[20] the significance of citizenship declined and the distinction between *cives* (full citizens) and *peregrini* (noncitizens) became less prominent.[21] To convey higher status, other divisions began to matter, in particular those between the elite *honestiores* and the humble *humiliores* (Jones 1963: 17; Garnsey 1970: 118; Mathisen 2006: 1013–15). A fiscal understanding of status fixated on people as resources, not least in their capacity as taxpayers. Hence, the place of one's *origo* (origin), rather than being designated by birth, was by then an administrative classification signifying where citizenship may be registered and enacted, and where taxes were paid (Thomas 1996: 187–90; Schmidt-Hofner 2017: 375). What mattered was a notion of belonging that was determined by one's function in the wider system of obligations and responsibilities. People had to have a reason for being in the place where they were, or conversely for being mobile, whether in their capacity as peasants, soldiers, merchants, clergy, or other professions. There was an interest in controlling those who did not have the right reasons, and new terms appeared, such as *transitor* or *vagus*, relating to vagrancy and loitering.[22] These terms come closest to today's colloquial use of *migrant* in the popular press,[23] as people began to be labeled negatively for their propensity to be mobile,[24] or more precisely, due to their being out-of-place of their designated system of obligations.[25]

The relational position of the body is key to understanding what displacement might signify in Late Antiquity, meaning that by being differently situated the nature of obligations and duties is either disrupted, transformed, or shifted to someone else. This manifests most extremely in contexts of captivity—whether through raids, kidnap, or conquest. Although physical removal into captivity could result in being transferred into territory under a different jurisdiction, what mattered was the loss of status, and thus removal from access to rights and protection, which were not territorially determined. For those who were in positions of authority or control, it provided opportunities to demonstrate their power potential by facilitating or denying the return of the body into place where rights and protection could be reinstated. Most directly, the captive body was of value for the captor in a monetary sense, through ransom or the sale of labor on the slave market—a practice attested throughout antiquity (Bielman 1994). For those who could be ransomed, unless the ransom was put up as a gift, it was considered a loan that needed repayment, the failure of which could lead to debt-servitude (Levy 1943: 160–63).[26] Furthermore, the institution of *postliminium*—the reinstatement of the status before capture—required that the debt was repaid.[27] Hence, despite redemption, the out-of-placeness could continue. For the redeemer, therefore, as for the captor, the most direct value was the worth contained in the laboring body of the *redemptus*—the ransomed person. More indirectly, captives were valuable in treaty negotiations and to exert political pressure: the more eminent or ransomable those captured, the more value they had.

The prominence of clauses concerning the ransom of captives in post-conflict exchanges is charted, for example, in Priscus's testimony of the delegation sent by emperor Theodosius II in 449 to negotiate terms with the Hun ruler Attila.[28] These records delineate how a person taken out of place, in relation to rights and protection, becomes dehumanized. Reduced to their body,[29] they are transformed into a bargaining tool, including by the ransomers themselves, who often act on behalf of others or the captives, without necessarily having any personal connection to those being ransomed. To explore the forms of value held in a body out of place, we will focus on this kind of ransomer—the intermediary—specifically in the figure of the Late

Antique bishop: a figure whose authority had steadily increased in this period, penetrating across the sociopolitical sector, and who sought out opportunities for enhancing jurisdiction and influence.

Bishops in Late Antiquity, being in positions of authority as well as having access to networks and resources, had the capacity to ransom captives. They did not only respond to requests for ransom but actively solicited this responsibility, so much so that it became an integral part of episcopal self-definition.[30] This phenomenon is of particular interest to us in understanding the value that captive out-of-place bodies had for the bishops. Here, we need to distinguish between the justifications that bishops gave to their communities and higher authorities in order to secure the funds necessary for ransom, and other personal motivations that lay behind such efforts. Funds were raised not only through donations but also by drawing on the church treasury, and more controversially through reappropriating *vasa ministerii*—sacred objects, along with other legacies and donations made by worshippers to ensure salvation, and were thus technically inalienable (Klingshirn 1985: 185).[31] Ransoming as an act of charity was not a given but had to be argued for to justify such use of church property, as Bishop Ambrose of Milan (340–97) had done when addressing his critics (Ambrose, *De Officiis*: 2.15.70).[32]

> Another great act of generosity is to ransom prisoners, to snatch people from the hands of the enemy, to deliver men from death and—especially—women from dishonour, to hand children back to their parents and parents back to their children, and to restore citizens to their own country.

The focus of redemption according to this reasoning is to restore the captured person to a place of protection and rights in the eyes of the community, thus rehumanizing the captive and giving their life value distinct from the worth held in their living body alone. It allowed for the public expression of the Christian ideal of *liberalitas* (generosity) and *caritas* (charity), and for bishops to enhance their reputation of generosity. But there were other enticements for investing energies into redeeming people. Captive persons' out-of-placeness afforded bishops lucrative possibilities to build their power base, extend their influence, and see off rivals, which made opportunities for ransom valuable enough to be sought out. We may gauge from the criticisms against bishops for such actions that they were widespread and also profitable.[33]

The great redeemer—Bishop Caesarius of Arles

By having access to ransomable captives, bishops could showcase their leadership and patronage networks to the community and the papal authority. This could be demonstrated through their ability to raise funds, gather local support, and successfully negotiate with the captors for a good price as well as with the higher powers of state and church to allow access to funds for the purpose. A bishop's jurisdiction could thus be augmented, and potentially extended territorially, if ways could be found to ransom those who were outside the lands of their diocese. The career of Bishop Caesarius of Arles (468/70–542), who was appointed the papal vicar of Gaul,[34] exemplifies the role of successful ransoming strategies, and how these can be positioned for maximum effect. Ongoing instability, which consequentially resulted in numerous people being captured, is what allowed him continuous access to captives.[35] His surviving sermons and the hagiography *Vita Caesarii*[36] situate his redemptions thoroughly within the sacred context, as in this example following the lifting of the siege of Arles in 508. It refers to the many captives taken by the Ostrogothic troops of King Theodoric—responsible for freeing the city—in the process of defeating the Franks and Burgundians (*Vita* I. 32):[37]

> In Arles, however, when the Goths had returned with an immense number of captives, the sacred basilicas were filled with a dense crowd of unbelievers (*infideles*), as was the bishop's residence. On those in great need the man of God bestowed a sufficient amount of food and clothing alike, until he could free them individually with the gift of redemption. When he had spent all the silver which his predecessor, the venerable Aeonius, had left for the maintenance of the church, he observed that the Lord had dipped bread into an earthen bowl and not a silver chalice, and had advised his disciples not to possess gold or silver.

> [...] The consecrated ornaments (*species*) of the church (*templum*) were sold for the redemption of the true church (*verum templum*). Even today the blows of the axes can be seen on the podiums and railings from which the silver ornaments [...] were cut away.

It would take some convincing to be thus empowered to strip the church so thoroughly and visibly.[38] Caesarius's efforts were not limited to helping Catholics but also Arians and Pagans (in hope that they too would become Catholics),[39] and his favors extended not only to citizens but also to enemies (Klingshirn 1985). Through such acts, he built his reputation for generosity, enhanced further by not requiring repayment for redemption, thus attracting even more captives to Arles seeking to be ransomed (*Vita* I. 44; 2.8, 23–24).[40] He did not stop there but also sought out captives from beyond the diocese of Arles, (*Vita* I. 38):[41]

> Meanwhile, in Italy, he discovered and redeemed all the captives he could from beyond the Durance (*de ultra Druentiam*), especially from the town of Orange (*Arausici oppidi*). This town had been completely enslaved, and he had already redeemed part of its inhabitants in Arles. Moreover, so that their liberty might be made more complete, he paid for horses and wagons for the journey, and by the relief and organization of his own, he arranged to return them to their homes.

These efforts were celebrated upon his return to Arles (*Vita* I. 43):[42]

> From here he returned home and entered the city of Arles. He was received with the singing of psalms, and having left as an exile, brought back with him from Italy, after he had redeemed the captives, 8,000 solidi.

Caesarius's actions, though carried out under the aegis of Christian virtues of generosity, hospitality, and love of enemies, did not go unnoticed for what else they achieved—the construction of a substantial power base. They contributed to disputes over jurisdiction with other equally ambitious bishops, especially Avitus the Bishop of Vienne, who was also a keen negotiator in ransoming transactions, as evident by his letters.[43] We know that Caesarius requested reaffirmation of boundaries between the diocese from Pope Symmachus in 513, who showed favor to the Bishop of Arles, fully aware that his efforts were used to further ambitions.[44] Through redemptive actions, Late Roman bishops such as Caesarius were able to present themselves as the protectors of the needy and patrons of their own communities, thus strengthening their ties of loyalty.

People who were out of place, and who could be ransomed, were valuable for the opportunity their condition enabled for bishops, among others, to conduct supreme acts of charity, and from which they substantially benefitted as a result. For the purpose of this chapter, I have considered the fate of the captives largely from the perspective of their captors or redeemers and said little about forms of agency, beyond that which has been termed "contingent." (Other modes are the focus of exploration in my forthcoming work.[45]) For those who were captured and who had the potential of being ransomed, their fate inevitably depended on their status and access to supporters and intermediaries. Although those in captivity are mainly discussed en masse and, except for certain elite figures, rarely do we get insight into their individual stories from the ancient record,[46] we can glean from wider reports, as those of Caesarius, just how diverse these groups were in terms of origin, status, religion, age, gender, and profession. What joined them was their suspended state due to their captivity and their removal, in a relational sense, from that place which had meaning in the ways discussed here. Their en-placement—whether to their original site of belonging or to a newly allocated one—depended not on permission from authorities of the state or its territorial jurisdiction, but on the authorities of the men who had the power to control and activate the laws that governed the relationships of obligation and responsibility. Among these, the bishops of Late Antiquity proved themselves keen operators.

Conclusion

By looking at two moments in history, some thousand years apart, this chapter has tried to unsettle our current understandings of migration, displacement, and belonging by highlighting the way that the categories used through time to articulate the meaning of communities, and individuals' position in relation to them, are unstable, as are the perceptions of those who move between. In focusing on the condition of being out of place, there has been an attempt here to capture the multiplicity of ways that place itself can be conceived, beyond its connection to a physically located site, but rather in its relational form.[47] Such an approach recognizes the fluidity of most communities, not only in the makeup of their populations but the way in which they choose to determine their boundaries and custodians. Encounters with people who are deemed on the outside, simultaneously materialize and destabilize such boundaries and reveal frictions in internal power dynamics. As we have seen, the position of people seeking asylum allowed the ancient Greek tragedians to use their predicament as an expression of the tensions within the nascent democracy between the *demos* and their leaders, as to who is the true representative of the people's will. Furthermore, people seeking asylum create a measure for society by providing the opportunity for acts of honor, heroism, and charity.

People's out-of-place condition takes on different forms of value in the Late Antique world, where state structures and people's relationship to them created diverse forms of obligation and responsibility. Groups of captives available for ransom provided one site of discourse of power, jurisdiction, and community formation. Successful redemption allowed for increased authority, influence, and moral superiority for the mediators. More tangibly, the utility of ransomable captives was economically contingent for both the captors and the redeemers. The value of captured people lies in their condition of being bodies out of place—physically and in terms of status. They reflect a phenomenon that appears to be shared across historic and societal contexts: the unique role of people in states of liminality in articulating inter- and intracommunity relationships. From a twenty-first-century perspective, two such contexts are particularly poignant. The first is the impasse of permanent temporariness, that is the condition of people seeking asylum who are in states of suspension, especially in refugee camps.[48] The second is the proliferation of captivity, for example in the United States, whose incarcerated community, one of the highest in the world, is being used to provide free labor, which has been positioned as akin to continuing slavery.[49] These disparate contexts are sites of discourse and negotiation that are intrinsically defined by bodies out of place, revealing an interdependence between the one who is excluded and the other who is incorporated.

Notes

1 The initial research for the second part of this paper on Late Antiquity was conducted during my Fellowship in 2018, at the *Kolleg-Forschergruppe Migration und Mobilität*, at the University of Tübingen. I am grateful to the colleagues and students I worked with there for creating such a dynamic environment with a rich and lively exchange of ideas. I owe my introduction to the challenges of looking at migration in this period to the generosity of Stephen Mitchell, and the

hospitality of Matina Mitchell, during my time in their haven in Berlin, and Stephen's astute comments on the paper.

2 This could be the result of a transformation of the physical site, causing a disjuncture between the memory-place and the material fabric that embodies the memory (over-writing it). Alternatively, de-placement could result from the transfer of people to in-between sites, such as refugee camps: Isayev (2017a: 394); Isayev (2021).

3 Based on critical analysis and figures by Erdkamp (2008: 419); Osborne (1991); Laslett (1977).

4 Scheidel (2004: 13–20); Scheidel (2006: 223–24), with adjustment for higher population estimates for Italy, see discussion in Isayev (2017a: 22–24).

5 This includes, for example, bioarchaeological and epigraphic studies: Isayev (2017a: 118–20).

6 Purcell (1990: 41); Horden and Purcell (2000: 380); Rathbone (2008); Archibald (2011: 42–65); Isayev (2017a: 225–27).

7 Translation of Cicero *De Officiis* 3.11.47, by the author, adapted from the 1928 translation of Cicero's *De Officiis* by Walter Miller. This critical comment was in reference to a series of laws passed by rival Roman politicians in the mid-first century BCE, who were keen to exclude "foreigners" from the city to ensure they did not interfere in voting for the bills. Implicit in this episode is the fact that the Gracchi and Julius Caesar were quite happy for these foreigners to stay in Rome and to also be part of its civic community, as they were trying to give them citizenship. For a discussion of the Ciceronian passage and the events in relation to expulsion of foreigners: Broadhead (2008: 466–67); Isayev (2017a: 35–38); Noy (2000: 37–44); Lintott (1994: 76); Purcell (1994: 652–53); Wiseman (1994: 344–45).

8 See for example: Horden and Purcell (2000); Isayev (2017a); Tacoma (2016).

9 On stoicism and cynicism: Gill (2013); Desmond (2008, chapter 5: 199–207). On stoicism, exile, cosmopolitanism, and wandering: Montiglio (2005: 183–87, 211–13); Gray (2015: 306–10); Schofield (1999, chapter 3: 69–32); Konstan (2009); Moles (1996); Isayev (2021).

10 For the way that people who are out of place subvert their condition of wandering and permanent temporariness, see Isayev (2021).

11 Nationality remains the basis of entitlement to rights, despite the guarantees offered for legal personhood to those deemed stateless by international human rights law (Article 2.1 of the International Covenant on Civil and Political Rights, and Article 14 of the Universal Declaration of Human Rights). Addressing perplexities of human rights, which, while promising equality irrespective of citizenship status, are still articulated within the framework of the nation-state: Arendt (1968); Gundogdu (2015).

12 For example, the earthquake and tidal wave that buried Helike in 373 BCE were perceived as a reprisal by the gods for crimes against suppliants sheltering in its sanctuary of Poseidon: Pausanias 7.25.1. The earthquake which hit Sparta was blamed on the ejection of the Helots from the Poseidon Sanctuary in 464 BCE: Thucydides 1.128.1.

13 This section draws on more detailed extensive studies of these issues in Isayev (2017b and 2018).

14 Translation by Alan H. Sommerstein, Loeb Classical Editions, 2009.

15 Just two of the many works that grapple with this theme: Hartog (1988); Said (1978).

16 The Chorus of Euripides, *Children of Heracles* lines 329–32, echoes the Athenian slogan of generosity. Translation for all cited Euripides from David L. Kovacs, Loeb Classical Edition, 2005.

17 Thucydides in Book 5 (84–116) of his *Peloponnesian War* exposes the disparity between Athenian adherence to democratic principles in conducting internal affairs, and acting akin to a tyrant in external dealings, at times forcing autonomous states into submission. He provides a stark dramatization of the dialogue between the men of Melos—a neutral state—and the Athenians who were poised to destroy it in 416 BCE.

18 Thucydides: 1.144.2, 1.67, 1.39.1, 2.39.1; Herodotus on Sparta being closed to strangers: 1.65.6–9; 1.69–79; with Garland (2014: 95–98, 126); Sinn (1993: 67–84, 71).

19 Late Antiquity is a period that spans from circa the fourth century CE through to the Early Middle Ages.

20 Citizenship was extended by emperor Caracalla's issuance of the Antonine Constitution. Its decline is noted by various studies, including for example: Brown (1992, 154); Garnsey (2004: 140). Mathisen (2006), however, challenges the extent of this decline, arguing that concepts of citizenship still continued to play a role in defining personal and legal identity after this period.

21 The Roman law—*ius civile*—by default now applied to all *cives* (full citizens) but also to legally less advantaged people, such as slaves and freedmen. For example, the judges could have on their council both "provincial citizens" or *peregrini*, according to a law of 400: *CTh* (Theodosian Code) 1.34.1 (400); See Mathisen (2006: 1018–20) for further discussion of continuity of peregrine status.

22 For example, *transitor*—as used by Ammianus 15.2.4 (Isayev 2017a: 9–10). The developing characterisation of vagrants as suspect is explored by Moore (1993); in relation to laws against vagrant beggars in the Theodosian Code: Caner (2002: 166). See wider discussion of the issues of control in Horden and Purcell (2000: 384); Pottier (2009).

23 Critique and debate of its use by Marsh (2015).

24 In earlier periods, there was no interest in categorizing all those on the move under one label such as *migrant*. Terms do exist for the foreigner/outsider in Ancient Greek: *xenos* (although initially the term could also be used to mean "host"), or enemy (*polemios*), and in Latin for the friendly outsider (*hospes*), and the one who is much less so, an enemy (*hostis*), originally the term was also used to mean stranger or foreigner. None of these express the same sentiment as the modern usage of "migrant." Instead, they focus on the specific relationship of the individual to the host community: Cicero, *De Officiis* 1.12.37; Varro, *Lingua Latina* 5.3, with discussion in Isayev (2017a: chapter 2).

25 Terms do exist for the foreigner/outsider in Ancient Greek: *xenos* (although initially the term could also be used to mean host), or enemy (*polemios*), and in Latin for the friendly outsider (*hospes*), and the one who is much less so, an enemy (*hostis*, originally the term was also used to mean stranger or foreigner). None of these express the same sentiment as the modern usage of "migrant." Instead, they focus on the specific relationship of the individual to the host community: Cicero, *de Officiis* 1.12.37; Varro, *Lingua Latina* 5.3, with discussion in Isayev (2017a, chapter 2).

26 The *redemptus* could also become a dependent and provide services more akin to a patron–client relationship: Klingshirn (1985: 202).

27 Levy (1943: 163–71); Watson (1991); in relation to recovery of status and marriage: Sessa (2012: 140–41). For Republican traditions governing *postliminium*: Leigh (2004: chapter 3, 57–97); Isayev (2017a: 224).

28 For example, Priscus 9.3, for text and translation of Prisucus's fragments: Blockley (1983): "After the battle between the Romans and the Huns in the Chersonese, a treaty was negotiated by Anatolius. The terms were as follows: that the fugitives should be handed over to the Huns, and six thousand pounds of gold be paid to complete the outstanding instalments of tribute; that the tribute henceforth be set at 2,100 pounds of gold per year; that for each Roman prisoner of war who escaped and reached his home territory without ransom, twelve solidi were to be the payment, and if those who received him did not pay, they were to hand over the fugitive; and that the Romans were to receive no barbarian who fled to them."

29 Gundogdu's (2015: esp. 16, 76, 116) reinterpretation of Arendt's analysis of state-lessness and human rights, points to emerging de-politicizing trends, especially with the humanitarian regime's increasing emphasis on the suffering body, which undermines the ability of displaced persons to make their actions and their speech relevant, hence excluding them from political community, which for Arendt equates to expulsion from humanity.

30 For an overview of such practices as exemplified in relation to bishop Caesarius of Arles: Klingshirn (1985: esp. 185–87).

31 Such actions initially required specific permission, but increasingly they formed part of directives of papal edicts, and eventually were written into the law codes: Codex of Justinian, Clause 1.2.21.2 (with Kearley 2007; *Codex of Justinian* [2007, 2016]): "For if the cause of the redemption of captives requires it, then we permit both sale of such property devoted to God as well as hypothecation and pledge thereof, since it is not improper to prefer human souls to vessels (vasis) and vest-ments of every kin. These provisions shall apply not only in future but also in pending causes."

32 Translation, Davidson (2002). He articulates this in his writings to counter the criticism against him by Milan's Arian clergy, for melting sacred church vessels for ransom. Criticisms by the Arian clergy: Ambrose, *De Officiis* 2.28.136. For the way that ransom fits into the tradition of helping the needy: Osiek (1981).

33 As noted above, for example: Ambrose, *De Officiis* 2.28.136. See also Klingshirn (1985: 185).

34 For an intricate study of the Bishop's career: Klingshirn (1985, 1994, 2004). For the the appointment of Casarius as the papal vicar of Gaul: Klingshirn (1985: 195).

35 This brings to mind the proliferation of captivity in the United States, for which see the final note at the end of the chapter.

36 The sermons and the *Vita* are compiled by Morin (1937–42). Summary of compi-lation and composition, Klingshirn (1985: 187–89).

37 Translation from Klingshirn (1985: 189).

38 By 506 Caesarius would have no longer needed permission from his clergy to alienate church property, instead, the Council of Agde required a bishop to obtain the agreement of two to three other bishops to make transactions valid: Canon 7, *Corpus Christianorum, Series Latina* 148 (1963): 195–96; see Klingshirn (1985: 189); Klingshirn (1994: 90, n. 36).

39 Caesarius could show how through acts of redemption, these people may be brought to Catholic Christianity. Daly (1970) traces how through such reasoning Caesarius anticipates the Medieval notion that all Christians are bound together by ties of fraternity and not divided by ethnic or geopolitical boundaries. In rela-tion to challenging restrictive citizenship, and Christian development of the classi-cal philosophical sense of being a cosmopolitan into the Christian concept of being citizens of a world community, as "City of God," see Mathisen (2006: 1017).

40 While he may not have always collected economic repayment, still such *beneficium* on his part would have created ties and obligations through duties owed, putting the bishop and *redemptus* into a form of patron-client relationship (Klingshirn 1985: 201–02).
41 Translation from Klingshirn (1985: 191).
42 Translation from Klingshirn (1985: 196).
43 Especially *Epistulae* 10, 12 and 35: Shanzer and Wood (2002); McCarthy (2016).
44 *Epistolae Arelatenses* 25 in *MGH* (*Ep.*) (= *Monumenta Germaniae Historica, Epistulae*) III, ed. W. Gundlach, 35–36.
    Symmachus's support and promotion of Caesarius was evident in other honours bestowed during his visit to Rome, where the Pope being "greatly moved by the worthiness of his good deeds (*meritorum eius dignitate*) and by reverence for his sanctity (*sanctitatis eius reverentia*)," (*Vita* I.42), gave him the right, among other things, to wear the *pallium*, which at that time was permitted only to the Bishop of Rome. Discussion in Klingshirn 1985: 195–97.
45 For forms of agency held by people who are displaced, especially in earlier periods, see Isayev (2017b). For other forms of agency, these are part of current research for a work on *Bodies Out of Place*.
46 Priscus 11.406 recounts an encounter with a Scythian-looking Greek merchant, who had been taken captive, enslaved, and eventually won his freedom.
47 For a discourse on relational perceptions of place: Lefebvre (1991); Massey (2004); Amin (2007); Darling (2010); Collins 2011. In relation to the ancient world: Isayev (2017a: 390–94); Thalmann (2011).
48 See Hillal and Petti (2018), with discussion in Isayev (2021).
49 See for example: the special issue on *Historians and the Carceral State* edited by Hernández, Muhammad, and Thompson (2015); Chase (2015); Peláez (2008); Simon (2012); Rockwell, Crocket, and Davis (2020).

References

Amin, Ash (2007). "Re-thinking the Urban Social." *City* 11: 100–114.
Archibald, Zosia (2011). "Mobility and Innovation in Hellenistic Economies: The Causes and Consequences of Human Traffic," 42–65. In Z. Archibald, J. K. Davies, and V. Gabrielsen, eds., *The Economies of Hellenistic Societies, Third to First Centuries BC*. Oxford: Oxford University Press.
Arendt, Hannah (1968). *The Origins of Totalitarianism* [1951]. New York: Harcourt.
Bakewell, Geoffrey W. (2013). *Aeschylus's Suppliant Women: The Tragedy of Immigration*. Madison: University of Wisconsin Press.
Bielman, Anne (1994). *Retour à la liberté. Libération et sauvetage des prisonniers en Grèce ancienne. Recueil d'inscriptions honorant des sauveteurs et analyse critique*. Athens and Lausanne: Ecole Française d'Athènes, Université de Lausanne.
Blockley, Roger C. (1983). *The Fragmentary Classicising Historians of the Later Roman Empire. Eunapius, Olymplodorus, Priscus and Malchus*. (*Text, Translation and Historiographical note*). (*Arca, Classical and Medieval Texts, Papers and Monographs 6*). Liverpool: Francis Cairns.
Broadhead, William (2008). "Migration and Hegemony: Fixity and Mobility in Second-century Italy," 451–70. In Luke De Ligt and Simon J. Northwood, eds., *People, Land, and Politics: Demographic Developments and the Transformation of Roman Italy 300 BC–AD 14*. Leiden: E. J. Brill.

Brown, Peter (1992). P*ower and Persuasion in Late Antiquity: Towards a Christian Empire*. Madison: University of Wisconsin Press.

Caner, Daniel (2002). *Wandering, Begging Monks. Spiritual Authority and the Promotion of Monasticism in Late Antiquity.* Berkeley: University of California Press.

Chaniotis, Angelos (1996). "Conflicting Authorities: *Asylia* between Secular and Divine Law in the Classical and Hellenistic Poleis." *Kernos* 1: 65–86.

Chase, Robert T. (2015). "We Are Not Slaves: Rethinking the Rise of Carceral States through the Lens of the Prisoners' Rights Movement." *Journal of American History* 102(1): 73–86.

Codex of Justinian (2007). Blume, Fred H. (2007). *Annotated Justinian Code.* Second edition, Timothy G. Kearley, ed. University of Wyoming. Accessed July 21, 2020, from http://www.uwyo.edu/lawlib/blume-justinian/

Codex of Justinian (2016). *The Codex of Justinian: A New Annotated Translation, with Parallel Latin and Greek Text Based on a Translation by Justice Fred H. Blume.* Bruce W. Frier, general ed. with Serena Connolly et al., 2016. Cambridge: Cambridge University Press.

Cole, Susan G. (2004). *Landscapes, Gender, and Ritual Space: The Ancient Greek Experience*. Berkeley: University of California Press.

Collins, Francis L. (2011). "Transnational Mobilities and Urban Spatialities: Notes from the Asia-Pacific." *Progress in Human Geography* 36(3): 316–35.

Daly, William (1970). "Caesarius of Arles, a Precursor of Medieval Christendom." *Traditio* 26: 1–28.

Darling, Jonathan (2010). "A City of Sanctuary: The Relational Reimagining of Sheffield's Asylum Politics." *Transactions of the Institute of British Geographers* 35: 125–40.

Davidson, Ivor J. (2002). *Ambrose: De Officiis: Edited with an Introduction, Translation, and Commentary (Two Volume Set)*. Oxford: Oxford University Press.

Derrida, Jacques (2000). *Of Hospitality: Anne Dufourmantelle Invites Jacques Derrida to Respond* [1997]. Rachel Bowlby, trans. Stanford, CA: Stanford University Press.

Desmond, William (2008). *Cynics*. Stocksfield: Acumen.

Desmond, William (2020). "Diogenes of Sinope." In D. Wolfsdorf, ed., *Early Greek Ethics*. Oxford: Oxford University Press.

Erdkamp, Paul (2008). "Mobility and Migration in Italy in the Second Century BC," 417–50. In Luke De Ligt and Simon J. Northwood, eds., *People, Land, and Politics: Demographic Developments and the Transformation of Roman Italy 300 BC–AD 14*. Leiden: E. J. Brill.

Garland, Robert (2014). *Wandering Greeks: The Ancient Greek Diaspora from the Age of Homer to the Death of Alexander the Great*. Princeton, NJ: Princeton University Press.

Garnsey, Peter (1970). *Social Status and Legal Privilege in the Roman Empire*. Oxford: Oxford University Press.

Garnsey, Peter (2004). "Roman Citizenship and Roman Law in the Later Empire," 133–35. In S. Swain and M. Edwards, eds., *Approaching Late Antiquity: The Transformation from Early to Later Empire*. Oxford: Oxford University Press.

Gill, Christopher (2013). "Cynicism and Stoicism," 93–111. In R. Crisp, ed., *The Oxford Handbook of the History of Ethics*. Oxford: Oxford University Press.

Gray, Benjamin (2015). S*tasis and Stability: Exile, the Polis, and Political Thought, c. 404–146 BC*. Oxford: Oxford University Press.

Gundogdu, Ayten (2015). *Rightlessness in an Age of Rights*. Oxford: Oxford University Press.

Hartog, Francois (1988). *The Mirror of Herodotus: The Representation of the Other in the Writing of History*. Berkeley: University of California Press.

Hernández, Kelly L., Khalil G. Muhammad, and Heather A. Thompson, eds. (2015). *Historians and the Carceral State*. Special Issue, *Journal of American History* 102(1).

Hillal, Sandi, and Alessandro Petti (2018). *Permanent Temporariness*. Stockholm: Art and Theory Publishing.

Horden, Peregrine, and Nicholas Purcell (2000). *The Corrupting Sea: A Study of Mediterranean History*. Cambridge: Cambridge University Press.

Isayev, Elena (2017a). *Migration, Mobility and Place in Ancient Italy*. Cambridge: Cambridge University Press.

Isayev, Elena (2017b). "Between Hospitality and Asylum: A Historical Perspective on Agency." *International Review of the Red Cross, Migration and Displacement* 99(904): 1–24.

Isayev, Elena (2018). "Hospitality: A Timeless Measure of Who We Are?" In M. Berg and E. Fiddian Qasmiyeh, eds., *Hospitality and Hostility Towards Migrants: Global Perspectives*. Special Inaugural Issue of *Migration and Society: Advances in Research* 1: 7–21.

Isayev, Elena (2021). "Ancient Wandering and Permanent Temporariness." In E. Isayev and E. Jewell, eds., *Displacement and the Humanities*. Special Issue of Humanities 10: 91 Accessed from https://doi.org/10.3390/h10030091

Jones, Arnold H. M. (1963). *The Later Roman Empire, 284–602: A Social, Economic and Administrative Survey*. Oxford: Blackwell.

Kasimis, Demetra (2013). "The Tragedy of Blood-based Membership: Secrecy and the Politics of Immigration in Euripides's Ion." *Political Theory* 41(2): 231–56.

Kasimis, Demetra (2018). *The Perpetual Immigrant and the Limits of Athenian Democracy*. Cambridge: Cambridge University Press.

Kearley, Timothy G. (2007). "Justice Fred Blume and the Translation of Justinian's Code." *Law Library Journal* 99: 525.

Klingshirn, William (1985). "Charity and Power: Caesarius of Arles and the Ransoming of Captives in Sub-Roman Gaul." *The Journal of Roman Studies* 75: 183–203.

Klingshirn, William (1994). *Caesarius of Arles: Life, Testament, Letters*. Liverpool: Liverpool University Press.

Klingshirn, William, ed. (2004). *Caesarius of Arles: The Making of a Christian Community in Late Antique Gaul*, 2nd edition. Cambridge: Cambridge University Press.

Konstan, David (2009). "Cosmopolitan Traditions," 473–84. In R. Balot, ed., *A Companion to Greek and Roman Political Thought*. Oxford: Blackwell.

Laslett, Paul (1977). *Family Life and Illicit Love in Earlier Generations: Essays in Historical Sociology*. Cambridge: Cambridge University Press.

Lefebvre, Henri (1991). *The Production of Space (Production de l'espace)* [Paris: 1974]. D. Nicholson-Smith, trans. Oxford: Wiley-Blackwell.

Leigh, Matthew (2004). *Comedy and the Rise of Rome*. Oxford: Oxford University Press.

Levy, Ernst (1943). "Captivus Redemptus." *Classical Philology* 38(3): 159–76.

Lintott, Andrew (1994). "Political History, 146–95 B.C.," 40–103. In J. A. Crook, A. Lintott, and E. Rawson, eds., *The Cambridge Ancient History: The Last Age of the Roman Republic, 146–43 B.C.*, vol. 9, 2nd edition. Cambridge: Cambridge University Press.

Marsh, David (2015). "We Deride Them as 'Migrants'. Why Not Call Them People?" *The Guardian*, August 28, 2015. Accessed February 11, 2021, from https://www.theguardian.com/commentisfree/2015/aug/28/migrants-people-refugees-humanity

Massey, Doreen (2004). "Geographies of Responsibility." *Geografiska Annaler: Series B* 86: 5–18.

Mathisen, Ralph M. (2006). "Peregrini, Barbari, and Cives Romani: Concepts of Citizenship and the Legal Identity of Barbarians in the Later Roman Empire." *The American Historical Review* 111(4): 1011–40.

McCarthy, Brendan (2016). "The Letter Collection of Avitus of Vienne," 358–66. In C. Sogno, B. K. Storin, and E. J. Watts, eds. *Late Antique Letter Collections: A Critical Introduction and Reference Guide*. Berkeley: University of California Press.

Moatti, Claudia (2000). "Le contrôle de la mobilité des personnes dans l'empire romain." *MEFRA* 112: 925–58.

Moatti, Claudia, ed. (2004). *La Mobilité des personnes en Méditerranée de l'antiquité à l'époque moderne. Procédures de contrôle et documents d'identification*. Rome: École française de Rome.

Moatti, Claudia, and Wolfgang Kaiser, eds. (2007). *Gens de passage en Méditerranée de l'Antiquité à l'époque moderne*. Paris: Maisonneuve et Larose.

Moatti, Claudia, Wolfgang Kaiser, and Christophe Pébarthe, eds. (2009). *Monde de l'itinérance en Méditerranée de l'antiquité à l'époque moderne: procédures de contrôle et d'identification: tables rondes, Madrid 2004-Istanbul 2005*. Bordeaux: Ausonius.

Moles, John (1996). "Cynic Cosmopolitanism" ["Cosmopolitanism"]," 105–20. In R. Bracht Branham and M.-O. Goulet-Cazé, eds., *The Cynic Movement in Antiquity*. Berkeley: University of California Press.

CTh (Theodosian Code): Teodor Mommsen, Paul M. Meyer, and Paul Krüger, eds., (1902). *Theodosiani libri XVI*. Berlin.

Montiglio, Silvia (2005). *Wandering in Ancient Greek Culture*. Chicago: University of Chicago Press.

Moore, Stephen (1993). *Perceptions of Vagrancy in Extant Legal Records throughout the Later Middle Ages, 11001400*. Ottawa: Carleton University (Canada), ProQuest Dissertations Publishing, MM89802. Accessed February 11, 2021, from https://search.proquest.com/dissertations/docview/304059252/135BD0FB03256F66FA6/283

Morin, Germain, ed. (1937–42). *Sancti Caesarii Episcopi Arelatensis Opera Omnia*, 2 vols. (Maredsous: Abbaye de Maredsous). Reprinted in *Corpus Christianorum, Series Latina*, 103, 104. Turnhout: Brepols, 1953.

Noy, David (2000). *Foreigners at Rome. Citizens and Strangers*. London: Duckworth with the Classical Press of Wales.

Osborne, Robin (1991). "The Potential Mobility of Human Populations." *Oxford Journal of Archaeology* 10(2): 231–52.

Osiek, Carolyn (1981). "The Ransom of Captives: Evolution of a Tradition." *The Harvard Theological Review* 74(4): 365–86.

Pelàez, Vicky (2008). "The Prison Industry in the United States: Big Business or a New Form of Slavery?" El Diario-La Prensa, *New York and Global Research*, March 10, 2008. Accessed February 11, 2021, from http://www.globalresearch.ca/the-prison-industry-in-the-united-states-big-business-or-a-new-form-of-slavery/8289

Pottier, Bruno (2009). "Contrôle et mobilisation des vagabonds et des mendiants dans l'Empire romain au IVe et au début du Ve siècle," 203–40. In C. Moatti, W. Kaiser, and C. Pébarthe, eds., *Monde de l'itinérance en Méditerranée de l'antiquité à l'époque moderne: Procédures de contrôle et d'identification: Tables rondes, Madrid 2004-Istanbul 2005*. Bordeaux: Ausonius.

Purcell, Nicholas (1990). "Mobility and the Polis," 29–58. In O. Murray and S. Price, eds., *The Greek City from Homer to Alexander*. Oxford: Clarendon Press.

Purcell, Nicholas (1994). "The City of Rome and the Plebs Urbana in the Late Republic," 644–88. In J. A. Crook, A. Lintott, and E. Rawson, eds., *The Cambridge Ancient History: the Last Age of the Roman Republic, 146–43 B.C.* vol. 9, 2nd edition. Cambridge: Cambridge University Press.

Rathbone, Domenic (2008). "Poor Peasants and Silent Sherds," 305–32. In Luke De Ligt and Simon J. Northwood, eds., *People, Land, and Politics: Demographic Developments and the Transformation of Roman Italy 300 BCA–D 14*. Leiden: E. J. Brill.

Rockwell, Casey C., David Crockett, and Lenita Davis (2020). "Mass Incarceration and Consumer Financial Harm: Critique of Rent-seeking by the Carceral State." *Journal of Consumer Affairs* 34: 1–20.

Said, Edward (1978). *Orientalism*. New York: Pantheon Books.

Scheidel, Walter (2004). "Human Mobility in Roman Italy, I: The Free Population." *Journal of Roman Studies* 94: 1–26.

Scheidel, Walter (2006). "The Demography of Roman State Formation in Italy," 207–26. In M. Jehne and R. Pfeilschifter, eds., *Herrschaft ohne Integration?: Rom und Italien in Republikanischer Zeit*. Frankfurt am Main: Verlag Alte Geschichte.

Schmidt-Hofner, Sebastian (2017). "Barbarian Migrations and the Socio-economic Challenges to the Roman Landholding Elites in the Later 4th C. AD." *Journal of Late Antiquity* 10(2): 372–404.

Schofield, Malcolm (1999). *The Stoic Idea of the City*. Chicago: University of Chicago Press.

Sessa, Kristina (2012). *The Formation of Papal Authority in Late Antique Italy: Roman Bishops and the Domestic Sphere*. Cambridge: Cambridge University Press.

Shanzer, Danuta, and Ian Wood, eds. and trans. (2002). *Avitus of Vienne: Letters and Selected Prose*. Liverpool: Liverpool University Press.

Shumsky, Neil L. (2008). "Noah Webster and the Invention of Immigration." *The New England Quarterly* 81(1): 126–35.

Simon, Jonathan (2012). "Mass Incarceration: From Social Policy to Social Problem," 23–52 In J. Petersilia and K. R. Reitz, eds., *The Oxford Handbook of Sentencing and Corrections*. Oxford: Oxford University Press.

Sinn, Ulrich (1993). "Greek Sanctuaries as Places of Refuge," 67–84. In N. Marinatos and R. Hagg, eds., *Greek Sanctuaries: New Approaches*. London: Routledge.

Tacoma, L. E. (2016). *Moving Romans: Migration to Rome in the Principate*. Oxford: Oxford University Press.

Thalmann, William G. (2011). *Apollonius of Rhodes and the Spaces of Hellenism*. Oxford: Oxford University Press.

Thomas, Yan (1996). *Origine' et 'commune patrie'. Étude de droit public romain (89av. J.-C. - 212 ap. J.-C.)*. Rome: Collection de l'École Française de Rome.

Thompson, Peter (2003). "'Judicious Neology': The Imperative of Paternalism in Thomas Jefferson's Linguistic Studies." *Early American Studies: An Interdisciplinary Journal* 1(2): 187–224.

Walbank, Michael B. (1978). *Athenian Proxenies of the Fifth Century B.C.* Toronto: Samuel-Stevens.

Watson, Alan (1991). "Captivitas and Matrimonium," 37–54. In *Studies in Roman Private Law*. London: The Hambledon Press.

Wilson, John-Paul (2006). "Ideologies of Greek colonization," 25–58. In G. Bradley and J.-P. Wilson, eds., *Greek and Roman Colonization, Origins, Ideologies and Interactions*. Swansea: Classical Press of Wales.

Wiseman, Timothy P. (1994). "The Senate and the Populares, 69–60 B.C.," 327–81. In J. A. Crook, A. Lintott, and E. Rawson, eds., *The Cambridge Ancient History: The Last Age of the Roman Republic, 146–43 B.C.* vol. 9, 2nd edition. Cambridge: Cambridge University Press.

Zeitlin, Froma (1992). "The Politics of Eros in the Danaid Trilogy of Aeschylus," 203–52. In R. Hexter and D. Seldon, eds., *Innovations of Antiquity*. London: Routledge.

3   Reimagining "refugee" protection

Beyond improving the status quo

Jennifer Hyndman

Introduction

The research and analysis presented in this chapter challenge the long-standing, state-sanctioned paradigm of managing migration and refugee displacement. Since World War II, throughout the Cold War, and until the early 1990s, refugees were regarded as proof of ideological superiority, but today the geopolitical valence of refugees has waned, and their protection has suffered across the world. *Non-refoulement* (no forced return to danger) remains the bare minimum provision of safety still recognized by many, if not all states, but one must still get into a country to claim such protection. Protecting borders against uninvited migration is a much greater priority than protecting people whose own governments cannot or will not.

One way to understand and interrogate the salient paradigm of managing migration is to examine the recent Global Compacts released in 2018: one "on refugees" and the other "for migration." Christina Clark-Kazak (2018) noted in reference to the Global Compact *on* Refugees (GCR), "on" is an unusual choice of preposition, given that in comparison the Global Compact *for* Migrants is so much less paternalistic. Two main approaches to challenging the paradigm emerge. First, for whom and to whom do the compacts apply? Are "refugees" and "migrants" discrete groups, and hence stable categories? Or do they overlap? And which of all 193 UN member states that signed the New York Declaration on Refugees and Migrants in 2016, a progenitor to the Compacts, are included in each Compact? (Hint: many are tacitly left out.) Another way to question this paradigm—one that quietly countenances refugee-migrant containment in "regions of origin" (Chimni 2008)—is to challenge the *scale* at which protection and safety from violence is forged and provided. International law and the refugee regime take *states* as the main unit of analysis, operations, and solutions to human displacement, as well as subsequent protection measures for those affected. This obscures protection or self-authorized security that is operating at other scales, such as refuge and livelihood in cities, or among faith groups sharing customary law in local encounters among people from different countries. Many of these strategies of finding security or providing protection are invisible from the salient state-centric perspective.

DOI: 10.4324/9781003170686-4

A good deal of support for those who are displaced, in the form of material assistance and protection of people who seek safety in other countries, flies under the radar of the international refugee regime, its member states, and related institutions operating at a global level. As Nick Gill (2018) contends, where states, or governments, are inclined to exclude migrant-asylum seekers and refugees from their territories, many grassroots efforts by civil society and local organizations have proven welcoming to refugee-migrants at or inside those same borders (Bagelman 2018; Sparke 2018). To complicate any simple rule, many governments—from South Africa to Sweden, Thailand to Germany, Lebanon to India to Canada—have taken steps to provide sanctuary to asylum seekers and refugee-migrants.

Refugee-migrant subjects are the authors of and decision-makers in these journeys, albeit under conditions not of their own making. The communities in which they arrive also shape their prospects, providing tacit information, resources, and relationships to sustain and facilitate their security and well-being in very immediate ways (Darling 2017). As Mona Fawaz and her colleagues in Lebanon demonstrate, refugees are "city-makers" (Fawaz et al. 2018) They are not passive recipients of instructions from others; rather, they decide and forge pathways by negotiating conditions that put their lives at risk so they may pass into safer contexts where they rebuild their lives. They are protagonists in shaping self-authorized notions of protection. Is *protection* even the right word? It implies something one party does for another, even *to* another.

In the following pages, I first briefly outline the Global Compact on Refugees and the Global Compact for Migration (2018) and assert that they are, at best, an affirmation of the current paradigm and status quo. Second, I show how the Global Compact on Refugees (UNGA 2018) leaves out significant world regions where sizeable groups of displaced persons live. Drawing on the concept of Orientalism (Said 1978)—which describes the West's (Occidental) representation of the East as an inferior and primitive, if exoticized, place and people—I transpose the concept of *legal orientalism* (Ruskola 2013) into the context of Refugee Studies and the protracted exile most refugees experience. Orientalism is the production of the East by the West, a representation or narration of a less civilized, even barbaric "Orient" (Said 1978). Drawing on Said's critical work about sociocultural constructions of other, Teemu Ruskola (2002) coins the term "legal orientalism" to show how the law of a country in the "Orient" (China in his example) is portrayed as weak and flawed in relation to the West, specifically the United States. Put in the context of the Global Compact on Refugees, which relies heavily on the 1951 Convention Relating to the Status of Refugees and its 1967 Protocol, an exclusionary "club" of countries and geography emerge as relevant to the GCR because they are part of this international treaty, generated in the aftermath of World War II Europe.

To avoid replicating the terms of the GCR, which succumb to legal orientalism in my view, I contend that rescaling protection to sites finer than the nation-state help to render visible tactics of staying safe and providing protections that are missing from the international relations and state-centric analyses around refugee-migrant protection. Faith practices, city policies, and quotidian routines can facilitate *self-authorized security of person* for many refugee-migrants who might not meet the official refugee definition as outlined in international law, or even national legislation.

The current paradigm and international system of laws, policies, and purported solutions do not address the intransigent and unequal distribution of refugee-migrants facing protracted displacement—not to mention the millions more who are called "migrants" because they are viewed as having less legitimate grounds for asylum. I could rattle off the usual statistics, but let it suffice to say that more than 80% of the world's refugees live in global South host countries (UNHCR 2012, 2019). The GCR advocates *more support* to these host countries through loans as the most compelling response to the situation of massive deaths at sea on the Mediterranean and Andaman Seas; this idea is underwhelming, to say the least. Inviting the World Bank to lend to already indebted states in order to make refugees more "self-reliant," and therefore less costly, is insulting and merely a grand band-aid that does nothing to change the current paradigm. The status quo must change. The "refugee" definition provides some safeguards and rights that "migrant" does not, but the refugee definition in international law and domestic legislation adopted by signatory states is very narrow, as I elaborate below (Betts 2013; Crawley et al. 2017).

As for terminology, I use the concept "refugee-migrant" as a shorthand here, an imperfect term that aims to capture the range of categories and monikers that states apply to people on the move, but also to destabilize the idea that there is any clear or precise distinction between them. These statuses may overlap and are often imposed on people, rather than claimed by them. A person may hold both statuses, either at once or sequentially over the course of a displacement journey. The debate and discussion about refugee-migrant terminology has been ongoing for years (Zetter 2007; Al Jazeera 2015; Kyriakides 2017; Crawley and Skleparis 2018). Beyond single term, the language of "solutions" in the refugee regime can itself limit the scope of change (Long 2011).

Refugees are often represented as bona fide and "deserving," while those fleeing structural violence, slow violence, or acute deprivation are seen as opportunistic. Authors Betts and Collier assert this questionable distinction in their 2017 book: "Migrants are lured by hope. Refugees are fleeing fear. Migrants hope for honeypots; refugees need havens" (Betts and Collier 2017: 30). Such false binaries and simplistic stereotypes do little more than obscure the multifaceted conditions and considerations that shape refugee-migrant journeys and the people who refuse the migrant/refugee distinction.

The definition of "refugee" in international law—outlining five grounds of persecution based on race, religion, nationality, political opinion, or membership in a social group—is drawn from the 1951 Convention (and its 1967

Protocol) but is highly Eurocentric, based largely on the experience of people in Europe displaced during World War II who were deprived of civil and political rights, including persecution and death (Hathaway and Foster 2014). As flagged above, the refugee definition is narrow, exclusionary, and not relevant to the postcolonial experiences of much of the world, and yet it remains a major basis for refugee determination. In addition to this overly narrow definition of "refugee," many states and world regions are not signatories to the 1951 Convention Relating to the Status of Refugees or its 1967 Protocol, and so are not bound by its provisions. Nonetheless, they and their civil societies do provide various kinds of support and protection for people forced outside the borders of their countries of origin.

Setting the stage

Not until 2016, when all member states of the United Nations General Assembly voted in favor of the New York Declaration for Migrants and Refugees, did governments across the world agree that the current paradigm of managing displacement was not working (UN 2016): all 193 member states declared that there was "the political will of world leaders to save lives, protect rights and share responsibility on a global scale." A similar consensus agreed on the importance of protection for those displaced during the World War II period, but not since. At that time, human rights instruments were drafted and the 1951 Convention Relating to the Status of Refugees written (Jubilut et al. 2019: 23). While the post-Cold War paradigm neither provided adequate protection nor addressed refugee insecurity in a meaningful way (Hyndman 2000; Castles 2008), European states did not directly face the acute consequences of this containment strategy, at such a high human cost, until the events of 2015 (Morrissey 2020).

What the Office of the UN High Commissioner for Refugees (UNHCR) calls "durable solutions" to fix the refugee problem—namely return to one's country of origin, local integration into a neighboring host country, and/or resettlement—are not available to all displaced persons (Aleinikoff and Zamore 2019). The tacit post-Cold War consensus among global North states has been that refugee-migrants should remain in their "regions of origin" (a favorite UN term). As early as the 1990s, the then UN High Commissioner for Refugees declared that refugees had the "right to remain" in their homes and that safety should come to them before they had to flee (UN High Commissioner for Refugees, Sadako Ogata cited in Hyndman 2000). This approach culminated in two UN Security Council Resolutions in the early 1990s, resulting in the dubious creation of "safe cities" inside Bosnia-Hercegovina, where civilians could flee within their country (a war zone) to urban areas protected by UN peacekeepers. The demise of this strategy had tragic human consequences when, in July 1995, more than 7,000 Bosniaks (all Muslim) men and boys were loaded onto buses by Serb paramilitaries as a small number of Dutch UN peacekeepers looked on, monitoring the situation but vastly outnumbered. The men and boys were never seen again, until their bodies were exhumed from mass graves, most just 30km away (Agence France-Presse 2001). The killings were officially deemed "a genocide" by the UN War Crimes Tribunal in The Hague. While the idea of protecting people "at home," literally *in* their own country, was abandoned after this genocide, the norm of "care and maintenance" for refugees close to home was not.

Instead, the globalized nexus of refugee-migrant exclusion has been galvanized, the politics and practices of which have been well documented (Amoore 2006; Huysmans 2006; Sparke 2006; Squire 2011; Crawley et al. 2017). Beginning in the 1990s when the political valence of the "refugee" was transformed from "strategic and valuable" in a Cold War context to "costly and avoidable," global North countries found myriad ways to outsource refugee care and maintenance to host countries in the global South that were adjacent to displacement (Hyndman 1995; Castles 2008; Urry 2017). As Alex Betts and James Milner observed many years ago, making a point that still remains as relevant as ever today, if "African states were to reduce their commitment to the principle of territorial asylum [in adjacent countries], thereby undermining access to effective refugee protection within the region, this would almost certainly exacerbate the likelihood of onward movement and global insecurity" (Betts and Milner 2006: 4). The authors note that European states are willing to pay for, but not host, refugees: for global North states "it doesn't matter where asylum is provided as long as it is provided" (2006: 4). Drawing on the example of Tanzania, they note that the

> approach of the European states has so far assumed that cooperative agreements can allow Southern states to be enticed or persuaded to improve their own protection standards in order to reduce the need for the onward movement of asylum-seekers to Europe.
>
> (2006: 3–4)

I would add that Southern states hosting refugees have become far more savvy, demanding compensation for this protection work, and are often reluctant to host large refugee groups on their territory. Tanzania is a good example, moving from hosting displaced Burundians in settlements where they had access to land and employment in the post-1972 period of displacement to later providing basic support for them in refugee camps in the 1990s. The assumption that the global South can be paid to host refugees and migrant groups is not news. While not a conventional "global South" state, Turkey has been paid $6 billion to host refugees who otherwise might make an onward journey to Europe, a questionable strategy from the outset.

Incentivizing asylum seekers' return to their regions of origin occurs through financial rewards to so-called transit states, such as the $2 billion "anti-migration" trust fund established by EU leaders in 2015 to stem the onward migration of people from North Africa (Rankin 2017). Despite the funds, development initiatives to prevent migration by tackling unemployment, insecurity, and poverty, remittances to family members in sub-Saharan African countries were estimated to be worth US$34 billion in 2017, counter-incentivizing the migration of Africans trying to reach Europe (Rankin 2017).

Unpacking the global compacts

Providing a comprehensive overview and analysis of the context and content of the two Global Compacts are beyond the scope of this chapter, but I present a brief sketch and analysis of each here. They are highly readable documents, each under 35 pages, that have been well-covered elsewhere (McAdam 2018; Turk 2018; Ferris and Martin 2019). Volker Turk, former UNHCR Director for International Protection and lead for UNHCR on the GCR, notes:

> The answer to the challenges posed by high numbers of refugees does not lie in draconian measures or revisiting the international refugee protection regime, which has proven to be good law and practice when the political will is there to implement it. Rather, the answer can be found in a more robust, comprehensive, and good-faith application of the tenets of protection. This requires that the international refugee protection system be better capacitated to absorb the growing pressures.
>
> (Turk 2018: n.p.)

The main message here is to improve the status quo by increasing capacity to manage refugees in at least the 10 countries that already host 60% of the world's refugees. The trend signaled in the GCR is toward integrating institutions and services that support refugees into more mainstream local or national structures: move from segregated refugee camps to more integrated settlements and infrastructure, where schools for example can be shared between local and refugee children. Turk promotes creation of "an architecture of support for the countries most affected" by the large number of displaced persons they host (Turk 2018 n.p.). These measures, designed to improve the status quo, are reasonable enough *if* one suspends knowledge that more than 85% of displaced people reside in the countries of the global South (UNHCR 2017 Global Trends). The business of hosting refugees is unequivocally unequal and disparate, with poorer countries doing most of the hosting. The improvements noted by Turk should be noted, but they are mere modifications to the prevailing paradigm of protection with containment. Protection for refugee-migrants cannot simply *fix* a broken system; the constellation of power relations that structures the international refugee regime and migration institutions has to change.

The very existence of *two discrete* compacts, documents that are pledges of action if not binding in law, must be acknowledged. The neat categorical separation of "refugees" and "migrants" defies a much fuzzier reality. One can be fleeing genuine violence that is not persecution and therefore may not meet a country's definition of "refugee" (Betts 2013; Crawley et al. 2017). "Irregular migrants" have replaced "asylum seekers" in most global North states, downgrading the status of many people on the move, and yet they are the same persons. These categories are highly politicized and not equally regarded in public opinion or government policies. While the Convention refugee definition is clear and applies to the GCR, it also a narrow and exclusionary definition as established above. The Preamble to the GCM acknowledges that refugees and migrants are entitled to the same human rights and fundamental freedoms (section 4), and yet distinguishes between the two groups as though their differences were clear and beyond debate.

The Global Compact for Migration, or GCM (International Organization for Migration/IOM, 2018), in contrast to the GCR, is a much more utopian document, one that affirms Occidental human rights law and principles since World War II, but one that also aspires to positive change and development for all world regions, especially decent work among labor migrants who are often economically, socially, and politically disadvantaged. It is in large part a compilation of existing commitments made by states, in particular through the process that led to the UN's Agenda for Sustainable Development 2030; hence, it includes sustainable development goals (SDGs) in its gambit. The GCM goes further, however, and includes migration related to climate change (or crisis, depending on where you sit) by affirming the UN Convention to Combat Desertification, the Paris Agreement, and the Sendai Framework for Disaster Risk Reduction among other accords. It identifies its guiding principles as "people-centred," "gender-responsive," and "child-sensitive" as part of its "cooperative framework." Its stated objectives, however, are more sobering and bureaucratic. They belie the same general thrust as the GCR in terms of improving the status quo incrementally, specifically to ensure migration is more regular, with increased data collection to improve its management.

The lead agency for the creation of the Migration Compact was the International Organization for Migration (IOM), a recent addition to the United Nations associated organizations in 2016 (Geiger and Pecoud 2020). Many of the objectives in the GCM are recommendations for actions that follow from the Compact, ostensibly to be executed by IOM or a designated partner. To this end, I find the IOM rather self-serving in relation to its lead role in organizing the drafts of the GCM. A majority of the 23 objectives outlined in the GCM are functions that IOM has been paid to execute by governments before its inclusion as a UN-related agency. Until it was incorporated as such, IOM lacked a stated mission or mandate and operated at the request of states who hired it to do a vast array of migration-related work, from accompanying resettled refugees to their destinations, as it does in Canada, to running detention centers for asylum seekers offshore, a function it was hired to do by Australia on the island state of Nauru. IOM personnel collect data and conduct applied research, but as scholars have argued, the organization also governs migration in myriad ways (Geiger and Pecoud 2020). In its new incarnation, the IOM now has a formal mission:

> IOM is committed to the principle that humane and orderly migration benefits migrants and society. … IOM acts with its partners in the international community to: Assist in meeting the growing operational challenges of migration management. Advance understanding of migration issues. Encourage social and economic development through migration. Uphold the dignity and well-being of migrants.
>
> (IOM 2020)

A careful reading of its strategic focus, however, suggests it does not condone "irregular migration": the IOM (2020) aims to "to support States, migrants and communities in addressing the challenges of irregular migration, including through research and analysis into root causes, sharing information and spreading best practices, as well as facilitating development-focused solutions." Many people seek asylum by taking what many states might deem "irregular" routes to arrive somewhere safe. How does one reconcile this tension?

I point out IOM's strategic focus to prevent irregular migration precisely because of the slippery discursive slope alluded to above, which witnessed the "disappearance of the refugee" (Macklin 2005) when governments renamed asylum seekers and refugees "irregular migrants," when or if they entered their territories without authorization. This is the moment at which the migrant-refugee categories blur. People who are today "irregular arrivals" in Canada were once called asylum seekers or refugee claimants. Their "irregularity" risks becoming criminalized, even if they are at risk of facing persecution or violence. If people reach the territory of a state that is a signatory to the 1951 Convention or its 1967 Protocol, they have the right to make a refugee claim, assuming that they are not intercepted before arriving. Efforts to *preclude* these "irregular migrants" have been well documented in the mounting scholarship on externalization of asylum seekers and efforts to prevent would-be refugees from even reaching the territory on which they could make a claim; I have referred to this geographical tactic as *neo-refoulement.* If *refoulement* is the forced return of refugee-migrants to their country of origin where they face persecution or violence, and non-refoulement is the principle of protection that safeguards against this, then *neo-refoulement* is return of the "irregular migrant" before they are even subject to *refoulement* or *non-refoulement.*

Returning to the GCM text, the development goals ensconced in the draft are laudable, and the pledges to improve working conditions and contracts for migrants are hopeful, but the funding to execute these goals and address the 23 stated objectives of the GCM remains less clear. The GCR text shares a commitment to engaging gender, and it goes further to advance an almost intersectional approach (although not called that), one that encompasses other axes of difference or diversity such as ability and age but also specifies experiences of sexual and gender-based violence. While both Compacts appear focused on being responsive and sensitive to gender—as well as to children, to those with disabilities, and to other groups who may be vulnerable—I am less convinced that protecting and assisting the most vulnerable migrants or refugees are the central concerns.

The GCR reiterates the durable solutions that UNHCR has long relied upon (voluntary repatriation to one's country of origin, local integration in host countries, and resettlement to a third country) and adds room for "complementary pathways" that might add more capacity for resettlement, especially in light of the dramatic decrease in refugee resettlement spaces since President Trump took office in 2017. Nonetheless, the GCR sticks closely to the existing refugee definition in *law*, specifically that included in the 1951 Convention Relating to the Status of Refugees, to which I turn briefly. As noted above, the *refugee* in international law is based largely on the figure of the "displaced person" in Europe after World War II, deprived of civil and political rights. This Occidental definition does not capture much of the violence wrought by colonial borders, or Cold War conflicts, that generate displacement beyond Europe's borders.

Legal orientalism meets refugee studies

In the 2000s, Teemu Ruskola, a scholar of comparative law, introduced the idea of legal orientialism (Ruskola 2002), a reference to the tacit superiority of Western, particularly American, legal regimes and the declared inferiority of non-Western regimes. He specifically focused on the "othering" of China's legal system, especially in relation to the US system. Ruskola's argument can be transposed to analyze the GCR. As Edward Said (1978) argued, Orientalism is the politicized cultural production of the "East" (as inferior) by the "West." Law is not necessarily culturally constructed, but Ruskola persuasively shows how it can be. The 1951 Refugee Convention definition of a refugee, reiterated and expanded in its geographical applicability in its 1967 Protocol, are indeed Occidential political constructions shaped by Cold War geopolitics, the Holocaust, and widespread human displacement in Europe after World War II. That the GCR focuses primarily on countries that have signed the Convention and Protocol instantiates its geopolitical antecedents, Cold War legacy architecture, and clear emphasis on civil and political rights at the expense of economic, social, and cultural rights—and thus is an expression of "legal orientalism." In a practical vein, how does this translate into "who counts" in the GCR? Moreover, "whose law counts?" So many major refugee host countries are left off the global map because they have not signed the Convention or Protocol, and yet they provide support and protection for refugees on their territory. Of the 13 case studies commissioned for the Comprehensive Refugee Response Frameworks (CRRFs) as part of the Global Compact on Refugees, all were signatories to the 1951 Convention or its 1967 Protocol. Who and which countries does this exclude? Does exclusion signal that countries that have not signed on as inferior?

The states and civil societies of Lebanon, Jordan, and Turkey, along with major parts of South and Southeast Asia are simply left out of the GCR and its companion CRRFs. The GCR tacitly subscribes to a highly exclusionary geography of protection. Signed after World War II, the 1951 Refugee Convention its 1967 Protocol leaves a number of world regions off the protection map, and the GCR reproduces these omissions, perhaps not an act of commission. And yet the reproduction of this Occidentalism in the GCR must be recognized in order to change it.

Adat: A silver lining?

In response to the query "whose law counts?," I want to highlight the presence of *local* legal regimes that are off the grid compared to international treaties among states. Instead, customary law is distinction from national legislation or international law, occurring at regional or even local scales. The example of Indonesia, specifically the region of Aceh on the island of Sumatra is a case in point (Jones 2018). *Adat* is one example of a practice ensconced in customary law, distinct from Islamic law, that prevails across much of Indonesia (Lukito 1997). Jones (2018) and Missbach (2017) discuss the ways in which very local, faith-based solidarity among Acehnese fisherman and stranded Rohingya refugees on the sea led to the former rescuing the latter and looking after them in their own communities. Such forms of protection are largely invisible because of the scale at and context in which they occur. Focusing specifically on the "torch and displace" tactics of the Burmese military in Rakhine state of Myanmar (also known as Burma) in 2015, Antje Missbach (2015, 2017) writes about the ways in which the Malaysian, Thai, and Indonesian governments turned their backs on Rohingya asylum seekers forced to leave Burma, first in spring and summer 2015 and later after the August 2017 exodus. Aceh, the northern province of Sumatra in Western Indonesia, has its own long history of conflict with the national government of Indonesia over resource revenue allocation and autonomy. Acehnese fishers ignored government orders not to help those in distress on boats by rescuing displaced Rohingya stranded on the Andaman Sea. "We have to help strangers because we have a heart. This is Acehnese culture" (local official in Banda Aceh, cited in Missbach 2017: 41). Missbach also notes that the hospitality extended to the Rohingya has political and social roots as well, citing findings that show Acehnese respondents were sympathetic to the Rohingya because they too had faced conflict with the Indonesian military and related displacement for almost three decades. These smaller scale humanitarian responses to refugees are missing, seemingly unrepresented within a framework that assumes a national scale of response. Because *adat* falls outside of international treaties or law legible to the West, it is invisible and uncounted, and yet it provided critical support and protection for the Rohingya fleeing Burma.

Examples of civil society providing safe haven, support, and protection for refugee-migrants at such local scales abound, but they tend to be overlooked by governments (Hyndman and Reynolds 2020). Having illustrated the exclusionary geography of protection and Occidental othering in the Global Compact on Refugees, the redemptive example of *adat* highlights the importance of finer scales of protection and illustrates how faith-based communities that transcend nationality and borders are salient in moments of humanitarian need. The conditions of their possibility and their enactment in everyday practices are largely "off the grid" and unrecorded in relation to the international refugee regime.

Thinking beyond international law and state-led protection

Johanna Reynolds and I (Hyndman and Reynolds 2020) convened a workshop in June 2018 to foreground the gaps in the global compacts, the GCR and GCM, and also to highlight promising practices at finer scales than the global or international. The city, the fishing community, and the university may be unexpected sites of self-authorized safety and sustenance, but they have proven workable places to mobilize communities to engage people at risk and on the move in as Aceh, Indonesia; Beirut, Lebanon; Toronto, Canada; and Johannesburg, South Africa. The practices highlighted here reframe state-centric "solutions" as *local and civil society responses, faith-based initiatives, and decisions made by refugees.* One of UNHCR's three official "durable solutions," resettlement, can be recast as a *strategy*, not simply a "durable solution" or fix, to protracted displacement. Some people displaced from their homes may choose it and others not; people who choose it may stay in Canada, while others may return after obtaining the protection of citizenship in Canada (Hyndman and Giles 2017).

One outcome of the Toronto workshop was the Humane Mobility Manifesto (2018), an open statement for all interested allies to sign, which recognizes the wide range of people affected by displacement and the narrowness of the legal Convention refugee definition. The Manifesto begins:

> A deep reimagining of migration is urgently needed. We are profoundly concerned about responses to human mobility, including the Global Compact on Refugees and the artificial separation from wider migration issues. It emerges from exclusionary drafting and decision-making processes that ignore the lived realities of the people and spaces most affected by displacement. It privileges state sovereignty over human beings. It reinforces unequal power relations and waters down existing commitments to human rights and dignity.
>
> (Humane Mobility Manifesto 2018)

Another important and related subaltern response to the Global Compact on Refugees is the Kolkata Declaration, signed in that Indian city—host to many Rohingya refugees—in November 2018. A juxtaposition with and response to the New York Declaration of 2016, the Kolkata Declaration demands that the displaced be included in wide-ranging dialogues about their futures: "nothing about us without us" is a common refrain that has emerged from refugee advocate groups during the writing of the Global Compacts. The Declaration captures the irrelevance of the Global Compact on Refugees for South Asian states that have not signed onto international refugee law. The Declaration makes eight statements of purpose paraphrased below, with emphasis added (Kolkata Declaration 2018):

1. The right to move is a universal human right, and any restriction on that right cannot be subject to policies and measures that violate the dignity of human beings.
2. Refugees, migrants, stateless, and other displaced persons are central figures in any protection system, legal regime, government, and societal institutions,
3. The idea of *a global compact must acknowledge the practices of protection at various regional, country, local, customary, city, and other scales.* Any global compact aiming at sustainable resolutions must be based on wide-ranging dialogues involving refugees, migrants, stateless persons, and groups defending them.
4. Any protection framework—global and local—must combat discrimination based on race, religion, caste, ability, sexuality, gender, and class that affect rights and dignity of all human beings.
5. Any redesign of the global framework of protection, perpetrators of violence, and displacement must be held accountable for their actions.
6. Refugees, migrants, and stateless persons working as informal laborers are entitled to social and economic rights.

The Declaration highlights instead the salient issues that are rendered invisible by legal orientalism (Ruskola 2013): statelessness and migrant worker rights especially, all in the absence of the narrow, Occidental "refugee" definition invented more than seven decades ago.

Conclusion

In this final section, then, I suggest we adopt a way of thinking and imagining change that goes beyond the status quo. In this chapter, I have probed alternatives to the Convention-centric paradigm of containment and protracted displacement. We must also look awry for what is not easily seen. As with Burmese refugees in Thailand over the past three decades; Rohingya in Bangladesh and Malaysia; and Afghans, Tibetans, and Tamils in India, displaced migrant-refugees receive support and freedom from *refoulement*, but these actions are largely illegible to the West, an expression of the same legal orientalism alluded to above.

Refugee-migrants are (already) knowledge makers (Arendt 1958), city-makers (Fawaz et al. 2018), neighbors, parents, and teachers living and working in cities across the globe. They are subjects with agency to be recognized; their protection must be self-authorized. *Adat*, for example, shows constructive (if invisible) approaches to protection are happening at scales finer than the nation-state. Potential change beyond the status quo is truncated by the language we use. Distinct and hierarchically ordered categories of "refugee" and "migrant" are particularly problematic, generating different degrees of worthiness in public discourse. Humane mobility demands we rethink these monikers and reimagine protection beyond the current paradigm for people on the move (Hyndman and Reynolds, 2020).

References

Aleinikoff, Alex T., and Leah Zamore (2019). *The Arc of Protection*. Stanford, CA: Stanford University Press.

Amoore, Louise (2006). "Biometric Borders: Governing Mobilities in the War on Terror." *Political Geography* 25: 336–51.

Arendt, Hannah (1958). *The Origins of Totalitarianism*. Cleveland, OH: Meridian Books.

Bagelman, Jen (2018). Who Hosts a Politics of Welcome? *Fennia* 196(1): 108–110.

Betts, Alex (2013). *Survival Migration: Failed Governance and the Crisis of Displacement*. Ithaca, NY: Cornell University Press.

Betts, Alex, and Paul Collier (2017). *Refuge: Rethinking Refugee Policy in a Changing World*. Oxford: Oxford University Press.

Betts, Alex, and James Milner (2006). "The Externalisation of EU Asylum Policy: The Position of African States. Centre on Migration." Policy and Society Working Paper 36, University of Oxford.

Castles, Stephen (2008). "The Politics of Exclusion: Asylum and the Global Order." *Metropolis World Bulletin* 8: 3–6.

Chimni, B. S. (2008). "The 'Birth' of a Discipline: From Refugee to Forced Migration Studies." *Journal of Refugee Studies* 22(1): 11–29.

Clark-Kazak, Christina (2018). "Re-centring Refugee Agency." *Presentation to the Alternative Solutions to Refugee Protection Workshop*, June 5–6, Toronto, Canada.

Crawley, Heaven, Frank Duvell, Katherine Jones, Simon McMahon, and Nando Sigona (2017). *Unravelling Europe's "Migration Crisis": Journeys over Land and Sea*. Bristol, UK: Policy Press.

Crawley, Heaven, and Dimitri Skleparis (2018). "Refugees, Migrants, Neither, Both: Categorical Fetishism and the Politics of Bounding Europe's 'Migration Crisis'." *Journal of Ethnic and Migration Studies* 44(1): 48–64. doi: 10.1080/1369183X.2017.1348224.

Darling, Jonathan (2017). "Forced Migration and the City: Irregularity, Informality, and the Politics of Presence." *Progress in Human Geography* 41(2): 178–98.

Kolkata Declaration (2018). "Statement of Conference Participants." In *The State of the Global Protection System for Refugees and Migrants*. Calcutta Research Group. Accessed February 2, 2021, from http://www.mcrg.ac.in/RLS_Migration/Kolkata_Declaration_2018.pdf

Fawaz, Mona, Ahmad Gharbieh, Mona Harb, and Dounia Salame (2018). "Introduction." In Fawaz, Mona, Ahmad Gharbieh, Mona Harb, and Dounia Salame, eds. *Refugees as City-makers*. Accessed February 2, 2021, from https://www.aub.edu.lb/ifi/Documents/publications/research_reports/2018-2019/20180910_refugees_as_city_makers.pdf

Ferris, Elizabeth E., and Susan F. Martin (2019). "Special Issue: The Global Compact for Safe, Orderly and Regular Migration and the Global Compact on Refugees."

International Migration 57(6). Accessed February 2, 2021, from https://onlineli-brary.wiley.com/toc/14682435/2019/57/6

Agence France-Presse (2001). "305 Bodies Exhumed by Bosnia Muslims from a Mass Grave." *New York Times*, November 23. Accessed February 2, 2021, from https://www.nytimes.com/2001/11/23/world/305-bodies-exhumed-by-bosnia-muslims-from-a-mass-grave.html

Geiger, Martin, and Antoine Pecoud (2020). *The International Organization for Migration: The New UN Migration Agency in Critical Perspective.* New York: Springer.

Gill, Nick (2018). "The Suppression of Welcome." *Fennia* 196(1): 88–98.

Hathaway, James C., and Michelle Foster (2014). *The Law of Refugee Status.* 2nd edition. Cambridge, UK: Cambridge University Press.

Humane Mobility Manifesto (2018). *Statement of the Alternative Solutions to Refugee Protection Workshop Participants*, June 5–6, Toronto, Canada. Accessed February 2, 2021, from humanemanifesto.net.

Huysmans, Jeff (2006). *The Politics of Insecurity: Fear, Migration and Asylum in the EU.* London: Routledge.

Hyndman, Jennifer (1995). "Solo Feminism: A Lesson in Space." *Antipode* 27(2): 197–207.

Hyndman, Jennifer (2000). *Managing Displacement: Refugees and the Politics of Humanitarianism.* Minneapolis: Minnesota University Press.

Hyndman, Jennifer, and Wenona Giles (2017). *Living on the Edge: Refugees in Extended Exile.* New York: Routledge.

Hyndman, Jennifer, and Johanna Reynolds (2020). "Beyond the Global Compacts: Re-imagining Protection." *Refuge* 36(1): 66–74. Accessed February 2, 2021, from https://refuge.journals.yorku.ca/index.php/refuge/article/view/40768/36573

IOM (International Organization for Migration) (2018). Global Compact for Safe, Orderly and Regular Migration. Final draft, July 11, 2018. Accessed February 2, 2021, from https://refugeesmigrants.un.org/sites/default/files/180711_final_draft_0.pdf

IOM (International Organization for Migration) (2020). Homepage. Accessed February 2, 2021, from https://www.iom.int/mission

Al Jazeera (2015). "Migrants or Refugees?" Aljazeer.com., August 21. Accessed February 2, 2021, from https://www.aljazeera.com/programmes/insidestory/2015/08/migrants-refugees-150821200514311.html

Jones, Martin (2018). "Developing a Law of Asylum in the Middle East and Asia: Engaging the Law at the Frontiers of the International Refugee Regime," *Alternative Solutions to Refugee Protection Workshop*, June 5–6. University of Toronto, Toronto.

Jubilut, Liliana Lyra, André de Lima Madureira, and Daniel Bertolucci Torres (2019). "Current Challenges to the International Protection of Refugees and Other Migrants: The Role of and Developments from the United Nations 2016 Summit," 19–33. In Stephen M. Croucher, Joao R. Caetano, and Elsa A. Campbell, eds. *The Routledge Companion to Migration, Communication and Politics.* New York: Routledge.

Kyriakides, Christopher (2017). "Words Don't Come Easy: Al Jazeera's Migrant-Refugee Distinction and the European Culture of (Mis)trust." *Current Sociology* 65(7): 933–52.

Long, Katy (2011). *Permanent Crises? Unlocking the Protracted Displacement of Refugees and Internally Displaced Persons—Policy Overview.* Oxford: Refugee

Studies Centre. Accessed February 2, 2021, from https://www.internal-displacement.org/sites/default/files/inline-files/201110-Permanent-crises-Unlocking-the-protracted-displacement-of-refugees-and-internally-displaced-persons-thematic-en.pdf

Lukito, Ratno (1997). "Islamic Law and Adat Encounter: The Experience of Indonesia." MA thesis, McGill University, Montreal Canada.

Macklin, Audrey (2005). "Disappearing Refugees: Reflections on the Canada-US Safe Third Country Agreement." Accessed February 2, 2021, from https://www.researchgate.net/publication/228160962_Disappearing_Refugees_Reflections_on_the_Canada-US_Safe_Third_Country_Agreement

McAdam, Jane (2018). "The Global Compacts on Refugees and Migration: A New Era for International Protection?" *International Journal of Refugee Law* 30(4): 571–74. doi: org/10.1093/ijrl/eez004.

Missbach, Antje (2015). "The Rohingya in Aceh: Solidarity and Indifference" Indonesia at Melbourne. Accessed February 2, 2021, from https://indonesiaatmelbourne.unimelb.edu.au/the-rohingya-in-aceh-solidarity-and-indifference/

Missbach, Antje (2017). "Facets of Hospitality: Rohingya Refugees' Temporary Stay in Aceh." *Indonesia* 104: 41–64.

Morrissey, John. (2020). "Introduction," In John Morrisey, ed., *Haven: The Mediterranean Crisis and Human Security*. Northampton, MA: Edward Elgar.

Rankin, Jennifer (2017). "$2b EU-Africa 'Anti-Migration' Fund Too Opaque, Says Critics." *The Guardian*, October 31. Accessed February 2, 2021, from https://www.theguardian.com/global-development/2017/oct/31/2bn-eu-africa-anti-migration-fund-too-opaque-say-critics

Ruskola, Teemu. (2002). "Legal Orientalism." *Michigan Law Review* 101(1): 179–234. Accessed February 2, 2021, https://repository.law.umich.edu/mlr/vol101/iss1/4

Ruskola, Teemu (2013). *Legal Orientalism: China, the United States, and Modern Law*. Cambridge. MA: Harvard University Press.

Said, Edward W. (1978). *Orientalism*. New York: Pantheon Books.

Sparke, Matthew B. (2006). "A Neoliberal Nexus: Economy, Security and the Biopolitics of Citizenship on the Border." *Political Geography* 25(2): 151–80.

Sparke, Matthew B. (2018). "Welcome, Its Suppression, and the In-between Spaces of Refugee Sub-Citizenship—Commentary to Gill." *Fennia* 196(2): 215–19.

Squire, Vicki, ed. (2011). *The Contested Politics of Mobility: Borderzones and Irregularity*. London: Routledge.

Turk, Volker (2018). "The Promise and Potential of the Global Compact on Refugees." *International Journal of Refugee Law* 30(4): 575–83. doi: 10.1093/ijrl/eez004.

UNGA (United Nations General Assembly) (2016). "New York Declaration for Migrants and Refugees." Accessed February 2, 2021, from https://www.un.org/en/development/desa/population/migration/generalassembly/docs/globalcompact/A_RES_71_1.pdf

UNGA (United Nations General Assembly) (2018). "Report of the High Commissioner for Refugees: Part II Global Compact on Refugees." Accessed February 2, 2021, from https://www.unhcr.org/gcr/GCR_English.pdf

UNHCR (United Nations High Commissioner for Refugees) (2012). *State of the World's Refugees: the Search for Solutions*. Oxford: OUP.

UNHCR (United Nations High Commissioner for Refugees) (2017). "Global Trends: Forced Displacement in 2017." Accessed on Reliefweb, February 2, 2021, https://reliefweb.int/report/world/global-trends-forced-displacement-2017

UNHCR (United Nations High Commissioner for Refugees) (2019). "Global Trends: Forced Displacement in 2018." Accessed February 2, 2021, from https://www.unhcr.org/statistics/unhcrstats/5d08d7ee7/unhcr-global-trends-2018.html

Urry, John (2017). *Offshoring*. Cambridge: Polity Press.

Zetter, Roger (2007). "More Labels, Fewer Refugees: Remaking the Refugee Label in an Era of Globalization." *Journal of Refugee Studies* 20(2): 172–92. doi: 10.1093/jrs/fem011.

4 Governance of migration in South Asia

The need for a decolonial approach

Sabyasachi Basu Ray Chaudhury

Introduction

The world has been witnessing frequent mixed and massive flows of population in recent times, when refugees, asylum-seekers, and migrants on the move together can hardly be differentiated. The New York Declaration 2016 (UNGA 2016), and its two Global Compacts—the "Global Compact on Refugees" (GCR) (UNGA 2018a) and the "Global Compact on Safe, Orderly and Regular Migration" (GCM) (UNGA 2018b)—raised expectations for better governance of international human migration by committing itself to securing the rights and protection of refugees and migrants in the context of human rights discourse and working toward sustainable development. But even this laudable initiative seems to have been triggered specifically by Europe's reaction to its 2015 "crisis" and the disconcerting images that circulated as migrants and refugees from different parts of the world crossed the Mediterranean to reach European shores, with many perishing in the sea. Like all other previous attempts at the international level to formulate rules and norms for the governance of the people on the move, this effort turned out to be predominantly influenced by developments in the Global North (Hyndman 2012; Oelgemöller 2011, 2021).

In this chapter, I thus explore the nature of migration governance in South Asia in postcolonial, neoliberal times and consider the efficacy of a decolonial approach to governing human migration in South Asia in particular and the Global South in general.

I start with a brief look at the process of decolonization of countries in this region, scrutinizing the British colonial rulers' hurriedly redrawn boundaries of the late colonial period. I address ways in which the process of mapping the would-be postcolonial states has been problematic. Using several examples, I indicate how subsequent policies of postcolonial states in South Asia, including new citizenship acts, or amendments to existing acts, impacted migration in the region (as well as the generation of new refugees) by rendering more and more human beings stateless.

My examples represent major trends addressing how inhabitants of South Asia were gradually and systematically dispossessed of their basic rights as human beings—and expelled to a world of unfreedom and precarity—with

DOI: 10.4324/9781003170686-5

the assistance of discourse from modern nation-states and citizenship parameters derived from the Global North. Finally, I assess to what extent these primarily Eurocentric universal refugee and migration policies would be suitable enough to deal with the population flows in South Asia.

I pose three research queries in this chapter. First: Is the colonial legacy responsible for redrawing borders and, therefore, for marginalizing and dispossessing several ethnic groups in the region? Second: Why do the universal principles and laws quite often fail to tackle the pluralities and complexities of the postcolonial South Asia, thus widening existing tensions and creating more ethnic fault lines? And third, to conclude: How do the existing universal principles and international laws on migration and refugees, like the two recent Global Compacts, exacerbate problems created by colonial rule in South Asia?

Colonial legacy in South Asia

The Indian borderlands host and generate millions of refugees, but they also serve as a home to populations whose movements influence (and are heavily influenced by) the livelihoods, commerce, and border management regimes working in these areas. These multiple systems create opportunities for (and obstacles to) movements within borderlands and across newly created borders of the concerned countries. The mobility of capital and technology has been deregulated during the last 40 or so years. But migration—whether "voluntary" in the form of economic migration or involuntary (forced)—has increasingly been shaped by national, regional, and global regulatory norms, on the one hand, and national and/or regional security-enhancing institutions, on the other.

The Westphalian template of international (in reality, interstate) relations, which gives each state sovereignty over its territory, is ill-equipped to deal with tensions between more open economies and ever-increasing human mobility. As postcolonial societies of the Global South largely inherited colonially demarcated (and sometimes arbitrarily partitioned) territories (as in South Asia), they legitimized and reproduced the colonial boundaries. This quite often led to the fragmentation of earlier economic and cultural networks in the ethnic communities. Exclusionary impulses gained strength in a frame of state-nation underscored as nation-state. Spatially bounded parameters of belonging, based on the ideas of nation-state, borrowed from the Global North, have only excluded more people from the "nation" in South Asia, dooming them to a "bare life" of unfreedom and precarity. Subjugating colonialism has given way to postcolonial nations passionate about securing territorial integrity of the states yet unconcerned for the people who inhabited those territories for ages.

The birth of the modern territorial nation-state perhaps necessitated the mapping of nations and demarcation of borders and boundaries in Europe, and these new ideas soon spread to the one-time European colonies. In fact, European colonizers had already started to demarcate borders of their respective colonies to distinguish their own zones of control and influence. In certain cases, when colonial rulers could not control frontier areas, they granted some autonomy to deter "outside" powers and buy peace in the region. Accordingly, British colonial governance in the Indian subcontinent was primarily based on a strategic appreciation that allowed for a lighter colonial administrative footprint and greater autonomy for indigenous communities. This period began the demarcation between "inside" and "outside" a colonial territory.

Decolonization of the Indian subcontinent also led to partition in 1947, giving rise to two separate states, India and Pakistan. As the British colonial rulers hastily left their colonies, colonial boundaries and frontiers became *de facto* boundaries of the postcolonial states following the principle of *uti possidetis juris*, a specific legal mechanism in international law that serves to legitimize colonial borders in postcolonial times.

Territoriality, "the ability to enforce a writ over a particular geographical space or spaces," plays a key role in this conception of modern statehood; state boundaries became testing grounds for sovereign rule, especially in the newly demarcated boundaries of the Global South (Leake and Haines 2017: 964). In fact, imperial histories have demonstrated the complex nature of relationships between colonial sovereignty and border making (Baud and van Schendel 1997; Hansen and Stepputat 2006; Leake and Haines 2017).

Drawing borders posed cartographical and territorial challenges in South Asia, particularly when freshly demarcated lines on maps only nebulously followed geographical and sometimes ethnographic contours. But it also raised issues concerning relations between new states and their citizens as subjects of colonies constitutionally turned into citizens of new postcolonial nations. Citizenship remains a contentious issue, even after seven decades of decolonization in the region. This is true when assessing newly enacted citizenship laws (as in Burma, now Myanmar, or in Bhutan in the early 1980s) or existing acts that were amended (as in India in 2019); both efforts were part of a nation-building process that involved the inclusion of certain communities within the fold of a postcolonial state and the simultaneous exclusion of the remaining communities from the same political space (Basu Ray Chaudhury 2021).

When empires end, they often leave behind colonial borders and the subsequent complicated nature of nation-building with its social and political production of borders and nationalized spaces (Leake and Haines 2017). Close scrutiny of South Asia indeed uncovers complicated stories of postcolonial nation-building and boundary-making that involve reconciling the inherited complexities of colonial borders while coping with the exigencies of newly demarcated boundaries (Adelman and Aron 1999; van Schendel 2002). Nation-building in this region has been intended to homogenize a plural society consisting of different religious and ethnic communities and historically sharing the same territorial space. It, thus, homogenizes a heterogeneous space into territories to be controlled by a monopolistic state-power, with territorialization of the postcolonial space a major precondition for the emergence of politics in the modern sense (Balibar 2004).

For the architects of nation-building in South Asia, it was nearly impossible to identify any single logic of nation-building in the region. Rather, nation-building and boundary-making highlighted far more emotive, troubled processes through which colonies eventually became postcolonial states with intent to become nation-states in the European style, while still opposing the European colonizers. Borrowing from the decolonial thinkers, I thus argue that the Global South is a product of the empire, conceived to express all forms of subordination brought about by the capitalist world system (Santos 2014).

Once new borders were drawn or redrawn in newly decolonized South Asia, major distinctions needed to be made at the borders between citizens and foreigners, insiders and outsiders, and between those who would be allowed entry and those who would not be. Therefore, the new state-nations (my preferred name for them) started to close their borders. Within a decade, the new borders became inseparable from the concept of passport—except in case of the borders between India and Bhutan, and India and Nepal, as a result of contested bilateral treaties between the concerned states signed in 1949 and 1950, respectively.

Meanwhile, communal riots following the partition caused the deaths of about two million innocent men, women, and children. One of the largest mass migrations in the twentieth century occurred between 1946 and 1957, when at least 15 million people were forced to flee from (or in fear of) bloody communal riots and seek refuge across the newly demarcated "international boundaries." Because "partition continues to influence how the peoples and states of postcolonial South Asia envisage their past, present and future" (Jalal 2013: 4), fresh waves of migration to India occurred on the eve of and during the 1971 liberation war of Bangladesh.

Examples from South Asia—the Bhutanese Lhotsampa of Nepali origin, Myanmar's Rohingya, and the Bangladeshi undocumented migrants in India—explain how the situation continued to be complex, even after the birth of Bangladesh, when mapping the new nations had seemingly concluded.

Evicted people

The Himalayan kingdom of Bhutan was never colonized, but it had a close brush with British colonial powers when it signed the Treaty of Punakha in 1910. This treaty ensured that the British Indian administration would not interfere in Bhutan's internal affairs as long as Bhutan was willing to accept advice from the colonial powers regarding its external relations.

In the late nineteenth century, the Lhotsampa—people of Nepali origin (primarily of Hindu background)—settled in southern Bhutan, a previously underdeveloped area of the Himalayan kingdom. By the late 1970s, the dominant Drukpa community in Bhutan, who mainly follow Buddhist ideals, became apprehensive about the influence this growing number of Lhotsampa had on Bhutan's "cultural identity." Lhotsampa were soon identified as "illegal immigrants" and apparent threats to Bhutan's "survival as a distinct political and cultural entity" (Ministry of Home Affairs, Government of Bhutan 1993: 41). The government's mistrust of Lhotsampa minorities intensified in the late 1970s and the early 1980s when it adopted a policy of national cultural "orientation," and in 1977 and 1985, it passed two new Citizenship Acts in Bhutan that drastically altered citizenship requirements in the country. Having already viewed the Lhotsampa as "illegal immigrants," Bhutan refused to recognize the citizenship identity cards issued after the last census.

A decree from the king of Bhutan in January 1989 introduced the Driglam Namzha (Thronson 1993)—the traditional Drukpa code of values, etiquette, and dress—as the cornerstone of its Bhutanization policy. In February 1989, the Nepali language was excluded from the school curriculum (Rizal 2004; Ferraro 2012). Subsequently, through violent repression, forcible eviction, or formal pressure, the Bhutan government compelled more than 120,000 Lhotsampa to leave the country in 1990 (Amnesty International 1994). They entered India to reach Jhapa and Morang districts in eastern Nepal and seek temporary shelter in the refugee camps there.

As the Lhotsampa were excluded from the Bhutanese citizenship, they were also dispossessed of their land (Hutt 2003). The emptied agricultural lands, developed by the Lhotsampa in southern Bhutan, were subsequently handed over to the landless people of northern Bhutan. These violations of minority rights drew international condemnation to the Government of Bhutan (Amnesty International 1994). Based on the UN's elaboration of minority rights in the 1992 Declaration on the Rights of Persons Belonging to National or Ethnic, Religious and Linguistic Minorities, the UN repeatedly questioned Bhutan's denial of rights and citizenship in the Commission on Human Rights, and then in the Human Rights Council, albeit in vain (UN General Assembly 1992). The Bhutan government, viewing the Lhotsampa from the perspective of "terrorism" and "criminality," thus argued, "The problem of the people in the refugee camps in eastern Nepal is not a typical refugee situation, but one of highly complex nature, with its genesis in illegal immigration" (Kingdom of Bhutan 2009; Meier and Chakrabarti 2016).

In 2008, following several failed negotiations with the Bhutan government regarding repatriation, the UNHCR (United Nations High Commissioner for Refugees) decided to end this ongoing situation by resettling Bhutanese refugees in the United States and other countries, including Australia, Canada, Denmark, the Netherlands, New Zealand, and Norway (UNHCR 2009). The resettlement of nearly all the UNHCR-registered Bhutanese refugees is considered complete (Kantipur 2014) but not without a cost, as many refugee families were divided through this artificial process of resettlement.

The rejected Rohingya

The Rohingya, one of the most severely persecuted minorities in the contemporary world, were once residents of Arakan State (now Rakhine State) in

Burma (now Myanmar). They were excluded from Burma in 1982 when the nation's military junta enacted a new repressive citizenship law; it eventually rendered the Rohingya stateless, primarily on the basis of their religion and ethnicity (Basu Ray Chaudhury and Samaddar 2018). In 1991 and 1992, about 250,000 Rohingya fled an intensified post-election clampdown in Rakhine State, marking the first exodus of Rohingya from Myanmar.

The subsequent large-scale persecution of the Rohingya led to their further exodus, primarily into neighboring Bangladesh, but also in smaller groups to Malaysia and Thailand. After the August 25, 2017 massacre, the world was shocked by images of hundreds and thousands of innocent Rohingya men, women, and children fleeing mass atrocities. The 2018 report of a fact-finding mission (FFM) on Myanmar mandated by the United Nations Human Rights Council (UNHRC) also concluded that the State of Myanmar was responsible for carrying out acts of genocide against the Rohingya people and called for the arrest and prosecution of some top generals (OHCHR 2018).

Ahead of the August massacre, a February 2017 report released by the Office of the United Nations High Commissioner for Human Rights (OHCHR) found that an ongoing military crackdown by Myanmar's armed forces and police on Rohingya caused extrajudicial killings, burning of homes, massive and systematic sexual violence against Rohingya women, and mass killing of children (OHCHR 2017). With the August 2017 escalation of violence came a humanitarian disaster that forced over 730,000 Rohingya to flee their native land and seek refuge in makeshift and already overpopulated refugee camps near Cox's Bazaar, Bangladesh, where the concentration of Rohingya refugees is one of the densest in the world. Many Rohingya who settled outside the camps, with little official assistance, have in effect been integrated among the host societies of Chittagong following their stay in Bangladesh, possibly because of cultural and linguistic similarities between local Chittagonian people and Rohingya refugees (Human Rights Watch 2020). But Myanmar's complete nonrecognition of the Rohingya as legitimate citizens makes the future for more than a million stateless Rohingya refugees ambiguous and insecure. Following the military coup d'état in Myanmar in February 2021, it is likely that they will continue to languish indefinitely in the densely populated camps of southern Bangladesh. As their expectations of being recognized by citizens with due rights and entitlements fade away, their prolonged situation of precarity and unfreedom will create an intractable "Palestine-style" (Lintner 2020) scenario in South Asia, with the Rohingya as the "new Palestinians" (Considine 2017).

The experience of the Rohingya in Bangladesh, Malaysia, and India reflects the situation many refugees face in South and Southeast Asia. In the absence of appropriate legal frameworks dealing with asylum—and given the corresponding reliance on inconsistent *ad hoc* institutional practices—the protection environment for refugees and asylum-seekers in different South and Southeast Asian nations is quite uncertain and precarious (Basu Ray Chaudhury 2018; Cheung 2001).

A recent judgment supporting this prediction, delivered by the Manipur High Court in India on May 3, 2021, rightly admits that Article 51 of the Constitution of India "casts a non-enforceable duty upon the 'State' to promote international peace and security, apart from fostering respect for international law and treaty-obligations in the dealings of organized peoples with one another." It further says that the far-reaching and myriad protections afforded by Article 21 of the Constitution of India, "would indubitably encompass the right of *non-refoulement*, albeit subject to the condition that the presence of such asylum seeker or refugee is not prejudicial or adverse to the security of this country." The Division Bench of the Manipur High Court concludes that

> though India may not be a signatory to the Refugee Convention of 1951, its obligations under other international declarations/covenants … enjoin it to respect the right of an asylum-seeker to seek protection from persecution and life or liberty-threatening danger elsewhere."
>
> (High Court of Manipur 2021)

Only occasionally do such judicial decisions appear as bright sparks in the dark and inconsistent legal horizon of South Asia.

Partitioned lives

The inconclusive partitioning in the east of the Indian subcontinent began with the hurried border drawn in 1947 by the Bengal Boundary Commission, under the leadership of Sir Cyril Radcliffe (Chatterjee 1999). As a consequence, the physical and geographical contours and social and economic linkages and histories of the undivided Bengal were practically overlooked by the commission (Ahmed 2011; Bose 2000; Butalia 2000, 2015; Chatterjee 2007; Raychaudhury 2004; Samaddar 1977, 2003; Singh 2016). While the famous jute mills, symbols of industrialization in Bengal during the colonial period, were on both sides of Hooghly River in West Bengal (the new state of India), the raw jute was still being mainly produced in the fields of East Bengal (later East Pakistan and subsequently Bangladesh).

Migration within South Asia must be assessed in this context of uncomfortable and uneasy partition in the east. The partition of Bengal, which led to the division of its economy and society, meant the loss of home and sudden disappearance of *desh* (country) for hundreds and thousands of Bengalis and other smaller ethnic communities. The situation became more complex after the December 1970 elections in Pakistan, as civil war broke out when poll results were annulled, causing a huge exodus of refugees to India from East Pakistan, now Bangladesh. Many refugees could trek back to their native land after the liberation of Bangladesh, however, a situation not possible for partition refugees of 1947.

But the refugee influx from East Pakistan had already started to change the demographic composition of India's bordering states. For instance, in

Tripura the Bengali-speaking refugees of 1947 and 1971 soon outnumbered the earlier indigenous majority. Similarly, the demographic composition of Assam changed significantly. As a consequence, while the ethnic Assamese retained their predominance in the Brahmaputra Valley of the state, the Bengali refugees arriving from across the international border soon overwhelmed the Barak Valley in Assam.

Cross-border migration and the incoming refugees, a major issue in Assam since the decolonization and partition of India, forced the government to enact the Immigrants (Expulsion from Assam) Act in 1950 and prepare a National Register of Citizens (NRC) based on the 1951 census in India. In the context of the overall changing demography of the state, young educated ethnic Assamese soon started to spearhead the antiforeigner movement in the state. When the All Assam Students' Union (AASU) led this movement during 1979 and 1985 with the participation of other organizations in Assam, cross-border migration soon turned into a major political issue in India's northeast. Foreigners were classified into three categories for identification and differential treatment under Clause 5 of the 1985 Assam Accord.

In spite of this agreement, Bangladeshi undocumented migrants, commonly referred to as "illegal immigrants," became one of the most sensitive and most talked about topics in India's northeast; the foreigner issue remained unresolved and the process of identifying foreigners in Assam did not materialize as expected. Subsequently, initiatives were established to introduce identity cards for "genuine" Indian citizens in India's northeast and thus contain the influx of migrants from Bangladesh.

The Citizenship Amendment Act (CAA) 2019 (*The Gazette of India* 2019), amending the Citizenship Act of 1955, has opened a path to Indian citizenship for persecuted religious minorities—Hindus, Buddhists, Parsis, Sikhs, Jains, and even Christians—from Pakistan, Afghanistan, and Bangladesh. But the CAA excludes thousands of Tamil refugees (mostly Hindus) fleeing the sites of civil war and genocide in Sri Lanka, hundreds of Rohingya (mostly Muslims) fleeing ethnic cleansing in Myanmar, and Muslims of Shia and Ahmadi stock facing persecution in Pakistan. The cut-off date for granting citizenship has shifted from March 24, 1971, to December 31, 2014.

Although the inflow of Hindus to India has been considered a "natural homecoming," the inflow of Muslims, even if they are refugees, has been treated as "infiltration" (Basu Ray Chaudhury 2021: 212) or "illegal immigration." Meanwhile, updating the NRC in Assam in August 2019 to determine citizenship of "suspicious" residents of this state has led to the exclusion of 1.9 million people already residing there, both Hindus and Muslims, probably for decades (Basu Ray Chaudhury 2021: 212).

Inadequacies of (post)colonial approaches

As a result of complex situations emerging from the movements of refugees, stateless people, and undocumented migrants across the redrawn boundaries of South Asia (the eastern part in particular), international instruments have been set up to deal with issues relating to nationality, citizenship, refugees, migrants, asylum-seekers, and stateless people. Are they capable of addressing the complexities relating to the movements of people in the Global South, in general, and South Asia, in particular? Have the leaders of these postcolonial states been able to decolonize their thinking during the last 70 years?

Migration seems to be a legitimate area of sovereign regulation and governance for the postcolonial states of South Asia, where policies are calibrated according to the respective political goals and imperatives and where sovereignty has precedence over the entry and residence of non-nationals. But international human rights and humanitarian laws restrict a state's powers to socially and economically exclude or deport a migrant who resides legally in its territory. International law also limits states' sovereignty to regulate entry, for instance, through the principle of *non-refoulement*. But in South Asia, the state has the final say about the entry, residence, or deportation of a person arriving at the border or sneaking through it.

The 1951 Refugee Convention obliges a state not to return or expel migrants who would otherwise suffer persecution, but it and many other international conventions and laws are largely perceived as Eurocentric: in spite of their universal character and applicability, they mostly emerged in the context of World War II, partly because postcolonial India and Pakistan accommodated millions of refugees in a fairly competent manner in the late 1940s, when there was no UN support or no UNHCR.

Given the diverse origin and nature of migrant, refugee, and stateless peoples in different geographies and histories, we must look at these issues from a pluriversal perspective that takes into account different legal geographies and anthropologies of accommodation. Recent decolonial thinking, in contradistinction to most modernist and postcolonial ideas, has been more inspired by the indigenous movements in the Global South, particularly in Latin America (Hutchings 2019), by attempting to provide an alternative place and time from which to mobilize resistance to modernity. These decolonial and pluriversal arguments have two basic premises: first, that dominant Western theory and practice depend on the particular ethical and political effects of a Euro-modern ontology and, second, that the Western ontology's claim to universality is false (Hutchings 2019).

In fact, the critique of modernity in the Global South has largely been influenced by its subordination to and dependency on a Western-dominated world even after decolonization (Khoury2012). Therefore, even in postcolonial societies, the production of knowledge has predominantly been subject to colonial blueprints and geopolitics that are Eurocentric. Most postcolonial accounts, like their colonial counterparts, tend to disregard the existence of other knowledges in connection to refugees and migrants (Khoury and Khoury 2013) as well as different minorities within the boundaries of the state.

As the world witnesses massive numbers of people on the move, it is difficult to distinguish between refugees and asylum-seekers (normally protected by international refugee law) and the economic migrants (who usually do not enjoy that level of protection); earlier classifications are gradually becoming blurred. Similarly, post-9/11, increased efforts to securitize migration is pushing humanitarian concerns to the backburner. The view of state sovereignty from an absolutely fresh global perspective, which was seeming to emerge from the Westphalian order, now includes a renewed notion of sovereignty. This is evident from recent situations within the European Union (EU) and from approaches of bigger and smaller states elsewhere who are beset by populist ideas of ultra-nationalism, xenophobia, and therefore, intolerant policies vis-à-vis refugees, asylum-seekers, and migrants.

South Asian countries, already obsessed with their newly acquired sovereignty and distraught by the nuanced impact of partition, have started tweaking their citizenship laws in a larger context of populist politics, thereby exacerbating the ethnic fissures and eventually generating more refugees and stateless people. In this scenario, the so-called universal principles and laws fail to address the pluralities and complexities of postcolonial South Asia, thus creating more ethnic fault lines and widening the existing ones.

I, thus, maintain that pluriversal arguments grounded in anthropologies of accommodation offer critiques to the universal ideas and legal positions emanating from Europe. The pluriversal approaches, after all, claim to recognize and appreciate other cultures and legal instruments emanating from other knowledges, histories, and experiences (Mignolo 2011). The truly international legal framework cannot and should not be trapped within the Euro-American framework. I also argue that while creating emancipatory international and universal frameworks, we should consider the geographies of reason and law (Mignolo 2007).

Conclusion: examining the "global" in global compacts

On September 19, 2016, the UN General Assembly adopted the New York Declaration for Refugees and Migrants, thus sending a powerful political message that refugees and migrants had become major issues in the global agenda. UN Member States recognized the need for a comprehensive approach and enhanced cooperation regarding human mobility at the global level. But this approach once again reiterated the hierarchic treatment of refugees, asylum-seekers, stateless people, and migrants by making way for two Global Compacts in 2018, one for refugees (GCR) and one for migrants (GCM), thus ignoring the inseparability of these categories, even in the Global North during the recent Mediterranean "crisis."

The Global Compacts have been drafted mainly to address the refugee and migrant crisis in the Global North, thereby neglecting the concerns in the Global South and exacerbating problems created by colonial regimes in South Asia. Just like the disturbing death of the Syrian–Kurdish toddler Aylan Kurdi and his family, who drowned in the Mediterranean Sea, many Rohingya refugee children succumbed to death while trying to cross the rough stretches of the Bay of Bengal in leaky boats to reach the shores of South and Southeast Asian countries (Sengupta 2018). Although the UN has identified the Rohingya as the most persecuted minority, the initiative to formulate Global Compacts again failed to take any special note of the refugees and migrants in the Global South, thereby ignoring the other legal geographies.

The expanding discourses of securitization and the "illegality" of migrants and refugees across the globe have kept the universal principles and international laws on migration and refugees focused on the Global North, through which neoliberal ideologies persevere to contain racialized and dangerous "others" outside the state's juridical and spatial confines.

Taking a cue from the Kolkata Declaration 2018, adopted after a conference organized by the Calcutta Research Group (CRG) on "The State of the Global Protection System for Refugees and Migrants" in November 2018, I question just how global these compacts are if they do not acknowledge different practices of protection at various regional, national, local, customary, city, and other scales. Any global compact aiming at sustainable resolutions has to be "based on wide-ranging dialogues involving refugees, migrants, stateless persons and groups defending them" (Kolkata Declaration 2018: 6). To overcome the existing inadequacies and address the complex issues generated by the overlapping of refugees, migrants, and stateless people now requires looking beyond international refugee law per se. A framework for governing global migration must take into consideration the international human rights and humanitarian laws, along with the other international, regional, and even national legal frameworks. An emancipatory blueprint of governance must also appreciate the pluriverse of legal geographies. We urgently need an alternative, decolonial and pluriversal approach to migration, based on the histories, experiences, and occurrences in different parts of the world.

References

Adelman, J., and S. Aron (1999). "From Borderlands to Borders: Empires, Nation-States, and the Peoples in between in North American History." *The American Historical Review* 104(3): 814–841.

Ahmed, Ishtiaq (2011). *The Punjab Bloodied, Partitioned and Cleansed: Unravelling the 1947 Tragedy through Secret British Reports and First-Person Accounts*. New Delhi: Rupa.

Amnesty International (1994). "Bhutan: Forced Exile." Index number: ASA 14/004/1994. Accessed February 8, 2021, https://www.amnesty.org/en/documents/asa14/004/1994/en/

Balibar, Étienne (2004). "Europe as Borderland." *Alexander von Humboldt Lecture in Human Geography*. Institute for Human Geography, Universiteit Nijmegen, The Netherlands. November 24. Accessed April 1, 2021, https://www.ru.nl/gpe/research/alexander-von-humboldt-lectures/past-lectures/

Basu Ray Chaudhury, Sabyasachi, and Ranabir Samaddar, eds. (2018). *Rohingya in South Asia: The People without a State*. Abingdon: Routledge.

Basu Ray Chaudhury, Sabyasachi(2018). *Statelessness, International Conventions and the Need for New Initiatives? Addressing the New Frontiers of Statelessness.*

Migration and Governance V, Policies and Practices 102. Kolkata: Mahanirban Calcutta Research Group.

Basu Ray Chaudhury, Sabyasachi (2021). "Dispossession, Un-freedom, Precarity: Negotiating Citizenship Lawsin Postcolonial South Asia." *The South Atlantic Quarterly* 120(1): 209–219.

Baud, M., and Willem van Schendel (1997). "Toward a Comparative History of Borderlands." *Journal of World History* 8(2): 211–242.

Bose, Pradip Kumar (2000). *The Refugees in West Bengal: Institutional Practices and Institutional Identities.* Calcutta: Calcutta Research Group.

Butalia, Urvashi (2000). *The Other Side of Silence: Voices from the Partition of India.* London: Hurst & Co.

Butalia, Urvashi, ed. (2015). *Partition: The Long Shadow.* New Delhi: Zubaan.

Chatterjee, Joya (1999). "The Fashioning of a Frontier: The Radcliffe Line and Bengal's Border Landscape, 1947–52." *Modern Asian Studies* 33(1): 185–242.

Chatterjee, Joya (2007). *The Spoils of Partition: Bengal and India, 1947–67.* Cambridge Studies in Indian History and Society, no. 15. Cambridge: Cambridge University Press.

Cheung, Samuel (2001). "Migration Control and the Solutions Impasse in South and Southeast Asia: Implications from the Rohingya Experience." *Journal of Refugee Studies* 25(1): 50–70.

Considine, Craig (2017). "*The Rohingya are the New Palestinians.*" *Foreign Policy,* September 26. Accessed May 5, 2021, https://foreignpolicy.com/2017/09/26/the-rohingya-are-the-new-palestinians/

Ferraro, Matthew F. (2012). "Stateless in Shangri-La: Minority Rights, Citizenship, and Belonging in Bhutan." *Stanford Journal of International Law* 48(2): 405–435.

Hansen, Thomas Blom, and Finn Stepputat (2006). "Sovereignty Revisited." *Annual Review of Anthropology* 35(1): 295–315.

High Court of Manipur at Imphal (2021). *Nandita Haksar vs. State of Manipur and Ors.* W.P. (Crl. No. 6 of 2021).

Human Rights Watch (2020). *Burmese Refugees in Bangladesh,* May 1. Accessed February 11, 2021, http://www.hrw.org/en/reports/2000/05/01/burmese-refugees-bangladesh-0

Hutchings, Kimberley (2019). "Decolonizing Global Ethics: Thinking with the Pluriverse." *Ethics & International Affairs* 33(2): 115–125.

Hutt, Michael (2003). *Unbecoming Citizens: Culture, Nationhood and the Flight of Refugees from Bhutan.* Oxford: Oxford University Press.

Hyndman, Jennifer (2012). "The Geopolitics of Migration and Mobility." *Geopolitics* 17(2): 243–255.

Jalal, Ayesha (2013). *The Pity of Partition: Manto's Life, Times and Work across the India-Pakistan Divide.* Princeton, NJ: Princeton University Press.

Kantipur (2014). "90,000 Bhutanese Refugees Resettled: UNHCR, IOM." May 23. Accessed February 8, 2021, http://www.ekantipur.com/2014/05/23/top-story/90000-bhutanese-refugees-resettled-unhcr-iom/389922.html

Khoury, Laura (2012). "The Novel as the Voice of Subaltern Arabs: Critique of Western Modernity in Arab Thought." *Humanities and Social Sciences Review* 1(2): 419–435.

Khoury, SeifDa'Na, and Laura Khoury (2013). "'Geopolitics of Knowledge': Constructing an Indigenous Sociology from the South." *International Review of Modern Sociology* 39(1): 1–28.

Kingdom of Bhutan (2009). *National Report Submitted in Accordance with Paragraph 15(A) of the Annex to Human Rights Council Resolution 5/1* (Report Submitted for the Universal Periodic Review). Accessed May 31, 2021, https://digitallibrary. un.org/record/666675?ln=en

Kolkata Declaration 2018: *Protection of Refugees and Migrants (2018)*. Kolkata: Mahanirban Calcutta Research Group.

Leake, Elisabeth, and Daniel Haines (2017). "Lines of (In)Convenience: Sovereignty and Border-Making in Postcolonial South Asia, 1947–1965." *The Journal of Asian Studies* 76(4): 963–985.

Lintner, Bertil (2020). "Rohingya Refugees becoming Palestinians of Asia." *Asia Times*, August 26. Accessed May 5, https://asiatimes.com/2020/08/rohingya-refugees-becoming-palestinians-of-asia/

Meier, Benjamin Mason, and Averi Chakrabarti (2016). "The Paradox of Happiness: Health and Human Rights in the Kingdom of Bhutan." *Health and Human Rights* 18(1): 193–207.

Mignolo, Walter D. (2007). "Delinking: The Rhetoric of Modernity, The Logic of Coloniality and the Grammar of De-coloniality." *Cultural Studies* 21(2–3): 449–514.

Mignolo, Walter D. (2011). *The Darker Side of Western Modernity: Global Futures, Decolonial Options*. London: Duke University Press.

Ministry of Home Affairs, Government of Bhutan (1993). *The Southern Problem: Threat to a Nation's Survival*. Thimphu: Government of Bhutan.

Oelgemöller, Christina (2011). "Informal Plurilateralism: The Impossibility of Multilateralism in the Steering of Migration." *The British Journal of Politics and International Relations* 13(1): 110–126.

Oelgemöller, Christina (2021). "The Global Compacts, Mixed Migration and the Transformation of Protection." *Interventions* 23(2): 183–190.

OHCHR (2017). "Interviews with Rohingyas Fleeing from Myanmar Since 9 October 2016." Accessed February 11, 2021, https://www.ohchr.org/Documents/Countries/ MM/FlashReport3Feb2017.pdf

OHCHR (2018). "Report of the Detailed Findings of the Independent Fact-Finding Mission on Myanmar" (A/HRC/39/CRP.2), September 18. Geneva, Switzerland, Accessed February 9, 2021, https://www.ohchr.org/en/hrbodies/hrc/myanmarffm/ pages/index.aspx

Raychaudhury, A. B. (2004). "Nostalgia of 'Desh,' Memories of Partition." *Economic and Political Weekly* 39(52): 5653–5660.

Rizal, Dhurba (2004). "The Unknown Refugee Crisis: Expulsion of Ethnic Lhotsampa from Bhutan." *Asian Ethnicity* 5(2): 151–177.

Samaddar, Ranabir, ed. (1977). *Reflections on Partition in the East*, New Delhi: Vikas.

Samaddar, Ranabir, ed. (2003). Refugees and the State: Practices of Asylum and Care in India 1947–2000. New Delhi: Sage.

Santos, Boaventura de Sousa (2014). *Epistemologies of the South: Justice Against Epistemicide*. Abingdon: Routledge.

Sengupta, Sucharita (2018). "Stateless, Floating People: The Rohingya at Sea," 20–42. In Sabyasachi Basu Ray Chaudhury and Ranabir Samaddar, *The Rohingya in South Asia: People without a State*. Abingdon: Routledge.

Singh, Khushwant (2016). *Train to Pakistan*. New Delhi: Penguin.

The Gazette of India (2019). December 12. Accessed May 31, 2021, https://egazette. nic.in/WriteReadData/2019/214646.pdf

Thronson, David (1993). *Cultural Cleansing: A Distinct National Identity and the Refugee from Southern Bhutan*. Kathmandu: INHURED International.

UNGA (UN General Assembly) (1992). Declaration on the Rights of Persons belonging to National or Ethnic, Religious and Linguistic Minorities, A/RES/47/135, adopted on 18 December 1992.

UNGA (2016). "The New York Declaration for Refugees and Migrants." Accessed May 31, 2021, http://www.un.org/en/ga/search/view_doc.asp?symbol=A/RES/71/1

UNGA (2018a). "Part II Global Compact on Refugees" Accessed May 31, 2021, https://www.unhcr.org/gcr/GCR_English.pdf

UNGA (2018b). "Final Draft of the Global Compact on Migration." Accessed May 31, 2021, https://www.refugeesmigrants.un.org/sites/default/iles/180711_inal_draft_0.pdf

UNHCR (2009). *Over 20,000 Bhutanese Refugees Resettled from Nepal*. Accessed February 8, 2021, http://www.unhcr.org/4aa641446.html

van Schendel, Willem (2002). "Stateless in South Asia: The Making of the India-Bangladesh Enclaves." *Journal of Asian Studies* 61(1): 115–147.

5 Lives on the move

Experiences of exclusion, vulnerability, and resilience of Venezuelan forced migrants in Peru

Luisa Feline Freier and Andrea Kvietok

Introduction

Displaced Venezuelans' experiences with exclusion and vulnerability begin back home in response to rising levels of violence, lack of food, hyperinflation, and the breakdown of the public health system (OAS 2019; Vivas and Paez 2017). As a result of these and other factors that have compromised individuals' safety and ability to continue making a living, over 5.5 million Venezuelans have left behind their homes (R4V 2021) in an attempt to pursue better livelihoods for themselves and for their families. Importantly, Venezuelan forced displacement now constitutes the fastest-growing displacement scenario in the world.

Adding layers of exclusion and vulnerability, their journeys have taken on a predominantly regional dimension, with over four million migrants and refugees moving to neighboring countries (R4V 2021). Men, women, and children move largely on foot, along treacherous roads, fearing thieves and human traffickers, enduring inhospitable climates, and encountering problems with border patrol agents regarding required documentation needed upon entry: Venezuelans on the move often endure assaults to their physical and psychosocial well-being (Carroll et al. 2020).

Peru is the second-largest recipient of Venezuelan forced displacement worldwide, with over a million Venezuelan migrants and refugees (R4V 2021). Upon their arrival in the country, Venezuelans have increasingly encountered stricter legal entry requirements, reduced integration efforts, and declining public perceptions toward Venezuelan immigration (Aron and Castillo Jara 2020; Freier and Castillo Jara 2020; Freier and Pérez 2021). Most participants in our interviews experienced some or all of these integration barriers, which converge to prevent their access to the formal labor market. Given the difficulties associated with the regularization of their migratory status, or the recognition of their professional titles or skills (Equilibrium CenDE 2020a), displaced Venezuelans often end up performing underpaid jobs (for which they are overqualified) without formal payroll benefits or security (World Bank 2019). They also endure instances of xenophobia and gender-based discrimination (Freier and Pérez 2021), all of which force them into a more precarious social, economic, and legal standing.

DOI: 10.4324/9781003170686-6

The Covid-19 pandemic has further exacerbated the mounting layers of exclusion and vulnerability experienced by Venezuelan migrants and refugees. The country's mandatory lockdown, and subsequent mobility restrictions, deeply impacted their working arrangements, increased their risk of eviction, and lessened their ability to provide for their families in Peru and those left behind in Venezuela (Equilibrium CenDE 2020b); this led some to embark on journeys to return back home (Salazar Vega 2020). Their precarious position was also heightened, as they were not considered for state-led social protection programs and had extremely limited access to the Peruvian healthcare system (Freier and Vera Espinoza 2021; Zambrano-Barragán et al. 2021). This scenario not only increased their risk of infection but also deepened depression and anxiety levels (Bird et al. 2020).

Forcibly displaced Venezuelans' experiences of exclusion and vulnerability need to be understood within a framework that addresses their lack of international protection (UNHCR 2019). More specifically, legal scholars have confirmed the applicability of the extended refugee definition of the 1984 Cartagena Declaration, which has been incorporated into the domestic legislation of 15 countries in the region, including that of Peru (Berganza et al. 2020; Freier et al. 2020). In practice, however, Peru has only recognized 1,282 Venezuelans as refugees, out of a population of more than one million (R4V 2021), leaving many with partial or nonexistent protection.

Amid these cumulative instances of exclusion and vulnerability, displaced Venezuelans employ a number of resilience strategies that build on transnational linkages with different actors and that renegotiate notions of belonging and community: such strategies include their faith in God as a source from which to garner strength and seek company in the face of uncertainties and the hardships they have experienced, keeping in touch with family members back home through phone and video calls, and relying on migrant networks as sources of information, support, and solidarity.

In this chapter, we seek to nuance and diversify extant literature on forced displacement in the so-called Global South by placing migrants' voices and narratives at the forefront of our analysis. In doing so, we not only recognize their agency as storytellers and as active social actors throughout the migration process, but we also present valuable insight into the myriad and complex ways they experience, and make sense of, exclusion, vulnerability, and resilience. Based on semi-structured interviews conducted between 2019 and 2020 with 61 Venezuelans at the Ecuador–Peru border and in three Peruvian cities, we argue that the mounting layers of exclusion and vulnerability marginalize Venezuelans into more precarious social, economic, and legal standings, and negatively impact their physical and psychosocial well-being. In spite of this precarity, however, forcibly displaced Venezuelans often demonstrate resilient strategies that build on multistranded, transnational connections with family members, friends, and compatriots. These ties allow them to cope with hardships associated with the migratory process and to renegotiate notions of belonging and community.

On voices, narratives, and qualitative research

Although it is impossible to deny the role of global labor regimes, citizenship and migratory status, border securitization efforts, and other structural factors in shaping migratory flows, people both impact and are impacted by migration. Of course, individuals' migratory experiences are not one and the same. Rather, they are time- and place-specific, and vary according to individuals' sociocultural markers of identity. Qualitative studies on migration, especially those that adopt an ethnographic methodology (Boccagni and Schrooten 2018; FitzGerald 2006), allow us to engage with the multidimensional ways people on the move navigate their transnational worlds. Their voices and narratives, thus, serve as a central piece for understanding mobility as a highly heterogeneous process and an intricate interplay between socioeconomic and political systems, as well as familial and individual aspirations.

Given its ability to produce nuanced analyses, as well as to highlight the voices of immigrant groups, qualitative research in migration studies is also useful in refining conceptual frameworks and redefining existing theoretical paradigms and concepts regarding migration dynamics (Zapata-Barrero and Yalaz 2018: 2). Rather than working under assumptions of migration as a unidirectional process in which migrants travel from countries of origin to countries of destination, human mobility "involves multiple physical, social and symbolic locations, whether simultaneously or over time" (Boccagni and Schrooten 2018: 220), as well as the multistranded social relations migrants forge and sustain across these spaces (Basch et al. 1994).

As the voices and narratives laid out in the next pages will show, the multiple locations of the migratory process, as well as the transnational connections migrants maintain, need to be understood as socially interdependent sets of phenomena. More specifically, amid the hardships experienced when uprooted from their home country, displaced Venezuelans often cited certain ties—material (e.g., remittances), informational (e.g., tips, guidance), and affective (e.g., phone calls, memories, nostalgia, love)—with family members, friends, and compatriots in different geographical locations (Cole and Groes 2016). Such ties contribute to the long-standing critique of the isomorphism between space, place, and culture (Appadurai 1996; Gupta and Ferguson 1992; Sassen 1991) and provide the framework through which to analyze the ways migrants renegotiate notions of belonging and community within a context of forced displacement.

On exclusion, vulnerability, and resilience

Venezuela's deepening crises have forced millions to leave behind their homes. The majority of those displaced have not been able to afford plane tickets, or, in some cases, bus fares, and have had no other option than to travel on foot to neighboring Colombia, Ecuador, and Peru. This section highlights participants' testimonies on their experiences of exclusion and vulnerability during the journey from Venezuela to Peru, placing a special emphasis on the added precarity of certain groups: pregnant women and families with children.

Uprooting, danger, and resilience

For our participants, the "journey" was not a straightforward trip but rather included different stations, moments of immobility, and decision-making processes along the way. Highlighting the material, informational, and affective transnational ties displaced Venezuelans maintain with family members, friends, and compatriots in distinct geographical locations, this section also reveals the resilient strategies they employ to renegotiate notions of belonging and community.

A 44-year-old criminal lawyer named Esther[1] decided to leave Venezuela for her safety, after receiving threats from public officials for denouncing acts of corruption and finding herself unable to provide for her two children and parents in her role as head of household. Her story vividly depicts the affective dimensions of separation from one's support system. First settling in Colombia, she worked tirelessly in short-term jobs for six months before being able to send back money to bring her 20-year-old daughter and teenage son to join her. But not long after their arrival, her firstborn returned to Venezuela, as she needed to receive medical treatment. The separation deeply affected Esther, as she emphasized:

> My family is a team, and as a team there are three of us, me and my two children. We have always maintained a balance, and wherever I am, my two children are there with me. The same needs, opportunities, strengths, and threats that I may have, they have, because we are a team, and when one leaves, the team is incomplete. This hurts, because I need the ideas of [my daughter] to strengthen the ideas of [my son] and to reinforce the security of all three.

After a year and six months in Colombia, Esther and her son decided to travel to Peru by bus, following an invitation they received from her son's friend. But the country's economy was slowly turning into a "Little Venezuela," meaning that it was becoming increasingly impossible to land a stable job that would allow her to make a living and take care of her family. Determined to reach Peru, they used their savings, as well as the money her son's friend had sent, to finance the trip. In Bogota, however, they were robbed of all their belongings. After a few days of asking around and assessing their options, they boarded one of the humanitarian buses provided by the International Organization for Migration (IOM) at the time, which offered safe transit for migrants and refugees in the region, mainly to cross Ecuador.

More often than not, displaced Venezuelans' perilous journeys compromise their physical and psychosocial well-being (Carroll et al. 2020). Eight months into her pregnancy, and traveling with her infant daughter, 24-year-old Maria Alejandra waited impatiently at the Bi-national Border Center

(CEBAF for its Spanish acronym) in Tumbes, Peru, in August 2019, for a humanitarian bus to take her to Lima, the country's capital. There, she would reunite with her partner, who had left two months earlier, searching for a way to sustain his growing family. Commenting on her journey through Colombia, Ecuador, and Peru, she highlighted that, ever since leaving Venezuela, life had been difficult. In addition to being pregnant and single-handedly caring for her daughter, rationing the little money she had on transportation and food, she recounted dangers she encountered on the road, as well as from inhospitable climates, all while lacking basic necessities. Bringing up a particular instance at the Colombia–Ecuador border, she shared:

> I spent three days there, sleeping on the floor, the bare floor! ... There weren't many tents available for you to sleep in, and one tried to stay inside, but the guards and police would kick you out. ... I had a suitcase, so I opened it and put my daughter to sleep there [and I slept on the floor]. One day it started pouring, so I moved her inside, but I got drenched.

Difficulties like these were exacerbated by the frustration of not being able to adequately care for her child, which took a toll on her well-being, and by the fact that she missed the company and support of her partner, despite communicating with him via phone on a regular basis.

Shifting policy requirements, designed and implemented to curb Venezuelan immigration, also hamper migrants' experiences on the road, leaving many stranded at the border without access to basic necessities. Although Esther's time spent at the Ecuador–Peru border in August 2019 was short because she held a passport and a clean criminal record—documents that, in theory, facilitated soliciting a humanitarian visa—many displaced Venezuelans do not possess the necessary documents to meet entry requirements, leaving them to wait at the border for days or weeks. In Maria Alejandra's case, she had left Venezuela with only an expired national identification card. Given the hardships she experienced at the Colombia–Ecuador border, she decided to cross without getting her Andean Migration Card stamped—a necessary step to certify her entry into and exit out of the country. This decision, as well as the fact that she is not officially married to her partner, hindered her ability to enter Peru regularly, despite her status as a pregnant woman with a partner who was working and residing in the country.[2] Initially rejected, she suffered extreme emotional distress. An NGO at the border eventually helped Maria Alejandra, and after four days the Peruvian border agents stamped her card.

Despite the arduous work of NGOs at the border, only a few Venezuelan migrants share Maria Alejandra's fate. Thirty-year-old Gabriel, who left Venezuela with his wife and three children because of the deteriorating economic situation, arrived at the Ecuador–Peru border in August 2019, after 10 days of walking and hitchhiking. Thinking they could cross with the Andean Migration Card, as he had done so earlier that year, they were now denied entrance by immigration agents, as they did not have a humanitarian visa.

Not in possession of the required documentation, and stranded at the border with his entire family, Gabriel shared, "if I could, I would [cross irregularly], because, really, Venezuela, [is not an option]." But he also added that he would be terrified of being caught and deported if he chose this path. His testimony showcases how the humanitarian visa requirement in June 2019 actually amplified this group's vulnerability (Freier and Luzes 2021), leading many to fall prey to additional perils, like human trafficking rings.

Amid these dangers, participants also employed resilient strategies to navigate situations of exclusion and vulnerability, which reveal how displaced Venezuelans renegotiate notions of belonging and community despite being uprooted from their home country and separated from their loved ones. In April 2019, waiting at the CEBAF with his wife, Alicia, and their infant son, Elias initially commented on the hardships they experienced, such as having to pay smugglers to cross the Colombia–Ecuador border, as well as alternating between walking and hitchhiking in frigid climates, which led to their son getting sick twice as a result of asthma complications. Recounting a particular experience, Elias shared:

> He got sick and, between Dorada and Honda, there are no nearby hospitals, no health centers, nowhere to take him. … I ran to a pharmacy and got him [an inhaler]. A man came and took us to a place in the road and left us somewhere dark and told me, "Do you see that walkway? Nearby, there is a hospital." We were travelling with a friend, my wife, the kid, and myself. We got to the hospital. Thank God, they treated us well. [We trust] in God. Whatever is God's will, we will follow.

Like Elias and his family, many displaced Venezuelans relied on their faith in God as a source from which to garner strength and seek security and safety in the face of multiple uncertainties and hardships experienced throughout their displacement. Their reliance on faith also reveals participants' need for belonging and community within a context of precarity and lack of institutional protection.

In a later part of the interview, Elias also discussed how Venezuelan migrant groups in different social media, such as Facebook, had helped him locate NGOs to secure humanitarian assistance. Recounting an interaction with IOM representatives in Rumichaca, Ecuador, he mentioned:

> [We] got there and there was a [tent] from IOM, and I asked them, "Look, what kind of help can you give us? Because they have told me about a [type of assistance]." Meaning, in Facebook, they had commented that there were giving out [humanitarian assistance] and I [said to] them, "I'm from the police sector in Venezuela and I'm traveling with my wife and son." They told me, "Yes, yes we can help you."

Thus, informational ties with migrant networks in different geographical locations proved instrumental to Elias and his family as they were able to access humanitarian buses from Ecuador to Peru through IOM's assistance. These networks also positioned them as members of a collective whose main purpose was to share similar experiences of displacement with fellow compatriots in order to ease their journeys.

Socioeconomic exclusion and resilience

Regarding their experience in Peru, participants often commented on feeling excluded from, or having limited access to, the formal labor market. This was partly because barriers blocked recognition of their professional education titles or regularization of their migratory status, which led to worsened economic conditions and precarity (Equilibrium CenDE 2020a). As a result, many joined the informal labor market or ended up performing underpaid, underqualified jobs, mostly in the service and commerce sectors, without the benefits of being on an official payroll (INEI 2018). Venezuelan forced migrants also brought up how their vulnerability was exacerbated because of nationality- and gender-based discrimination (Freier and Pérez 2021). Highlighting the informational and affective transnational ties Venezuelan migrants sustain with family members and compatriots in distinct geographical locations, this section also reveals the resilient strategies they employ to renegotiate notions of belonging and community.

Most, if not all, of our interviewees cited similar difficulties. Having fled Venezuela to join his siblings residing in Peru, 19-year-old Andrés, who traveled by bus with his mother, shared his work routine:

> If I tell you about this Friday, it would have to do with doing nothing, because there we work with merchandise, sometimes we have [it], and sometimes we don't. That is my concern about job stability, because sometimes you can earn a lot, but sometimes you can't. And this Friday, there was no merchandise, so they gave us the day off, which they are not going to pay for … because it's a job without payroll, well, it is more informal. So I rested. … I have to wake up at 6:30 a.m. and leave by 7:30 to be there by 8. … [I stay] until 8 p.m., and sometimes I stay until 9 or even 10.

Like other Venezuelan migrants, Andrés's precarious job situation included minimal rights, such as long working hours and unstable remuneration; this limited his integration and contributed to his socioeconomic marginalization and vulnerability. Participants' testimonies also signaled how these unregulated working conditions took a toll on their physical and psychosocial well-being. Andrés, who hopes to one day continue his studies in psychology and open his own clothing store, mentioned:

> My body is sore, super tired, that all it wants is to sleep, sleep, eat, and sleep, and that frustrates me because I want to do activities, I simply want to draw, to start working on my designs. But the time is not enough, the fatigue takes over the desire and that is what frustrates me and causes me anxiety.

Recently arrived migrants often maneuver the difficulties and injustices of the Peruvian labor market by drawing from the experiences and contacts of more solidified migrant networks. Having studied computer science back in Venezuela, albeit not finishing his degree, Jordan ran a promising business with his wife. But given the deteriorating economic situation in Venezuela, he decided to migrate to Peru in search of a better future. After receiving an invitation from his godchild's father, who was residing there, Jordan shared how he was initially able to get a job: "He brought me to where he works. It is an informal job, because it is a company that employs Venezuelans, as well as Peruvians, but more Venezuelans, selling soft drinks at traffic lights." Although Jordan was not working in his area of expertise, which affected him constantly, he recognized the importance of having a source of income to pay for food and rent, as well as being able to send some money back home. Thus, by revealing the material and informational ties migrants maintain with family members in different geographical locations, Jordan's testimony showcases the role migrant networks play as sources of information and support that help displaced Venezuelans, like Jordan, renegotiate notions of belonging and community.

Participants also emphasized the importance of keeping in touch with family members back in Venezuela, via phone or video calls, as a way to overcome the hardships and relieve the stress of everyday life in Peru. Having arrived in the country with her partner, in October 2017, Gisella mentioned that her main reason for migrating was to get a job to be able to take care of her two young daughters, who were back in Venezuela with her ex-husband. Soon after she arrived, she secured work as a waitress at a local restaurant near her place of residence, although making less than minimum salary while working 10-hour shifts. In addition, Gisella held a tense relationship with her boss, as they would usually complain about each other's accents. Recounting her routine after work, she mentioned:

> I go back to the room [I rent], shower, and start cooking dinner. And nothing, stay there, call and talk to my girls. … I talk to them every day. … If not, every other day, but there must always be communication. … Sometimes it depresses me, but sometimes it gives me strength, you know? That if I'm here, it's because I'm doing it for them, to be able to go back and have something to give them.

Like Gisella, many participants cited the importance of keeping in touch with family members back in Venezuela as they continued their collective pursuits of better livelihoods, despite the many hardships displaced Venezuelan's experience daily. Participants also cited that not being able to keep in touch with family back home had a negative impact on their well-being, citing feeling stressed, worried, and even depressed. In addition to showcasing the affective transnational ties migrants sustain with their family members back home, this resilient strategy demonstrates how uprooted Venezuelan migrants use technology to renegotiate notions of belonging and community.

Sexual discrimination, assault, and resilience

Venezuelan women face additional challenges and threats, both during the journey and in Peru. Hypersexualized within a rampant culture of sexism in Peru (Oxfam 2019), they are at risk of being attacked or harassed in different spheres of social life, but particularly in the workplace (Pérez and Freier forthcoming). Based on 72 in-depth interviews conducted with Venezuelan women across five Peruvian cities in 2019, the authors found that 29% had experienced gender-based discrimination, 31% hypersexualization, and 25% sexual harassment. Indexing these findings, 28-year-old Rosanna described a sexual harassment case with her boss while working as a nanny:

> I was making food and my boss grabbed me by the waist and I said, "Hey, sir, what's wrong?" [and he said] "Come here ... I need you to take care of me." And I said, "No, what's wrong, sir, respect yourself. I come to take care of the girls. I do not come [to work here] for that." And he said, "How come you don't come for that. You know what it means when we hire a maid to sleep at the house. You put the girls to sleep and later you come with me." And I said, "What? You are crazy, right?" ... I ran out and ... went to the police and said, "There was a man in that house who wanted to abuse me," and he says, "Oh, what were you doing?" And I told him, "What do you mean by what was I doing? I'm telling you that I was taking care of the girls because they hired me to take care of them." [He] asked me, "Are you Peruvian?" I said from Venezuela ... and he said, "What were you waiting for? You work as an overnight maid, what do you think?" I left.

While Rosanna was strong-willed enough to immediately leave the employer's house, such cases have repercussions for assault survivor's psychosocial well-being. Sexism in Peru, as demonstrated by the male employer and police officer, allows for these cases to go unregistered, especially in the case of Venezuelan women. Nevertheless, Venezuelans forced migrants to practice resilience by speaking out against, and reporting, such abuse.

Covid-19 and resilience

The Covid-19 pandemic further exacerbated the mounting layers of exclusion and vulnerability experienced by forcibly displaced Venezuelans. A common narrative among interviewees was that the country's mandatory lockdown and subsequent mobility restrictions impacted their working arrangements, increased their risk of eviction, and lessened their ability to provide for their families in Peru and back in Venezuela (Equilibrium CenDE 2020b), with women being more negatively impacted. Venezuelans were not considered in Peru's state-led social protection programs (Freier and Vera Espinoza 2021), which led many Venezuelan migrants into further socioeconomic marginalization. This scenario even led some to embark on journeys back home (Salazar Vega 2020). Highlighting the informational and affective ties Venezuelan migrants forge with migrant networks, mainly through social media groups, this section once again reveals the resilient strategies they employ to renegotiate notions of belonging and community, specifically regarding access to healthcare services.

In response to their precarious economic situation, Venezuelans saw themselves forced to disregard nation-wide lockdowns and endure more exploitative working conditions, in order to pay for rent, provide food and basic necessities for their families, and avoid the risk of eviction. Reporting on the above-mentioned hardships and stressors, Miguel, who was living in Lima with his wife, shared:

> My wife previously worked at a restaurant but lost her job due to the pandemic. I was called back to work and am going two to three times a week. For three months, we couldn't pay rent, but luckily we were able to reach an agreement with the owner's family to schedule payments on a monthly basis. Before the pandemic, we had a good income, we ate well, and we even sent money to Venezuela on a weekly basis. Now, we are eating twice a day—we get up late so we are not hungry in the morning and eat lunch and dinner—and we send money to Venezuela every month or month and a half.

Echoing similar experiences—but highlighting the substantial toll of the pandemic on her psychosocial well-being—Gladys, who lived in Trujillo with her husband, four children, brother, brother-in-law, and sister, often repeated how nervous and worried she was about making a living during the pandemic. Her husband and sister had lost their jobs, and the only ones chipping in to pay for rent, food, and utilities were herself, her brother, and brother-in-law. Commenting on her own mental health, she added:

> I have not been able to sleep for 15 days and am sleepwalking. I live under current stress, as having four children is not easy, especially now that they cannot go out and they are screaming all the time. Sometimes I tell my husband, as a joke, that I want to run out into the street and scream to vent.

The testimonies of both Miguel and Gladys are indicative of the disproportionate burden of unpaid domestic work that fell on women both during and after the lockdown period, further limiting their access to job opportunities. More specifically, participants frequently mentioned more traditional gender roles regarding work-home dynamics, with the men often leaving the house to work and do the groceries, while the women stayed at home taking care of the children. This is consistent with Castro's (2020) regional findings indicating that, between March and April 2020, around 70% of Venezuelan women in Colombia and Ecuador declared spending most of their time in the first quarantine period on unpaid domestic work, compared to 44% of their male counterparts. In line with the findings of Matthew Bird and colleagues (Bird et al. 2020), Gladys's testimony, as well as those of other participants (both male and female), reflect the substantial toll of the pandemic on this group's mental health.

Adding layers of vulnerability and exclusion, and despite their overproportional exposure to the virus, most Venezuelans lack access to the public or private healthcare system in Peru, as a result of legal and financial obstacles, information asymmetry, and fear of being refused medical attention because of their nationality and migration status (Zambrano-Barragán et al. 2021). In spite of these obstacles, however, many of the participants cited alternative strategies to access healthcare services. Having left Venezuela seven months pregnant to be able to afford the costs of giving birth, Lisbeth, who was living in Trujillo with her husband and infant twins, discussed the help she received from a licensed Peruvian healthcare professional, to treat her husband outside his professional practice:

> In the case of my husband, he doesn't have access to SIS [public health insurance]. When he's had health complications, due to kidney colic problems, we go to the Regional Hospital and there, a Peruvian doctor, who helps out Venezuelans a lot, sees him in the parking lot. We found out about him through WhatsApp groups [of Venezuelan migrants]. The consultation is free and he sometimes gives him free medicines.

Lisbeth also mentioned using the services of Venezuelan migrant healthcare professionals, who offer unofficial services because they are not yet licensed in Peru, for minor health issues for the twins. Further elaborating on these services, she said:

> I've used the services of Venezuelan doctors, which I have found through Venezuelan migrant groups in WhatsApp. I've used them mostly for the twins. The consultations are free and, depending on the case, they can be either virtual or in-person. If it is something serious, they will refer you to a doctor and a health establishment. The advantages of using these services are that we are already familiarized with the working method of Venezuelan doctors, there is trust, and we understand their dialect and vice-versa. Especially in these times of Covid-19, one can use these services that do not require going out, which means that there is less of a chance of getting infected with the virus.

Amid a context of partial (or nonexistent) access to the public and private health systems in Peru for the majority of this population, both quotes reveal how informational ties through social media groups to migrant networks in Trujillo, as well as in other parts of Peru, proved instrumental for Lisbeth and her family to gain access to virtual or in-person medical services of Peruvian or Venezuelan doctors. Within this framework, Lisbeth's testimonies also showcase the importance of virtual migrant networks as sources of information, support, and solidarity for fellow forcibly displaced compatriots, thus renegotiating notions of belonging and community.

Conclusion

Displaced Venezuelans' experiences of exclusion and vulnerability begin upon being uprooted from their home country as a result of the pernicious effects of the political, humanitarian, and economic crises on their safety and ability to continue making a living. Throughout their journeys, Venezuelans on the move continue to endure complex hardships and dangers, which encompass, but are not limited to, separation from family members, exposure to inhospitable climates, traversing extensive distances, falling prey to robberies, and being denied entrance at borders. In Peru, Venezuelans' precarious legal status, their limited access to the formal labor market, and increasingly negative perceptions of Venezuelan immigration, heighten their exclusion and vulnerability. The Covid-19 pandemic further exacerbated Venezuelan migrants' precarity, as the country's mandatory lockdown, and subsequent mobility restrictions, impacted their working arrangements, increased their risk of eviction, and lessened their ability to provide for their families in Peru and back in Venezuela. Their precarious position, and risk of infection, was also heightened as a result of their exclusion from state-led social protection programs, as well as their limited (or nonexistent) access to public and private healthcare systems. A common narrative throughout participants' testimonies was the impact of these hardships and barriers on their physical and psychosocial well-being.

Nonetheless, it is essential to avoid reproducing monolithic assumptions of migration and one-sided narratives of despair regarding forced migrants' experiences. In this chapter, we thus highlight the resilient strategies that displaced Venezuelans adopt throughout the migratory process, which build on material, informational, and affective linkages forged with multiple actors in different geographical locations; these strategies draw upon their faith in God as a source of strength and company in the face of uncertainties and hardships; depend upon keeping in touch with family members back home through phone and video calls; and rely on migrant networks as sources of information, support, and solidarity. In establishing these transnational connections, Venezuelans on the move cope with these experiences of exclusion and vulnerability, and renegotiate notions of belonging and community.

Notes

1 We use pseudonyms to refer to interviewees to preserve their privacy and anonymity.
2 Some sub-groups are exempted from the humanitarian visa requirement: children, pregnant women, people over age 60, those with medical issues, and those with nuclear family members in Peru, as long as those family members entered legally and have regular status.

References

Appadurai, Arjun (1996). *Modernity at Large: Cultural Dimensions of Globalization*. Minneapolis: University of Minnesota Press.

Aron, Valeria, and Soledad Castillo Jara (2020). Reacting to Change within Change: Adaptive Leadership and the Peruvian Response to Venezuelan Immigration. *International Migration*. Special Issue: 1–20.

Basch, Linda G., Nina Glick Schiller, and Cristina Szanton-Blanc (1994). *Nations Unbound: Transnational Projects, Postcolonial Predicaments, and Deterritorialized Nation-states*. Philadelphia. PA: Gordon and Breach.

Berganza, Isabel, Cécile Blouin, and Luisa F. Freier (2020). El elemento situacional de violación masiva de derechos humanos de la definición ampliada de Cartagena: hacia una aplicación en el caso venezolano. *Revista Chilena de Derecho* 47(2): 385–410.

Bird, Matthew, Luisa F. Freier, and Marta Luzes (2020). For Venezuelan Migrants, COVID-19 Is Fueling a Mental Health Crisis. *American Quarterly*. Accessed February 22, 2021, https://www.americasquarterly.org/article/for-venezuelan-migrants-covid-19-is-fueling-a-mental-health-crisis/

Boccagni, Paolo, and Mieke Schrooten (2018). Participant Observation in Migration Studies: An Overview and Some Emerging Issues, 209–225. In Ricard Zapata-Barrero and Evren Yalaz, eds., *Qualitative Research in European Migration*. New York: Springer International Publishing.

Carroll, Haley, Marta Luzes, Luisa F. Freier, and Matthew Bird (2020). The Migration Journey and Mental Health: Evidence from Venezuelan Forced Migration. *SSM—Population Health* 10: 1–11.

Castro, Marta (2020). *Migrantes y COVID-19: ¿Qué tienen en común Perú, Colombia y Ecuador? Similitudes en la respuesta institucional y lecciones aprendidas para el escenario postpandemia*. Lima: Equilibrium CenDE.

Cole, Jennifer, and Christian Groes (2016). Introduction: Affective Circuits and Social Regeneration in African Migration, 1–26. In Jennifer Cole and Christian Groes, eds., *Affective Circuits: African Migrations to Europe and the Pursuit of Social Regeneration*. Chicago: University of Chicago Press.

Equilibrium CenDE (2020a). *La Calidad Migratoria Humanitaria y su relación con los derechos de la población Venezolana en el Perú*. Lima, Perú: Equilibrium CenDE.

Equilibrium CenDE (2020b). *Segunda Encuesta Regional. Migrantes y Refugiados Venezolanos. Octubre 2020*. Lima, Perú: Equilibrium CenDE.

FitzGerald, David (2006). Towards a Theoretical Ethnography of Migration. *Qualitative Sociology* 29(1): 1–24.

Freier, Luisa F., and Soledad Castillo Jara (2020). El Presidencialismo y la "Securitización" de la Política Migratoria en América Latina: un Análisis de las Reacciones Políticas frente al desplazamiento de Ciudadanos Venezolanos. *Internacia* 1(1): 1–27.

Freier, Luisa F., and Marta Luzes (2021). How Humanitarian Are Humanitarian Visas? An Analysis of Theory and Practice in Latin America. In Liliana L. Jubilut, Gabriela Mezzanotti, and Marcia Vera Espinoza, eds., *Latin America and Refugee Protection: Regimes, Logics and Challenges*. Oxford: Berghahn.

Freier, Luisa F., and Leda M. Pérez (2021). Nationality-Based Criminalisation of South-South Migration: The Experience of Venezuelan Forced Migrants in Peru. *European Journal on Criminal Policy and Research* 27: 113–133.

Freier, Luisa F., and Marcia Vera Espinoza (2021). COVID-19 & Immigrants' Increased Exclusion: The Politics of Immigrant Integration in Chile and Peru. *Frontiers* 3(1): 1–10.

Freier, Luisa F., Isabel Berganza, and Cécile Blouin (2020). The Cartagena Refugee Definition and Venezuelan Displacement in Latin America. *International Migration.* Special Issue: 1–19.

Gupta, Akhil, and James Ferguson (1992). Beyond "Culture": Space, Identity, and the Politics of Difference. *Cultural Anthropology* 7(1): 6–23.

INEI (2018). *Condiciones de vida de la población venezolana que reside en el país. Resultados de la "Encuesta dirigida a la población Venezolana que reside en el país" ENPOVE 2018.* Lima, Perú: INEI. Accessed February 22, 2021, from: https://www.inei.gob.pe/media/MenuRecursivo/boletines/enpove-2018.pdf

OAS (2019). *Final Report of the OAS Working Group to Address the Regional Crisis Caused by Venezuela's Migrant and Refugee Flows.* Washington, DC: Organization of American States.

Oxfam (2019). *Sí, pero no aquí: Percepciones de xenofobia y discriminación hacia migrantes de Venezuela en Colombia, Ecuador y Perú.* Oxford: Oxfam.

Pérez, Leda M., and Luisa F. Freier (forthcoming). Of Prostitutes and Thieves: The Hyper-sexualization and Criminalization of Venezuelan Migrant Women in Peru. *Journal of Ethnic and Migration Studies.*

R4V (2021). Respuesta a los Venezolanos. Accessed February 22, 2021.

Salazar Vega, Elizabeth(2020). El doble éxodo: la pandemia fuerza el retorno de venezolanos al país que dejaron. *Ojo Público.* Accessed April 22, 2021, https://ojo-publico.com/1919/el-doble-exodo-la-pandemia-fuerza-el-retorno-de-venezolanos

Sassen, Saskia (1991). *The Global City: New York, London, Tokyo.* Princeton, NJ: Princeton University Press.

UNHCR (2019). Guidance Note on International Protection Considerations for Venezuelans—Update I. Accessed February 19, 2021, https://www.refworld.org/docid/5cd1950f4.html

Vivas, Leonardo, and Tomas Paez (2017). *The Venezuelan Diaspora: Another Impending Crisis?* Washington, DC: Freedom House.

World Bank (2019). *An Opportunity for All: Venezuelan Migrants and Refugees and Peru's Development. Technical Report.* Washington, DC: World Bank. Accessed January 18, 2021, https://openknowledge.worldbank.org/handle/10986/32816?show=full

Zambrano-Barragán, Patricio, Sebastián Ramírez Hernández, Luisa F. Freier, Marta Luzes, Rita Sobczyk, Alexander Rodríguez, and Charles Beach (2021). The Impact of COVID-19 on Venezuelan Migrants' Access to Health: A Qualitative Study in Colombian and Peruvian Cities. *Journal of Migration and Health* 3: 1–8.

Zapata-Barrero, Ricard, and Evren Yalaz (2018). Introduction: Preparing the Way for Qualitative Research in Migration Studies, 1–8. In Ricard Zapata-Barrero and Evren Yala, eds., *Qualitative Research in European Migration.* New York: Springer International Publishing.

Part II

Drivers of displacement

6 War and forced migration in medieval Iberia (1085–1266)

Between al-Andalus and the feudal world

J. Santiago Palacios

Introduction

In 1309, an "old Moor," upon being expelled from his land by the Christian armies, complained about his fate in a fictional dialogue held with King Ferdinand IV, as recounted in the *Crónica de Fernando IV*. The Muslim used to live in Seville, but in 1248 King Ferdinand III conquered Seville and threw him out of the city. He then went to live in Jerez, but King Alfonso X conquered Jerez and expelled him from there as well. After that, he moved to Tarifa, further south in the Iberian Peninsula. When King Sancho occupied that city and drove him out yet again, he finally came to Gibraltar. The Muslim thought that this place, al-Andalus, was the safest place in the whole land of the Moors on this side of the sea. But eventually, King Ferdinand IV conquered Gibraltar, and he forced the "old Moor" to travel to North Africa, where he settled in a secure place (O'Callaghan 1975).

The history of al-Andalus since 1085 reflects the desperation and rootlessness encountered by all emigrants throughout history who have been made refugees because of war and conquest. The memento of home lived in their collective memory, as it had in the minds of many other peoples who lost their home. (*Nihil novum sub sole.* Nothing new under the sun.) But such recall is scarce in the contemporary historiography that has recounted those events. It mattered little what happened to that "old Moor" and the rest of the co-religionists, in comparison to the achievements of the kings. And part of the historiography was even more interested in transmitting the fact that, still after the conquests, the Muslims were able to continue living in their lands, thanks to the generosity of the Christian rulers.

In medieval Spain, from the eleventh to the thirteenth century, the northern feudal states conquered vast amounts of land from the Muslims, triggering many situations similar to the one described above. Spanish traditional historiography has defined this process *Reconquista* (reconquest), which led to the progressive reduction of the political space of al-Andalus under the military pressure of the Christian powers. From the standpoint of the indigenous population (called *andalusíes*), this had two consequences. First was the forced emigration of the Muslim dwellers (that is, Muslims who were descendants of the Arab and Berber invaders, as well as Christians who had

DOI: 10.4324/9781003170686-8

converted to Islam generations ago). Second was their replacement by new settlers from the north—a process defined with another specific term by historians of medieval Spain: *Repoblación* (repopulation) (MacKay 1977). The Christian colonists called to repopulate the conquered lands of al-Andalus settled on property owned by the former indigenous population. Consequently, a new society arose, one that implemented a sort of medieval colonialism, based precisely on the dominance and exploitation of the "different" (Torró 2008). The victors' community was, in part, the result of a process of segregation and control of the vanquished.

The Muslims were forced to choose. Some of them stayed behind voluntarily and were subjected to several economic and legal conditions, although they kept their religious beliefs (called *Mudéjares, mudaŷŷan*, namely submissive, domesticated) (Burns 1990; O'Callaghan 1990). Others, less "fortunate" were enslaved and diminished in number. And still others, as we know, migrated.

This chapter focuses on those refugees, drawing examples from medieval Spain between the conquest of Toledo in 1085 up to the Mudejar revolt in 1266. I first consider under what conditions the emigration of the Muslims took place after the Christian conquest. I address the conditions imposed by the victors on the vanquished in order to discuss the consequences of how loss of property and land caused a demographical and socio-cultural impact. I next detail who these people were, the places they emigrated to, and the feelings that overwhelmed them. In summary, I call attention to immigrants who were not visible in history because, as in many other cases, they were not able to write such pages themselves.

Feudal conquest of al-Andalus: stay or emigrate

The Islamic conquest of the Iberian Peninsula in the eighth century was conducted by a small contingent of Arab and Berber conquerors. They asserted their authority on the indigenous population by spreading the Arab culture and language (as well as the Islamic religion), although by no means did that exclude the use of violence and coercion. At that time, only a few Hispano-Goths went into exile (García Sanjuán 2013).

In contrast, the processes of feudal conquest of al-Andalus, from the eleventh century onward, materialized on the basis of a total or partial replacement of the indigenous Muslim population by the new Christian colonists. Thus, Muslims were displaced, expelled, or held captive by the latter. But there were also attempts to assimilate or integrate the defeated population into the society of the victors, albeit under difficult and often unbearable conditions.

I chronicle these episodes by outlining the different shapes that the conquest of the various territories of al-Andalus could take. These circumstances determined the terms and conditions of the Muslim surrender. Also, I speculate whether the continued presence of these indigenous populations in the conquered territory was really possible.

The prolonged resistance or the conquest by force, in a radicalized context, entailed the possibility of extermination, slavery, or mass deportation. This occurred from the pre-Pyrenees city of Huesca in 1096 to Seville in 1248, where a complete expulsion of the Islamic population took place as a condition imposed after the Christian victory (Sénac 2000; García Sanjuán 2017). Despite these events and other examples, the extermination of the defeated Muslim population was not widespread (García Fitz 2008). General enslavement was not always a viable option, and mass deportations only occurred under specific conditions.

In cases of covenants or obeisance pacts with the Muslim leaders, many times following some military resistance in the face of the conquest, Andalusi communities could retain part of their properties and remain in their towns without suffering violence. This might have been the case for the inhabitants of the kingdom of Toledo following the occupation of the city in 1085 (Molénat 1997). Nevertheless, intermediate situations also ensued. Despite the signing of a formal capitulation with conditions set to render the surrender effective, the conquerors finally claimed their right to the spoils of war. They imposed harsh obligations on the Muslim populations, and hence emigrating was no longer a voluntary act but rather the only option. This occurred in major cities such as Toledo in 1085 or Tortosa in 1148 (Molénat 1997; Virgili 2001), and in many Andalusian cities and territories throughout the thirteenth century (González Jiménez, 1994; García Fitz 2002). These agreement breaches set the conditions for a mass rebellion, the so-called Mudejar revolt, which took place between 1264 and 1266, and led to the definitive expulsion or exile of most of the Muslim population after the Christian victory.

Therefore, even when the Andalusi population could remain on their lands, it was just a provisional status quo, which could be revoked at short notice and with the most trivial excuses. This precarious balance neither lasted long nor was particularly common, and it did not consider the possibility of conversion, except for some attempts (Kassis 1990). The expansion of Latin Christianity during the eleventh, twelfth, and thirteenth centuries generally involved, with the blessing of the Church, the destruction of pagans, heretics, or non-believers that stood in its way (Iogna-Prat 1998). In the best-case scenario, the presence of the vanquished was understood provisionally, and they were segregated from the rest of the dwellers.

The conditions for staying are evident in the capitulations signed between the conquerors and the Andalusis, which account for a number of scenarios (Burns 1973; O'Callaghan 1990). The vanquished could be freed and, thereby, leave their places of origin once they received the king's safe-conduct (*amān*) designation within a certain time frame. Emigrants were allowed to bring their portable properties along with them, but they lost their ownership of real estate properties (land and dwellings). Nevertheless, Christians occasionally conceded some time to sell and liquidate them before leaving.

Those who voluntarily wished to remain could do so but had to pay high duties, such as a type of tithe (*diezmo*) (Sénac 2000). This payment allowed them to hold certain legal and political rights, among other freedoms, for the organization of the local communities (*aljamas*). In any case, they could be forced to move to segregated areas within their own cities (*morerías*), such as neighborhoods or ghettos located in the outskirts beyond the city walls (Burns 1973; Virgili 2001).

Under these circumstances, despite the initial pacts between the conquering Christian powers and the Muslim communities, most of the indigenous population that endured feudal pressure left their homes after the conquest. The city of Toledo, for example, surrendered after several years of military campaigns in the surrounding areas. This did not prevent the terms of surrender imposed by Alfonso VI from being advantageous for the Muslims, at least in the beginning (Molénat 1997). Nevertheless, some Muslims left the city even before its fall, according to the account of the Arab historian Ibn Bassām. According to Kristine Vlaminckx, he chronicles that "most of the Toledans were massacred or fled" at the same time that the military pressure on Toledo's territory was progressively being lifted (Vlaminckx 1985: 187).

On the other hand, after the conquest of Andalusia and Murcia in the thirteenth century, the kingdom of Castile and León ruled the most densely populated region of al-Andalus. Many Muslims then emigrated to the Maghreb or to the kingdom of Granada. Until 1266, however, most of the population remained in their properties. It was undoubtedly neither wise to leave those highly productive lands lying fallow, nor feasible to improvise an alternative colonization to replace Muslims with new Christian immigrants from the north. For that reason, they initially signed apparently advantageous vassalage agreements and pacts. In these cases, the Muslim population was allowed to remain only under strict conditions of fiscal subjugation.

The ubiquitous expansion of Latin Christianity and the constitution of strong monarchies, however, imposed the eradication of any resistance and set an expiry date for these covenants. Thus, war and feudal conquest caused the forced emigration of thousands of Muslims in medieval Iberia. The exile was dramatic and caused a social rupture in many parts of the peninsula, mainly in the cities, which were emptied, according to Arab and Christian chroniclers (Torres Balbás 1955).

Not all historians agree on the extent and scope of this exodus (Catlos 2004). But most of them agree with those accounts of expulsions (Sénac 2000). Carlos Laliena Corbera, for example, speaks about an "Andalusian society that has been conscientiously demolished," and Muslims "evicted and expelled" and forced into a shift "on a remarkably large scale, either through coercive procedures or through the intensification of ethnic segregation" (Laliena Corbera 1998: 200, 209).

The emigrants

Classical Islamic law, based on the Quran and the Sunnah, dictates that Muslims living in an infidel country have to leave it and emigrate. The Maliki legal school, the dominant one in al-Andalus, also imposed the obligation on all Muslims to emigrate if the Christian authorities conquered their lands (Fierro 1991; Abou El Fadl 1994; Aldeeb Abu-Sahlieh 1996; Masud 2001; Molénat 2001; Verskin 2013; Verskin 2015; Hendrickson 2020). But not all Muslims could afford the *hijra* (migration) in the same conditions, meaning that there must have been a wide gap between the legal rhetoric and the believer's reality. The first ones who could emigrate were the ruling class within Andalusian society. We know they preferably escaped to other regions of al-Andalus still in Muslim hands: from Toledo (1085) to Badajoz, Seville, Cordoba, or Valencia. In turn, when the last city fell under the control of El Cid (1094), the refugees headed for Almería, Córdoba, or Jaén. After losing Zaragoza (1118), some went to Islamic territories in the south, to the Levante region (Sharq al-Andalus), or to Marrakech. And when the Christians definitively conquered Valencia (1238) many scholars traveled to Tunisia, Algeria, or Egypt, while others headed for Andalusia (Molina 1987). But, as the "old Moor" narrated at the beginning, the more land the Christians conquered, the greater the pressure to emigrate, and the less safe was land they went to (López de Coca 1988; Vallvé 1988; Valencia 1992; Marín 1995; García Sanjuán 2016; Torró 2019).

Although there were also peasants among them, the conquests first resulted in the flight of the administrative, military, religious, and intellectual classes (García Sanjuán 2021). In short, these were the vital forces who had resources or supporting networks, or could find accommodation and ultimately continue their occupations in other Muslim countries. Along with them, but with more difficulties, many members of the underprivileged classes left their homes carrying only a few belongings. Some of them ended up settling in Christian cities in northern Castile, where they formed larger and more prosperous communities (Echevarría 2006), places from which their ancestors were expelled some generations before.

Unsurprisingly, the King of Seville, al-Mu'tamid, wrote: "better to be cameleer in Africa than a swineherd in Castile." The dilemma of emigration emphasized the feeling of precariousness that characterized Islamic society in al-Andalus. It was a community aware of living on the edge of Islam, surrounded by the sea and threatened by Christians. Muslims knew that, sooner rather than later, the Christians would destroy them, as some millennialist traditions predicted (Fierro 1997; Stearns 2009). Moreover, the feelings of defeat and the nostalgia for the lost homeland were evident. These are common traits in history that often characterize the exile of the defeated populations forced to emigrate.

Having outlined these situations, I can summarize the elements that caused the social rupture and marked the path of emigration for many Andalusis. Undoubtedly, the military pressure and fear that preceded the feudal domination of al-Andalus was a key factor driving emigration. After the conquest, either agreed or by force, the main rupture from the previous social structure occurred because portions of the defeated populations were held captive. They could remain as settler-slaves, forming part of the loot, or be sold on other markets. All the same, other types of social segregation and fiscal pressure also instigated emigration. This refers to forced relocations of the population to new sites, the seizure of property, or the high taxes, which resulted in unsustainable living conditions.

The loss of freedoms, properties, and lands; the impossibility of taking all their properties with them; the fees they had to pay, and the expenses they faced; all impoverished the Muslim emigrants. But the common thread running through all these elements has not only a material but also an emotional and socio-cultural implication.

al-Andalus: lost paradise and homeland

Muslims from the Iberian Peninsula primarily lost their *home*: al-Andalus. For Arab geographers, it was a territory with specific geographical limits and a defined shape. But there are also references to al-Andalus as the exclusive part of the Iberian Peninsula under Muslim political-administrative control. That is, not as a territorial entity but in the sense of belonging to a cultural, religious, and even emotional community (García Sanjuán 2003).

When the expulsion or denaturation of the Andalusian population occurred, the chance to find a home within al-Andalus was still possible. Although territorially reduced, a part of Andalusi community still existed, and there was always the option of moving and settling in the kingdom of Granada, which underwent a remarkable population growth due to these migrations. Although such a possibility existed, the awareness of living under a permanent threat, the feeling of disunity and weakness in the face of feudal armies, and, of course, the constant territorial mutilations led to a climate of general fear and despair.

Thus, from the capture of Toledo, the poets sang of the progressive loss of Andalusi cities until the fall of Granada in 1492, and they launched appeals for help to their co-religionists (Rubiera 1988; Garulo 1998; Martínez de Francisco 2003).

> Save al-Andalus.
>
> Come with your army, the army of God to al-Andalus
> because their chances of salvation are fading away.
> Grant her the help that implores you
> for the honor of your help is always implored ...

> Call from Seville.
>
> The land of the enemies announced the reconquest,
> Doesn't the morning come after the night?
> They didn't stop to stir up the lion of war
> to devour them, being them wolves.
> They lit the fuse of war ...
>
> (Ibn al-Abbār, c. 1238, in Martínez de
> Francisco 2003: 57, 85)

Lamentations about the country's decline began very soon, when civil war that destroyed the Umayyad Caliphate (and especially Córdoba and Madīnat al-Zahrā') broke out at the beginning of the eleventh century. In these poems, the metaphor of the ruins is always present in relation to al-Andalus and its people, a poetic symbol that would later become recurrent in the face of the Christian threat.

> Oh, garden scourged by stormy winds,
> which drove away your inhabitants,
> destroyed, like you devastated [...]
> There is no person among the ruins
> that of our friends give news,
> who will we ask how they are?
> (Ibn Šuhayd, c. 1035, in Garulo 1998: 50)

The elegies for the decline of al-Andalus in the face of Christian threat sometimes mixed mentions of the defeated battles, the brave combatants who defended their land, or the ephemeral joy for the recovered lands, such as the reconquest of Gibraltar by Yūsuf III in 1414, sung by the King of Granada himself in his *dīwān*.

> Whose red banner trembles triumphantly
> surrounded by the warriors of the Naṣries?
> He is going to al-Jabal to conquer it, and to confirm his
> omen, because, after the successive difficulties
> will come, no doubt, the joys [...]
> A city that is already a joy for the eyes,
> and before which no more can be waited for:
> How could it not be so if our dynasty
> only aspire to the greatest!
> A Naṣri is always a torch for the road!
> (Yūsuf III, 1414, in Moral 1987: 94–96)

Wherever the Andalusis were expelled, they remembered the Andalusi refinement, associating their lands with Paradise or with the mythical Arabia and remembering the noble origin of its inhabitants, just as they were reminded of their opulent cities and beloved relatives lost in the war launched by the Christians against them (Elinson 2009; Barceló 2012; Foulon 2013; García Sanjuán 2021).

> What a shame I can't sleep,
> and have lost friends and loved places!
> The hand of misfortune broke my group [...]
> The distance from my people is unbearable for me,
> and I am overcome with nostalgia for my homeland.
> (Ibn al-Abbār, c. 1238, in Martínez de Francisco
> 2003: 57, 85)

In medieval sources, it is not easy to find the voice of the common people. Only poetry could express those feelings, and thanks to it we know what the Andalusian emigrants felt when they lost their homes. The poets composed elegies in which they featured several themes: laudatory literature of merits (*faḍā'il*), the call for help to other Muslim rulers, the search for explanations for the defeats, and nostalgia and the memory of the lost home. The authors' memories of their youth revived the image of al-Andalus as the lost paradise, land of *jihād* with similarities to Jerusalem, and the homeland where they used to live and which they fought for to the end.

> From the homeland where we spent our youth and where we had the most beautiful loves, we have said goodbye forever because it has fallen on her the ruin, the Islam has abandoned it and there is disorder and uprooting everywhere [...]
>
> They captured Denia and its vineyards, they moved away even though it was the next one. And Xàtiva and its valley, victim of the injustice of time! What a misfortune about Tudmir and its hills, about Cordoba and its river, about Seville and yours! All of them have seen their fields devastated, their population dispersed: the majority suffered the siege and the infidels destroyed their sources.
>
> Elvira [Granada] is about to be lost and Rayya [Málaga] is like a throat fenced with a bracelet. There is no doubt that Almeria has next the fall.
>
> (Ibn al-Abbār, c. 1238, in Rubiera 1988: 43)

21. Yet under the Prophet Muhammad's religion we
    used to oppose the governors of the Cross with
    our inner intentions,

22. Facing grave dangers in Holy War because of killing
    and capturing, hunger and dearth.

23. But the Christians attacked us from all sides in a vast
    torrent, company after company

24. Smiting us with zeal and resolution like locusts in
    the multitude of their cavalry and weapons.

25. Nevertheless, for a long time we withstood their
    armies and killed group after group of them.

26. Though their horsemen increased every moment,
    whereas ours were in a state of diminution and
    scarcity,

27. Hence, when we became weak, they camped in our
    territory and smote us, town after town.
    (Anonymous Morisco after the conquest of
    Granada, 1492, in Monroe 1966: 296)

But in the end, where once the voice of the muezzin was heard, "the bells are ringing and crosses are standing" (O'Callaghan 1975: 356–57).

> Christians share as spoils
> the beautifully demure wives.
> How many taverns where there were holy places,
> how many churches where there were bans!
> Woe to mosques, made synagogues by the enemy
> while the call to prayer was supplanted by the bells!
> What a pity that we cannot recover his past,
> and the Koranic schools now ruined ...
>> (Ibn al-Abbār, c. 1238, in Martínez de
>> Francisco 2003: 57–58)

The Andalusian Garden of Eden transformed into a ruin, wretched and devastated; its population walked scattered, was enslaved, or was forced to convert to Christianity (Pérès 1983; Rubiera 1988; Kassis 1990; Stearns 2009).

> Valencia was nothing but a Paradise
> full of beauty, through which the rivers ran.
> Their sunsets were scented
> by the aroma of the daffodils,
> and their trees were embalmed with their zephyr ...
>> (Abū-1-Muṭ arrif b. 'Amīra, c. 1238, in
>> Stetkevych 1993: 106–07)

> Where is Valencia and its houses, and its chants and cooing of its doves?
> Where are its shady places that spilled welfare, and its bright irrigated land?
> Where are their overflowing streams and their leafy trees?
> Where are their fragrant gardens and their delightful spots?
> The necklaces of their flowers that have fallen from their necks are
> undone, and they have lost the shining light that their Albufera and
> their sea had!
>> (Ibn al-Abbār after de conquest of
>> Valencia, 1238, in Terés 1965: 296)

96. [...] your majesty, we complain to you, what we have encountered is the worst form of estrangement.

97. Could our religion not be left to us as well as our ritual prayer, as they swore to do before the agreement was broken?

98. If not let them allow us to emigrate from their land to North Africa, the homeland of our dear ones, with our belongings.

99. For emigration is better for us than remaining in
    unbelief enjoying power but having no religion.
                        (Anonymous Morisco after the conquest of
                                    Granada, in Monroe 1966: 302)

What makes the history of al-Andalus unique is that, from the nineteenth century and even today, Arab writers evoke that lost home (Martínez Montávez 1992; Noorani 1999), despite the fact that they have little to do with today's Spain. In different contemporary cultural products, the image of the Moor submitted to a cruel fate still emerges, characterized by the loss and final expulsion from his land (Civantos 2017). For many Muslims, al-Andalus means a golden age of Islam, epitomized by cultural splendor and tolerance. But even among other social or racial communities, not necessarily related to Islam, Muslim Spain is a symbolic place to yearn for because of Moorish Spain's multiracialism and *convivencia* (Aidi 2003).

In spite of this, the hegemonic discourse in Spanish historiography has not paid enough attention to the phenomenon of forced emigration of Muslims after the Christian conquest. The vanquished remain hidden in a historiography more concerned with the glorious Reconquest, rather than with the Andalusi population that suffered the consequences of the feudal war. This phenomenon is partly explained by a perverse idea: the people from al-Andalus were strangers in the land where they lived and were not their legitimate owners (Maíllo Salgado 2011; Fanjul 2000). Consequently, not only was their expulsion from the peninsula legitimized, but also was their expulsion from history itself, almost as though the history of al-Andalus had been an undesirable parenthesis in the history of Spain, considered from the old paradigm of the Reconquest.

As we see, however, al-Andalus was an Arab and Islamic country shaped in the Iberian Peninsula during the Middle Ages, whose inhabitants created strong identities tied to their land. The cohesive factors were many: political, religious, and cultural. The sense of belonging to a distinct community was also evident. The solution adopted for them could only be devastating to their sense of unity.

Conclusions

For 200 years, a religious war waged on the Iberian Peninsula, resulting in the loss of al-Andalus and the displacement or disappearance of most of its population. Large groups of people were either condemned or forced to leave their homes and emigrate, being dispossessed of most of their belongings and suffering the uncertainties and fears inherent to all wartime migrants. Some left their homes by their own will, though moved by fear. But there were also other possible situations.

Some historians point out that conversion was the path chosen by some to avoid that fate. Others contest the chronicles of mass emigration offered by certain sources. In some cases, historians insist on the exceptional nature of the expulsions, whereby a majority of Muslims stayed on their lands and Christian repopulation had a scarce impact. And some scholars have even benevolently pondered the agreements signed with the Muslims at the time of the conquest, pointing out that the Christian generosity and tolerance toward the defeated illuminated a new stage of tolerance and coexistence in medieval Iberia.

Whether the Andalusians remained in the early stages of the conquest as "a faceless sea around the colonial [Christian] atolls" (Burns 1973: 25), the result at the end of the period is completely unambiguous. The Islamic population in the Christian-dominated areas was drastically reduced after the massive Castilian and Catalan-Aragonese conquests during the thirteenth century. There is clear evidence of the extensive movement of people toward the kingdom of Granada and the Maghreb, or even their resettlement in new *aljamas* in northern Christian cities. Only those who had been deprived of their property and of a large part of their rights and freedoms remained, as subjugated Mudejars. We can, thus, only imagine that all the emigrants who left their homes felt the same way as the inhabitants of Cordoba in 1236 who, as expressed in the words of the chronicler Juan de Osma (thirteenth century), "fainting from hunger, left their settlement crying, screaming and moaning because of the anguish of their spirit" (Juan de Osma 2010: 116).

References

Abou El Fadl, Khaled (1994). "Islamic Law and Muslim Minorities: The Juristic Discourse on Muslim Minorities from the Second/Eighth to the Eleventh/ Seventeenth Centuries." *Islamic Law and Society* 1(2): 141–87.

Aidi, Hisham (2003). "Let Us Be Moors: Islam, Race and 'Connected Histories'." *Middle East Report* 229: 42–53.

Aldeeb Abu-Sahlieh, Sami (1996). "The Islamic Conception of Migration." *The International Migration Review* 30(1): 37–57.

Barceló, Carmen (2012). "Endechas por la pérdida de al-Andalus en dos zéjeles de Cútar." *Al-Qanṭara* XXXIII(1): 169–99.

Burns, Robert I. (1973). *Islam under the Crusaders. Colonial Survival in the Thirteenth-Century Kingdom of Valencia*. Princeton, NJ: Princeton University Press.

Burns, Robert I. (1990). "Muslims in the Thirteenth-century Realms of Aragon: Interaction and Reaction," 57–102. In James M. Powell, ed., *Muslims Under Latin Rule, 1100–1300*. Princeton, NJ: Princeton University Press.

Catlos, Brian A. (2004). *The Victors and the Vanquished: Christians and Muslims of Catalonia and Aragon, 1050–1300*. New York: Cambridge University Press.

Civantos, Christina (2017). *The Afterlife of Al-Andalus. Muslim Iberia in Contemporary Arab and Hispanic Narratives*. Albany: State University of New York Press.

Echevarría, Ana (2006). "La 'mayoría' mudéjar en León y Castilla: Legislación real y distribución de la población (Siglos XI–XIII)." *En La España Medieval* 29: 7–30.

Elinson, Alexander E. (2009). "The View from Al-Andalus: Looking East, West, and South for Andalusī Identity," 117–50. In Alexander E. Elinson, ed., *Looking Back at Al-Andalus. The Poetics of Loss and Nostalgia in Medieval Arabic and Hebrew Literature*. Leiden-Boston: E. J. Brill.

Fanjul, Serafín (2000). *Al-Andalus contra España*. Madrid: Siglo XXI.

Fierro, M. Isabel (1991). "La emigración en el Islam: conceptos antiguos, nuevos problemas." *Awrāq* 12: 11–41.

Fierro, M. Isabel (1997). "Christian Success and Muslim Fear in Andalusi Writings during the Almoravid and Almohad Pèriods," 155–78. In Uri Rubin and David J. Wasserstein, eds., *Dhimmis and Others: Jews and Christians and the World of Classical Islam, Israel Oriental Studies 17*. University Park: Pennsylvania State University Press.

Foulon, Brigitte (2013). "Paysage et nostalgie dans la poésie andalouse." *L'écriture de la nostalgie dans la littérature árabe*. París: l'Harmattan.

García Fitz, Francisco (2002). *Relaciones políticas y guerra. La experiencia castellano-leonesa frente al Islam. Siglos XI–XIII*. Sevilla: Universidad de Sevilla.

García Fitz, Francisco (2008). "¿'De exterminandis sarracenis'? El trato dado al enemigo musulmán en el reino de Castilla-León durante la plena Edad Media," 113–66. In Maribel Fierro and Francisco García Fitz, eds., *El cuerpo derrotado. Cómo trataban musulmanes y cristianos a los enemigos vencidos (Península Ibérica, siglos VIII-XIII)*. Madrid: Consejo Superior de Investigaciones Científicas.

García Sanjuán, Alejandro (2003). "El significado geográfico del topónio al-Andalus en las fuentes árabes." *Anuario de Estudios Medievales* 33(1): 3–36.

García Sanjuán, Alejandro (2013). *La conquista islámica de la península ibérica y la tergiversación del pasado*. Madrid: Marcial Pons.

García Sanjuán, Alejandro (2016). "La conquista de Andalucía y el destino de la población musulmana. La aportación de las fuentes árabes," 33–58. In Manuel González Jiménez and Rafael Sánchez Saus, eds., *Arcos y la frontera andaluza (1264–1330)*. Sevilla: Universidad de Sevilla.

García Sanjuán, Alejandro (2017). "La conquista de Sevilla por Fernando III (646 h/1248). Nuevas propuestas a través de la relectura de las fuentes árabes." *Hispania* LXXVII(255): 11–14. doi: 10.3989/hispania.2017.001.

García Sanjuán, Alejandro (2021). "Retroceso territorial y declive musulmán en la obra de al-Qurṭubī (m. 671/1273), un ulema andalusí del exilio." In Carlos Ayala, Francisco García-Fitz, and Santiago Palacios, eds., *Memoria y fuentes de la guerra santa peninsular (ss. X–XV)*. Gijón: Ediciones Trea.

Garulo, Teresa (1998). "La nostalgia de al-Andalus, génesis de un tema literario." *Qurtuba: Estudios andalusíes* 3: 4–63.

González Jiménez, Manuel(1994). "Alfonso X y Andalucía," 69–82. In Manuel González Jiménez," *Andalucía a debate y otros estudios*. Sevilla: Editorial Universidad de Sevilla.

Hendrickson, Jocelyn (2020). *Leaving Iberia. Islamic Law and Christian Conquest in North West Africa*. Cambridge, MA: Harvard University Press.

Iogna-Prat, Dominique (1998). *Ordonner et exclure. Cluny et la société chrétienne face a l'hérésie, au judaisme et a l'islam, 1000–1150*. París: Aubier.

Juan de Osma (2010). *Crónicas hispanas del Siglo XIII*. Luis Charlo Brea, Juan A. Estévez Sola and Rocío Carande Herrero, eds., Corpus Christianorum in translation, 5. Turnhout: Brepols.

Kassis, Hanna E. (1990). "Muslim Revival in Spain in the Fifth/Eleventh Century. Causes and Ramifications." *Der Islam* 67(1): 78–110.

Laliena Corbera, Carlos (1998). "Expansión territorial, ruptura social y desarrollo de la sociedad feudal en el valle del Ebro, 1080–1120," 199–228. In Carlos Laliena Corbera and Juan Utrilla Utrilla, eds., *De Toledo a Huesca: Sociedades medievales en transición a finales del siglo XI (1080-1100)*. Zaragoza: Institución Fernando el Católico.

López de Coca, J. Enrique (1988). "Granada y el Magreb: la emigración andalusí (1485-1516)," 409–51. In Mercedes García-Arenal and Mª Jesús Viguera, eds., *Relaciones de la Península Ibérica con el Magreb (siglos XIII–XVI)*. Madrid: CSIC.

MacKay, Angus (1977). *Spain in the Middle Ages: From Frontier to Empire, 1000–1500*. New York: Macmillan.

Maíllo Salgado, Felipe (2011). *De la desaparición de al-Andalus*. Madrid: Abada.

Marín, Manuela (1995). "Des migrations forcées: les 'ulamā' d'al-Andalus face a la conquete chrétienne," 41–59. In Mohammed Hammam, ed., *L' Occident Musulman et l' Occident Chrétien au Moyen Age*. Rabat: Faculté des Lettres et des Sciences Humaines.

Martínez de Francisco, Santiago, ed. (2003). *Ibn al-Abbar. Salvad Al-Andalus y otros poemas*. Madrid: Huerga y Fierro.

Martínez Montávez, Pedro (1992). *Al-Andalus, España, en la literatura árabe contemporánea: La casa del pasado*. Madrid: MAPFRE.

Masud, Muhammad Khalid (2001). "The Obligation to Emigrate: The Doctrine of *Hijra* in the Islamic Law," 29–49. In Dale F. Eickelman and James Piscatori, eds., *Muslim Travellers Pilgrimage, Migration, and the Religious Imagination*. London: Routledge.

Molénat, Jean-Pierre (1997). *Campagnes et monts de Tolède du XIIe au XVe siècle*. Madrid: Casa de Velázquez.

Molénat, Jean-Pierre (2001). "Le problème de la permanence des musulmans dans les territoires conquis par les chrétiens, du point de vue de la loi islamique." *Arabica* 48(3): 392–400.

Molina, Emilio (1987). "Algunas consideraciones sobre los emigrados andalusíes," 419–32. In *Homenaje al Prof. Darío Cabanelas Rodríguez, O.F.M., con motivo de su LXX aniversario* (vol. I). Granada: Universidad de Granada.

Monroe, James T. (1966). "A Curious Morisco Appeal to the Ottoman Empire." *Al-Andalus* 31(1): 281–303.

Moral, Celia del (1987). "El Dīwān de Yūsuf III y el sitio de Gibraltar," 79–96. In *Homenaje al Prof. Darío Cabanelas Rodríguez, O.F.M., con motivo de su LXX aniversario* (vol. II). Granada: Universidad de Granada.

Noorani, Yaseen (1999). "The Lost Garden of al-Andalus: Islamic Spain and the Poetic Inversion of Colonialism." *International Journal of Middle East Studies* 31: 237–54.

O'Callaghan, Joseph F. (1975). *A History of Medieval Spain*. Ithaca, NY & London: Cornell University Press.

O'Callaghan, Joseph F. (1990). "The Mudejars of Castile and Portugal in the Twelfth and Thirteenth Centuries," 11–56. In James M. Powell, ed., *Muslims Under Latin Rule, 1100–1300*. Princeton, NJ: Princeton University Press.

Pérès, Henri (1983). *Esplendor de al-Andalus. La poesía andaluza en árabe clásico en el siglo XI*. Madrid: Hiperión.

Rubiera, María J. (1988). "La Conquista de València per Jaume I com a tema literari en un testimoni de l'esdeveniment: Ibn al-Abbar de València." *Aiguadolç*, 7: 33–44.

Sénac, Philippe (2000). *La frontière et les hommes (VIIIe-XIIe siècle). Le peuplement musulman au nord de l'Ebre et les débuts de la reconquête aragonaise*. Paris: Maisonneuve et Larose.

Stearns, Justin (2009). "Representing and Remembering al-Andalus: Some Historical Considerations Regarding the End of Time and the Making of Nostalgia." *Medieval Encounters* 15: 355–74.

Stetkevych, Jaroslav (1993). *The Zephyrs of Najd. The Poetics of Nostalgia in the Classical Arabic Nasib*. Chicago: University of Chicago Press.

Terés, Elías (1965). "Textos poéticos árabes sobre Valencia." *Al-Andalus* 30 (2): 291–307.

Torres Balbás, Lepoldo (1955). "Extensión y demografía de las ciudades hispanomusulmanas." *Studia Islamica* 3: 35–59.

Torró, Josep (2008). "Colonizaciones y colonialismo medievales. La experiencia catalano-aragonesa y su contexto," 91–118. In G. Cano and A. Delgado, eds., *De Tartessos a Manila. Siete estudios coloniales y poscoloniales*. València: Publicacions Universitat de València.

Torró, Josep (2019). "*Expellere sarracenos*. Expulsions, reassentaments i emigració dels musulmans del regne de València després de la conquesta cristiana (1233–1348)," 71–103. In Flocel Sabaté (coord.), *Poblacions rebutjades, poblacions desplaçades (Europa medieval)*. Barcelona: Pagès editors.

Valencia, Rafael (1992). "La emigración sevillana hacia el Magreb alrededor de 1248," 323–27. In *Actas del II Coloquio Hispano-Marroquí de Ciencias Históricas. Historia, Ciencia y Sociedad (Granada, 6-10 noviembre 1989)*. Granada: Agencia Española de Cooperación Internacional.

Vallvé, Joaquín (1988). "La emigración andalusí al Magreb en el siglo XIII (despoblación y repoblación de al-Andalus)," 87–129. In Mercedes García-Arenal and Mª Jesús Viguera, eds., *Relaciones de la Península Ibérica con el Magreb (siglos XIII–XVI)*. Madrid: CSIC.

Verskin, Alan (2013). *Oppressed in the Land?* Fatwās *on Muslims Living Under Non-Muslim Rule from the Middle Ages to the Present*. Princeton, NJ: Markus Wiener Publishers.

Verskin, Alan (2015). *Islamic Law and the Crisis of the Reconquista. The Debate on the Status of Muslim Communities in Christendom*. Leiden-Boston: E. J. Brill.

Virgili, Antoni (2001). *Ad detrimentum Yspanie: La conquesta de Turtusa i la formació de la societat feudal (1148–1200)*. Collecció Oberta, Barcelona: Universitat Autònoma.

Vlaminckx, Kristine (1985). "La reddition de Tolède (1085 A.D.) selon Ibn Bassām Šantarīnī'." *Orientalia Lovaniensia Periodica* 16: 179–96.

7 Migration and modern slavery

Perspectives in Africa to Europe migration

Olayinka Akanle

Introduction

Studies abound on international migration to examine causes, moderators, and consequences of the global phenomenon (Akanle et al. 2019; Ikuteyijo 2020). Although international migration is as old as humans, recent experiences, particularly exemplified in the European migration crises of 2015 and persistent modern slavery, suggest the need for more nuanced and continuous research into international migration, given its dynamics, ramifications, and trajectories—especially from Africa (Baker 2019a). Understanding modern slavery is important when assessing current migratory flows between Africa and Europe. For clarification, modern slavery as discussed here refers to processes and consequences of facilitation, sourcing, recruitment, transporting, keeping, and using people by open and/or subtle force, deception, and other means—usually and most explicitly for exploitation (Anti-Slavery International, UK 2020; IOM 2019). According to the US Department of State (2020), modern slavery has become an umbrella term for both sex trafficking and compelled labor.

Modern slavery is usually hidden, and as a result many concealed slaves exist in contemporary societies even though slavery is now generally considered illegal. Unlike historical slaves during the transatlantic slave trade, modern slaves do not usually carry an open tag for identification. Nevertheless, they operate under restricted conditions and do not freely complain to third parties or authorities. They live and work in less-than-human conditions and earn below legal wages. They are systematically exploited and maltreated (US Department of State 2020). Individuals entangled in modern slavery sometimes enter into it willingly, while many are deceived or coerced into it. Some modern slaves can work and struggle themselves out of slavery, but many can remain permanently in slavery depending on the character of the enslaving syndicates, the nature of agreements and contracts, and the vigilance of host countries' law enforcement authorities and institutions.

Generally acceptable theories are lacking on how migrant brokers connect modern slavery, yet migrant brokers and networks are key actors in the context of international migration and modern slavery (Triandafyllidou 2018; Deshingkar et al. 2019). Theoretically, relationships between migrant brokers

DOI: 10.4324/9781003170686-9

(or networks) and migrants sometimes appear mutual, but trust is essential. Real and/or perceived mutual benefits are assumed and imagined until real distrusts and betrayals set in. Migration brokers leverage migrants' goals against the brokers' experiences and exploitive behaviors, even when some migrants are aware of risks involved in the journey or of the migration brokers' earlier failures, as in the case clearly demonstrated by Libyan routes' brokers and networks (Triandafyllidou 2018). The experience is not entirely different in the migration corridors of the Horn of Africa, where migrants and brokers relate and work together as coworkers and partners in the migration space. Migration brokerage and networks are very complex at origin and become more precarious at destination.

Intending migrants count their risks and take them with little attention to long-term danger. The illegality of brokers and challenges at destination do not count for much when intending migrants deal with brokers. To understand how modern slavery sustains itself in relation to migration brokers and networks, it must be appreciated that in theory, little or no distinction exists between il/legality of brokers and il/legality of migration. It is also difficult for migrants to ascertain the genuine motivations of brokers, their ultimate intentions, and their real capacities to facilitate movements—or whether migrants are dealing with modern enslavers. According to Adugna Tufa Fekadiu and colleagues, migration brokers play a crucial role in organizing and sustaining irregular migration, as exemplified among Ethiopians (Fekadiu et al. 2019). Such efforts can lead to modern slavery as brokers and smugglers assist migrants in circumventing layers of migration controls, and even as they navigate complex and risky mobility landscapes in the context of the governments' organized campaign to stop "illegal migration." Although migration brokers help migrants navigate complex migration systems at the origin, the brokers' profit appropriation often extends to the destination. As demonstrated in the case of Bangladeshi construction workers by Priya Deshingkar and colleagues, brokers, in relation to employers and even state actors, create multiple, precarious, and unpredictable risks and opportunities through which migrants must navigate to reach the future they imagine and desire (Deshingkar et al. 2019). Sometimes, however, the journey to fulfill these dreams leads to modern slavery. Migration and modern slavery exist in precarious, complex, and multilayered entanglements, and brokers and migrants are often engaged in dialectical relationships within complicated actions in various contexts and scenarios.

Modern slavery is a profound element of international migration, particularly in destination countries of the European Union (EU) (Hashimu 2018). Modern slavery as a consequence of international migration has gained considerable attention among scholars as it continues to shape relations among (and have great impact on) people and countries (IOM 2019). Modern slavery within the corridor of international migration is common as people move with or without required documents and credentials (IOM 2019; Ikuteyijo 2020). It is also an outcome of ir/regular migration papers or credentials facilitated within migrant networks at origin, transit, and destination

(Triandafyllidou 2018). Migration networks lure and sometimes coerce people at origin and transit and adopt many strategies to perfect their trades. Modern slavery not only makes migration management and integration difficult, but it also finances criminal networks across the value chain at home/origin, transit, and destination countries.

Economic migration, now the most common type of migration from Africa, is mainly motivated to negate insecurity, to find better jobs, and to access better education and good health systems (Ikuteyijo 2020, Akanle et al. 2019). Economic migration is born out of the desire to escape poverty and underdevelopment at origin by moving into more developed countries (Omobowale et al. 2019b; Triandafyllidou et al. 2019). Economic migration can be as desperate as other types of migration and can be as irregular as any other type (Omobowale et al. 2019b; Triandafyllidou et al. 2019). Regardless of the type of international migration from Africa, economic considerations are ultimately involved, and the economics of migration is important for irregular migration (Baker 2019a). Desperation is a major element of migration from Africa that has led many migrants to risk perilous movements; as a result, many who reach their destination end up in modern slavery.

Migrant agency is key to understanding modern slavery. Contextually, migrant agency of Africa is very dynamic and related to survival. The context of migrant agency is mostly belied by social, political, economic, and underdevelopment experiences. Due to survival pressures and existential challenges of Africans, many are easily deceived and lured into migrations that lead to modern slavery. There are, however, individuals who knowingly enter into these highly exploitative relationships. Such individuals enter modern slavery agreements because they think that no matter the slavery abroad, it will ultimately be better than remaining in Africa with widespread underdevelopment and poverty (Olaniyi 2009; Akanle 2018). Many of the willing individuals sign agreements and swear oaths through voodoo in shrines (Carling 2006; Comino et al. 2016). While there are cases where migrants enter modern slavery because they already lack human rights at home, this kind of motivation is not the norm.

Most migrants from Africa are economic migrants who enter exploitative relationships out of desperation to escape poverty and underdevelopment in Africa (Akinyemi et al. 2005; Omobowale et al. 2019b,). Many migrants remain in exploitative positions because they are bound by oaths and agreements. They have no legal paperwork, as their travel documents are usually confiscated upon arrival at their destination; as a result, they fear deportation and imprisonment. They also fear being killed by the migration syndicates or fear that the syndicates will harm their kin and family members back home if they do not keep to terms of the agreements at destination.

Unfortunately, modern slavery is among the least researched subjects in migration studies. This chapter, therefore, examines international migration from Africa and modern slavery relative to sustainable migration managements at origin, transit, and destination countries. Migration and modern slavery are emerging—and related, sustained phenomena are manifesting—as patterns of social relations develop between Western countries and underdeveloped countries, mostly dependent on countries in the north, for constructed and real livelihood (United Nation Population Division 2002; Alemazung 2010; Ikuteyijo 2020; Omobowale et al. 2019a). Issues engaged in this chapter are worthy of examination to further understand the dynamics and nuances of international migration from Africa particularly as the EU struggles to understand and cope with migrations from Africa (Akanle 2018).

Narratives of international migration and modern slavery

Slavery is a historical fact and manifests in various forms. Although several declarations have been made with the aim of wiping the slate clean of both past and present manifestations of slavery, elements of slavery subsist in contemporary international migration spaces (IOM 2019). Slaves suffer inhuman treatment from slave masters, and slaves (having little or no rights) are vulnerable to exploitation and abuse. The dimension slavery takes today within international migration is modern and complex, and sometimes systemic to the extent that slaves may not consider their status as such. The essence of this method is to conceal the act of modern enslavement by convincing migrants that it is better abroad in the context of modern enslavement than it is to live back home, even if that is not always the case. Modern slavery in migration circles involves diverse inhumane acts like human trafficking, sex trafficking, forced labor, domestic servitude, debt bondage, voodooism, unlawful recruitment, and visa/document racketeering. In the context of international migration, modern slavery includes engaging in activities that dominate and embondage migrants. These migrants might have entered the bondage and enslavement either willingly/knowingly or unwillingly/unknowingly as they commit to emigrate at all cost regardless of the terms and conditions. Many migrants entered oaths through voodoo, by incurring debt, by selling investments at origin or disposing of kin and community properties, and by signing bonds to emigrate. They, thus, remain illegally and legally in slavery abroad and must not return to their origin for whatever reason, as long as the often-impossible conditions attached to their migration are in place.

Modern slaves within international migration are made or forced to work, and often to sell their bodies, for little or nothing. The International Labour Organization (ILO) estimates that more than 40 million people are trapped in modern slavery; the human trafficking sector (HTS) of modern slavery alone is worth up to $150 billion a year across the world (Baker 2019b). This is a sector rich enough to conceal, rejuvenate, re-create, and protect itself from law enforcement structures. Capabilities to harness and manipulate people and environments are key in migration processes. Generally, geography, social relations, climate change, perceptions of migrations, and benefits and mindsets are important to development and migration issues, including modern slavery. Against the background that most African migrations are ultimately meant to be permanent, especially for those fighting to survive, many tolerate modern slave conditions at their destination (Akanle 2018). Given that migration is a major component of population dynamics characterized by deliberate decision-making processes of the migrants, many African migrants rationalize the conditions of slavery that accompany their migration as long as it takes them north away from the socioeconomic crises at home in Africa. This is particularly noteworthy because most migrants from Africa are relatively young, educated, and more achievement oriented (Todaro 1976; Akanle et al. 2019).

As a result, the young, energetic, educated, and ambitious population that are supposed to drive the development processes of the original location—Africa—are enslaved and entrapped in migration chains and networks in the north, thereby keeping the continent perpetually underdeveloped and the destination region perpetually endangered. Hence, it's important to note here that in the long run, these migrants become problematic to the ultimate development of both the origin and host nations. As they become enslaved, their capacity for optimal self-development is compromised. They turn bitter against themselves as well as their origin and host countries: they lack the capacity to add value to the place they migrated from as they bridge the functional systems at their destination; they service illegal networks and operate at the fringe of society just as they negate formal development structures. Unwholesome migration systems that sustain enslavement are, therefore, an ultimate loss-loss to all within migration and development systems in general; this result must be well recognized and understood. Just as the era of historical slavery and slave trade experienced serious development and economic downturn in the long run, the consequences will be dire if modern slavery within migration is not sufficiently understood and managed.

Migration offers both positive (celebratory) and negative (pessimistic) effects (Akanle 2018). On the positive side, migration enables the migrants to learn new skills, broaden their intellectual and social horizons, and earn a living—as well as help left-behind households through remittances sent back home for welfare enhancement and national/continental development. Migration also stimulates cultural innovations and technological changes (Akanle 2018). The negative effects of migration to the place of origin include brain drain and diplomatic backlash in the case of irregular migration and criminal tendencies of desperate migrants at destination countries (Dodani and La Porte 2005; Akanle 2018).

Modern slavery is an insufficiently studied negative effect of migration on transit and destination countries. Migration is a key context of modern slavery, and yet modern slavery is commonly overlooked in migration studies because enslavement itself is obscured, especially when all migrants are collapsed under one homogenous category, as though all are free to move from oppressive conditions. But such freedom is not always the case. Even with the strict regulations and the growing securitization of borders, and even with more rigorous requirements and processes, some migrants have devised ways to evade the rules of migration, as evident in innovative reversed rules that ultimately ensure the systematic survival of migrants' enslavement.

International migration: experiences in modern-day slavery

Development crises, especially poverty and conflicts, are among the most prominent reasons people migrate; these reasons also determine where migrants go (Alemazung 2010; Austin 2010; Toromade 2018). According to Simon Toromade (2018), widespread poverty all over the world drives international migration, human trafficking, prostitution, illegal recruitment, and other factors leading to modern slavery. At home and abroad, various interactions, negotiations, and tensions between migrant agencies and migrant brokers and networks exist to lure individuals (who sometimes enter these relationships knowingly) into these exploitative relationships. The exploiters and recruiters who introduce Africans into modern slavery are sometimes part of the same community. Although such recruitments can involve betrayals of kinship and community trusts, they are usually constructed or seen as a way of strengthening community-driven chain migration to increase remittances. Not all migrants in slavery have been lured into it. Many Sub-Saharan Africans even intentionally walk into slavery in Europe to escape poverty and brighten their chances for a better life. Many believe they will later fulfill the terms of their migration enslavement and become subsequently free to enjoy life in Europe by relocating to other European nations far from where they were previously enslaved. Recruiters target victims by sourcing (un)suspecting migrants to paint excellent pictures of their own migrations abroad. Such recruitments are part of a chain process; the layers and complexity at the start, middle, and end of the chain depend on the nature, size, and reach of the syndicate networks.

Nigeria, Senegal, Ethiopia, Cameroon, Mali, Guinea, Libya, and Burundi are among the nations with the largest number of slaves/victims globally (Walk Free Foundation and ILO 2016). Networks of migrant establishments, chain migration systems, and the preference among Africans for the Schengen visa—which allows a holder to enter and travel in a specified zone covering most of Europe for 90 days—drive migration from Africa toward Europe and sustain modern slavery. Migrants and their networks can move freely across the Schengen Zone until they finally settle in country of choice within the EU. According to the IOM (2019), contexts of vulnerability to modern slavery include private dwellings, border crossings, irregular migration routes, private businesses, displacement sites, refugee camps, natural disasters, rural areas, commercial sex establishments, and conflict zones.

Modern slaves experience severe exploitation by other people for personal or commercial gains (Anti-Slavery International, UK 2020). Although modern slaves can be seen providing services in public (for instance in stores, restaurants, factories, and farms), they are not usually recognized as members of an established workforce, given the clandestine nature of their slavery. Modern slaves are usually trapped in clandestine migration webs run by strong international syndicates (IOM 2019; Anti-Slavery International, UK 2020). The perpetrators linking international migration and modern slavery, therefore, are migration gangs and networks in Africa and Europe.

People in Africa have found themselves trapped in slavery abroad work in various factories manufacturing laptops, phones, fashions, or other products, or they provide cleaning and sanitation services; and young girls and women are vulnerable to sexual harassment by their hosts and bosses (Ikuteyijo 2020). Slave migrants suffer various forms of maltreatment in Europe, although this abuse is not generally confined to Europe. Most migrants who ended up in slavery experience and suffer crimes in their destination countries, including but not limited to verbal and mental abuse, rape, harsh labor, exploitation, extortion, and torture. According to Stefano Comino and colleagues, there is a greater possibility for migration slaves to become victims of a crime, despite laws aimed at promoting the rights and dignity of mankind across the world (Comino et al. 2016). In every country, atrocities and cruel treatment continue to manifest, but victims are not bold enough to report the situation, whether for fear of losing their migration status, or because of deep-seated beliefs, superstitions, and ignorance.

Broader-based and more inclusive migration management strategies are necessary to root out modern slavery because aggressive and stricter migration laws and policies generally empower migration syndicates to be more exploitative of the migration systems and Africans (Green and Grewcock 2002). As a consequence, while slave-generating syndicates and networks become more innovative, daring, and brutish, migration slaves become more compliant, timid, and driven further underground. So, despite more stringent migration policies and laws, destination countries of Europe continue to harbor growing numbers of slaves existing outside tax and legal migration nets who would benefit from better and more accountable migration management.

Information technologies, social media, international migration, and modern slavery

Information technologies (ITs) and social media are central to social relations in the twenty-first century. ITs and social media are even more crucial in effective human interactions as people move across borders. Social media is particularly pervasive, serving complex roles involving positive and negative elements in international migration. ITs and social media are thus double-edged swords within international migration; they affect how modern slavery is both perpetrated and perpetuated. According to Stephen Gelb and Aarti Krishnan, the relationship between technology and migration is central to achieving an agenda for sustainable development by 2030. Highly skilled migrants contribute substantially to technology innovation, research and development in destination countries, and the diaspora transfer technology to origin countries; digital connectivity impacts every aspect of migration, just as state or government migration management relies heavily on technology (Gelb and Krishnan 2018).

A high-profile issue in the last few years concerns how migrants and other key networks use ITs during migration journeys (McAuliffe 2018). Migrants and migration brokers use different applications (apps) and platforms to share information in real time, whether by supporting clandestine border crossings or by consolidating social media platforms to connect geographically dispersed groups with common interests. This raises valid questions concerning the extent to which technology has been used to support irregular migration as well as abusive and exploitative relationships (McAuliffe 2018). According to Jean-Yves Hamel, there is a strong relationship between information and communication technologies (ICTs) and migration. ICTs have not only positive but also negative implications for migration as ICTs are used to facilitate regular and irregular migrations and associated normativities and non-normativities, including modern slavery (Hamel 2009).

Hannah Thinyane has identified three cases in which ITs can help in understanding the relationship between migration and modern slavery: (1) recruitment, (2) identification of victims, and (3) individual reporting (Thinyane 2017). The number of cases can, however, be extended to five: (4) identification of actors and criminals in the networks of modern slavery and (5) monitoring of modern slavery systems. Modern slavery networks, gangs, syndicates, and systems utilize ITs and social media on a massive scale to perpetrate their trades and activities in key and strategic ways. Facebook, WhatsApp, Skype, Zoom, Twitter, Instagram, emails, and websites are commonly used to recruit, retain, monitor, and exploit modern slaves. In Africa, unsolicited emails are sent to potential recruits promising breakthroughs, help, and lucrative job offers in Europe. Such emails detail fantastic opportunities and assistance in Europe and overseas.

Examples abound to show how social media and other platforms are used to recruit African migrants into modern slavery. The case studies of Sunday Iabarot and Gladys from Benin Nigeria are relevant (Baker 2019a). Many Africans become entangled in modern slavery after hearing from friends in communities and on Facebook that life in Europe will bring them prosperity. Based on information from Facebook friends that jobs were plentiful in Europe, Sunday Iabarot, for instance, emigrated from Nigeria with a plan to go to Europe by paying for his passage on a migrant smuggler's boat (Baker 2019b). He was deceived, sold and resold, scared and dehumanized within the modern slavery syndicate. As Sunday explained his treatment in the syndicate, "It was as if we weren't human" (Baker 2019a, 2019b). For Gladys, from Nigeria's impoverished rural southwest where a generation of young people are seeking their fortunes abroad, modern slavery recruiters pose as families and friends, either on the ground or online via Facebook, Twitter, and Instagram, to deceive young women and men with unrealistic promises of money to be made in Europe (Baker 2019a, 2019b).

Deceptive email or spam messages are also common, especially prevalent among people who use yahoo!mail, the most common email server in Sub-Saharan Africa. Unfortunately, yahoo!mail has a weak spam filtering system, but it is free and Africa is poor—and usually unaware of dangers associated with free email accounts. Many people trust electronic messaging to the extent that they easily believe any message that comes to them via email. Those who are partially aware of the risks may decide to take a chance anyway. Interestingly, international migration processes that lead to modern slavery are sometimes free and portrayed as risk-free. Facebook messages that offer fantastic job and education opportunities in Europe, America, and Canada commonly circulate in Africa. People are enticed through social media, especially Facebook, and are discretely enslaved and circulated within the networks across countries. ITs are integral to this process: internet, webcams, and mobile phones are frequently used to monitor modern slaves and network members to prevent infiltrations by law enforcement institutions. ITs and social media make both text messaging and video meetings and conversations easy, allowing people and syndicates to manipulate messages, voices, and locations to lure people into the modern slavery net. Money is transferred and wired through ITs to sell goods and export online without physical contact and the profits used internationally to finance further human trafficking and modern slavery. It is also possible to launder the proceeds of modern slavery through virtual and cryptocurrencies.

In resolving the existential and analytical tensions in the context of entering modern slavery, it is noteworthy that two complementary categories of modern slaves exist in migration systems moving from Africa toward Europe: there are those who are easily duped and lured into exploitation within modern slavery, and those who enter modern slavery intentionally believing that they can survive, legitimize their status, and ultimately free themselves to live normal lives when they reach their destination. This category of modern slave believes living anywhere in Europe is better than living in Africa no matter if the way they migrate out of Africa involves conditions of slavery. They believe no condition in Europe can be as bad as in Africa.

ITs also allow human traffickers, kidnappers, and facilitators of illegal migration to alter encrypted codes or change strategies, routes, and agents with the speed of light to evade arrest and avoid detection (Lewin 2019). Such online activities make it difficult for modern slaves to escape and easy for criminal gangs and syndicates to outwit them. With ITs, the tricks of facilitating modern slavery are dynamic and difficult to comprehend by non-gang members. It is even possible to start modern slavery and international trafficking processes online and finish online, to start online and finish offline, to start offline and finish offline, or to start offline and finish online. The combination depends on the sensitivity of the transaction and the degree of perceived vulnerability and the level of infiltration by the networks. Information technologies are further obscuring the existence of modern slavery by the day and keeping syndicates mileage ahead of law enforcement apparatuses in terms of skills, techniques, and capabilities—especially relative to online/virtual operations (Lewin 2019). ITs and social media enable syndicates to reach further and faster and dominate legal and illegal spaces more effectively.

Even when leading companies in the technology market across the world claim to be waging war against modern slavery and slave networks, their efforts seem to be yielding significant fruits. For example, in 2017, Facebook, Google, Microsoft, and Uber met during the annual hackathon for development and testing of tools to counteract online trafficking (Heuty

2018), but substantial practical impacts are yet to be visible in the human trafficking and modern slavery sectors as a result of such actions, even though some progress may have been made.

Conclusion

The relationship between international migration and modern slavery is a general indication of underdevelopment of Africa. Europe has experienced consequences and outcomes of migration, and it still receives large number of immigrants with peculiar challenges concerning refugee issues, security matters, sociopolitical and economic criticisms, and integration. Interests and welfare of African migrants who fall into slavery in transit and destination remain very important scholarly, policy, and practical issues. It is important to better interrogate and understand the trajectories and ramifications of modern slavery in the contexts of movements from Africa to Europe. Now is the time for Europe to look deeper into the unusual spaces of migrants' livelihood and existences across migration chain; attention should be given to modern slavery and its processes and outcomes. Community-based targeting in Africa and Europe is important to support reporting, monitoring, and migration management, particularly with better appropriation of ITs. The tide of dangerous migration to the north, particularly to Europe, that proliferates modern slavery, thus, must be short-circuited and stopped. African leaders must be more accountable; their information and migration systems should be improved beyond just imposing migration (visa) restrictions in Europe. Migration and stricter visa processes will only lead to more desperation among Africans, make illegal routes more attractive, and ultimately make European migration management more tedious, disorderly, and expensive as migration continues to be more problematic.

Underdevelopment of Africa is a major problem not only to Africa but even to the developed nations, including Europe. Underdevelopment of Africa is, thus, dangerous to global development. Desperate migration from Africa to Europe and consequent slavery are critical and real signs of bigger problems. Against this background, perilous migration from Africa and underground slavery in Europe are inevitable and will remain a concern and stress on immigration as well as on refugee and integration management unless immigration and support systems are implemented within broader development strategies of Africa. Information technologies and social media are important in international migration management and in stopping human trafficking and modern slavery. Even though technology is neither a magic wand nor a quick fix, it offers important tools to prevent modern slavery and to set those already enslaved free. More strategic information and preventive education through ITs and social media in Africa are key to empowering Africans with the right information and warnings about modern slavery in Europe. Hence, more investments are needed in research on international migration and modern slavery to develop sufficient understanding and bridge knowledge gaps in modern slavery and African migrations to Europe.

References

Akanle, Olayinka (2018). "International Migration Narratives: Systemic Global Politics, Irregular and Return migrations." *International Sociology* 33(2): 161–70.

Akanle, Olayinka, Olufunke Fayehun, Gbenga Sunday Adejare, and Augustina Otomi Orobome (2019). "International Migration, Kinship Networks and Social Capital in Southwestern Nigeria." *Journal of Borderlands Studies*. doi: 10.1080/08865655.2019.1619475.

Akinyemi, Akanni S., Olawale Olaopa, and Olufunmi Oloruntimehin (2005). "Migration Dynamic and Changing Rural-Urban Linkages in Nigeria." Accessed January 27, 2021, from https://iussp2005.princeton.edu/papers/50208

Alemazung, Joy A. (2010). "Postcolonial Colonialism: An Analysis of International Factors and Actor Marring African Socio-economic and Political Development." *Journal of Pan-African Studies* 3: 16–25.

Anti-Slavery International, UK (2020). "What Is Modern Slavery." Accessed November 28, 2020, from https://www.antislavery.org/slavery-today/modern-slavery/#:~:text=Modern%20slavery%20is%20the%20severe,as%20cooks%2C%20cleaners%20or%20nannies

Austin, Gareth (2010). "African Economic Development and Colonial Legacies. 50 years of Independence Review: Major Development Policy Trends." *International Development Policy*. doi: 10.4000/poldev.78.

Baker, Aryn (2019a). "'It Was as if We Weren't Human': Inside the Modern Slave Trade Trapping African Migrants." *TIME*, March 14.

Baker, Aryn (2019b). "The Best Way to End Modern Slavery? Enable Legal Migration, Experts Say." *TIME*, December 2.

Carling, Jørgen (2006). *Migration, Human Smuggling and Trafficking from Nigeria to Europe*. Migration Research Series, 23. IOM: Geneva.

Comino, Stefano, Giovanni Mastrobuoni, and Antonio Nicolò (2016). *Silence of the Innocents: Illegal Immigrants' Underreporting of Crime and their Victimization*. IZA Discussion Papers 10306, Institute of Labor Economics (IZA).

Deshingkar, Priya, C. R. Abrar, Mirza Taslima Sultana, Kazi Nurmohammad Hossainul Haque, and Md Selim Reza (2019). "Producing Ideal Bangladeshi Migrants for Precarious Construction Work in Qatar." *Journal of Ethnic and Migration Studies* 45(14): 2723–38.

Dodani, Sunita, and Roland La Porte (2005). "Brain Drain from Developing Countries: How Can Brain Drain Be Converted into Wisdom Gain?" *Journal of the Royal Society of Medicine* 98(11): 487–91.

Fekadiu, Adugna Tufa, Priya Deshingkar, and Tekalign Ayelew (2019). *Brokers, Migrants and the State: Berri Kefach "Door Openers" in Ethiopian Clandestine Migration to South Africa*. Working Paper. Migrating out of Poverty, Brighton.

Gelb, Stephen, and Aarti Krishnan (2018). *Technology, Migration and The 2030 Agenda for Sustainable Development*. Briefing Note. London: ODI.

Green, Penny, and Mike Grewcock (2002). "The War Against Illegal Immigration: State Crime and the Construction of a European Identity." *Current Issues in Criminal Justice* 14(1): 87–101.

Hamel, Jean-Yves (2009). *Information Communication Technologies and Migration*. UNDP: Human Development Research Report.

Hashimu, Shehu (2018). *African Migration to Europe and the Emergence of Modern Slavery Africa: Who Is to Blame: Nigeria in Perspectives. Proceedings of International Academic Conferences 7208491*, International Institute of Social and Economic

Sciences. Accessed January 28, 2021, from https://ideas.repec.org/p/sek/iacpro/7208491.html

Heuty, Antoine (2018). *How 'Responsible Supply Tech' Is Helping Tackle Modern Slavery and Human Trafficking*. London: The Freedom Fund.

Ikuteyijo, L. O. (2020). "Irregular Migration as Survival Strategy: Narratives from Youth in Urban Nigeria." In McLean M. eds., *West African Youth Challenges and Opportunity Pathways. Gender and Cultural Studies in Africa and the Diaspora*. Cham: Palgrave Macmillan. doi: 10.1007/978-3-030-21092-2_3.

IOM (International Organization for Migration (2019). *Migrants and Their Vulnerability to Human Trafficking, Modern Slavery and Forced Labour*. Geneva: IOM.

Lewin, Ed (2019). "Technology Can Help Us End the Scourge of Modern Slavery. Here's How." *World Economic Forum*. Accessed January 27, 2021, from https://www.weforum.org/agenda/2019/04/technology-can-help-us-end-the-scourge-of-modern-slavery-heres-how/

McAuliffe, Marie (2018). "The Link Between Migration and Technology Is Not What You Think." *World Economic Forum*. Accessed January 27, 2021, from https://www.weforum.org/agenda/2018/12/social-media-is-casting-a-dark-shadow-over-migration/

Olaniyi, Rasheed (2009). "'We Asked for Workers But Human Being Came': A Critical Assessment of Policies on Immigration and Human Trafficking in the European Union," 140–61. In Akanmu G. Adebayo and Olutayo C. Adesina eds., *Globalisation and Transnational Migrations: African and Africa in the Contemporary Global System*. Cambridge: Cambridge Scholars Publishing.

Omobowale, Ayokunle Olumuyiwa, Olayinka Akanle, and Clement Akinsete (2019a). "Scholarly Publishing in Nigeria: The Enduring Effects of Colonization," 215–33. In James N. Corcoran, Karen Englander, and Laura-Mihaela Muresan eds., *Pedagogies and Policies for Publishing Research in English: Local Initiatives Supporting International Scholars*. London and New York: Routledge Taylor and Francis Group.

Omobowale, Ayokunle Olumuyiwa, Olayinka Akanle, Samuel Falase, and Mofeyisara Oluwatoyin Omobowale (2019b). "Migration and Environmental Crises in Africa," 1–10. In Cecilia Menjívar, Marie Ruiz, and Immanuel Ness eds., *The Oxford Handbook of Migration Crises*. London: Oxford University Press.

Thinyane, Hannah (2017). "Role of ICTs in Safeguarding Migrant Workers." UNU Institute of Computing and Society, Macua SAR. Accessed January 28, 2021, from http://collections.unu.edu/eserv/UNU:6207/ECDG17_212.pdf

Todaro, Michael P. (1976). *Internal Migration in Developing Countries*. Geneva: ILO.

Toromade, Simon (2018). "Nigeria Ends 2018 with 90.8 Million People Living in Extreme Poverty." *Pulse*, December 28. Accessed January 28, 2021, from https://www.pulse.ng/news/local/nigeria-ends-2018-with-908-million-people-living-in-extreme-poverty/nljxqtj

Triandafyllidou, Anna (2018). "The Migration Archipelago: Social Navigation and Migrant Agency." *International Migration* 57(1): 5–19.

Triandafyllidou, Anna, Laura Bartolini, and Caterina F. Guidi (2019). *Exploring the Links Between Enhancing Regular Pathways and Discouraging Irregular Migration*. IOM: Geneva.

United Nation Population Division (2002): *Population Database*. New York: United Nations.

US Department of State (2020). *What Is Modern Slavery*. Office to Monitor and Combat Trafficking in Person. Accessed October 23, 2020, from https://www.state.gov/what-is-modern-slavery/

Walk Free Foundation and International Labour Organisation (ILO) (2016). *Forced Labour and Forced Marriage: Executive Summary*. Accessed February 23, 2022, from https://www.ilo.org/wcmsp5/groups/public/@dgreports/@dcomm/documents/publication/wcms_575540.pdf

8 The anxious integration of former enclave or "new" citizens in North Bengal, India

Nasreen Chowdhory and Shamna Thacham Poyil

Introduction

An enclave, by definition, encompasses a territory whose borders are in turn enclosed by the mainland territory of another sovereign state; the borders of this irregular cartography are formally legalized, but the enclave exists outside the sovereign administration in which it sits. As a result, access to the enclaves requires officials of one country to traverse the border and sovereign land of another state to administer and manage its own citizens (Jones 2009; Cons 2013). But the enclave's ambiguous territoriality, when compared to the clearly demarcated frontiers of the nation-state system, effectively precludes its dwellers from completely realizing their rights as citizens.

How enclave dwellers experience *belonging* to the nation-state depends on issues of mobility/immobility, their citizenship status, and subsequent anxieties caused by integrating with their new societies. Scholars have addressed precarity according to various views and circumstances: from a statist perspective (Whyte 2002; Cons 2013); statelessness and national identity (van Schendel 2002); the territorial discontinuity of enclaves and its impact on sovereign authority of the home and host state (Jones 2009; Dunn and Cons 2013); and citizenship, abandonment, and vulnerability of enclave dwellers (Shewly 2012, 2013). In this chapter, we examine the ways in which enclave dwellers challenge the immobility within the contested and ambiguous territories of enclaves located in different nation-states: the *Chitmahal* people—caught amid the territories of two sovereign nation-states, India and Bangladesh—who have navigated a precarious existence (see Figure 8.1).

The term "nation-state" derives from the incongruous hyphenation of "state" as the sovereign institutional entity within a given territory, and "nation," which symbolizes the essence of its peoples' common culture through shared ethnicity, religion, language, or history. The association between the cultural entity of the nation and the institutional entity of the state is not organic. The problems caused by an artificial conflation between nation and state have been acknowledged (Shapiro 2000). A prevalent assumption about the "territorial contiguity" of national space nevertheless prevails, although the precarity of enclave dwellers remains anchored within the noncontiguous nature of both the territorial borders of the state and

DOI: 10.4324/9781003170686-10

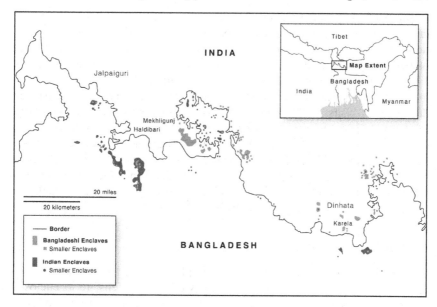

Figure 8.1 Enclaves on the India–Bangladesh border.

Data & Sources: ISCGM/Survey of India; Natural Earth Data; "Diagrametic Sketch Map of Cooch Behar" (Banerjee 1966). Cartographer: Gregory T. Woolston.

cultural borders of the nation (van Schendel 2002: 115). But nation-states strive to preserve and uphold their integrity by maintaining a coherent narrative as a nation while consolidating the state's territory in terms of "symbolic maintenance" (Shapiro 2000: 80). The emergence of the nation-state as the tacit, basic unit of world order has linked the "symbolic management of nation-state" to the ways in which the state itself undertakes the "symbolic management of citizen" (Shapiro 2000: 81).

The spatial component of this symbolic management has involved configuring and demarcating the territory of the sovereign state and securing its borders. These spatiotemporal alterations of the nation-state have made it necessary to reconfigure the terms of citizenship as well. The 2015 Land Boundary Agreement (LBA) ratified between India and Bangladesh, for instance, involved consolidating the otherwise discontinuous territory of enclaves to the mainland. This "symbolic maintenance of [the] nation-state" has an impact on citizens but more so on enclave dwellers in these ambiguous territories. The "new" citizenship these enclave dwellers acquired by choice after the 2015 LBA should be seen as a reclamation of their belonging to the nation-state. Citizenship is essentially a legal validation based on an individual's right to belong in the territory of the state. Realizing citizenship substantively requires a conscious re/claiming to belong to the territory of state.

Stories of enclave dwellers challenge the prevalent binary view of the mobility spectrum as "sedentary versus immobile," especially for forced migrants. The disproportionate focus on how vulnerable people are rendered "immobile" overlooks the ways in which they contest and exist within this immobility. Here we examine the anxiety of "new" citizens who are conditioned by the perpetually liminal citizenship as they navigate the shared sense of belonging to the nation in terms of everyday life in resettlement camps. The first section of the chapter draws on the im/mobility experienced by former enclave dwellers whose narratives we source from fieldwork undertaken during three visits, in 2016 and 2017, to resettlement camps of Haldibari, Dinhata, and Mekhligunj in North Bengal. We also draw from fieldwork to share narratives from former enclave dwellers attempting to realize their mobility as new citizens. In the second section, we look into the notion of belonging that conditioned their choice of citizenship, which in turn facilitated the realization of other substantive citizenship rights and brought on anxieties about integrating with their immediate societies.

Enclave dwellers and their im/mobility

Scholars have attributed the problematic implications of mobility across borders of the nation-state, as well as the entry of "noncitizens" to its enclosed and secure borders, as the primary catalyst for reconfiguring frameworks of citizenship. Noncitizens' immobility was a structural constraint initiated when the nation-state invoked territorial sovereignty by creating a stringent legal definition of who constitutes a citizen within its territory. Bryan S. Turner (2007: 288) points to debates from the 1980s and early 1990s regarding the "character of nation-state citizenship" and how it might be transformed to accommodate transnational reverberations on multiple fronts resulting people moving across borders. Mimi Sheller and John Urry (2006) examine a view from the social sciences of a "new mobilities paradigm" as a result of complex social relations that emerge when issues concerning movement take a central stage. Aihwa Ong (1999) asserts how the idea of "transnationality" has caused citizenship to "mutate" (Ong 2006) and be "flexible" (Ong 2003).

Mobility often helps to achieve and materialize the rights that citizenship provides. Conversely, having citizenship doesn't guarantee an individual's mobility. Enclave dwellers are "citizens" of the country to which the enclave belongs, but they are precluded from realizing the substantive rights and privileges accrued by citizenship because of their unique territorial location. Though the state accords rights to mobile populations, these rights are never as substantial or full as rights enjoyed by citizens. Mimi Sheller and Tore Sager (2008: 28) suggest how the state, in exercising sovereignty by granting its citizens freedom of movement, is committing a "mobility injustice" because such movement is conditioned by the "denial of [an]other's mobility." As a result, a "mobility gap" (Shamir 2005: 199) persists, reflecting an unequal capacity to exercise "freedom of movement."

Figure 8.2 depicts how im/mobility is experienced within the fenced borders of the Karela enclave, located in the Dinhata region. India's Border Security Force (BSF) guards the typical barbed-wire fencing that demarcates

Figure 8.2 The fenced borders of Karela enclave.
Copyright: Nasreen Chowdhory.

borders and, in this instance, the India–Bangladesh boundary. This enclave is located outside of that boundary yet within Indian territorial limits. The nine families residing, bearing the brunt of that ambiguity, have refused to move, and their sense of mobility as citizens depends on kind cooperation and help from the BSFs. The precarious nature of their everyday life is captured by the lament that they love being "half citizens." Muhammed Jamal (MJ), who lives with his four family members, and Abdul Huq (AH), who has school-age children, both narrate the difficulties of navigating such precarity. MJ pointed out "amader obostha anekta ordek nagarik … na jete pari na aste pari" (our existence is like a half citizen, cannot move around). AH, on the other hand, stated "kosto holeo eta amar desh" (despite all hardship this is my country).

Abdul Huq's children go to school in Dinhata, which necessitates the BSF officials to open the border fencing for access to education. Says AH, "anok upakar koren BSF … na hole jibon cholto na" (BSF officials cooperate, otherwise life would have been more difficult). Officials guarding the fence maintain a roster to record when dwellers cross to the mainland. Dwellers must plan their routines to align with when the gate opens, three times daily before 4 p.m. Officials acknowledge that "ora nogarik r amader kotobbo saharjo kora" (they are citizens and it is our duty to help them). But the sovereign state lacks formal administration powers in the enclaves to which the enclave dwellers ultimately belong; the lack of legal documentation prevents them from adequately accessing their rights as citizens.

The unique mode of state formation in South Asia dictated a particular trajectory for citizens' rights that excluded noncitizens yet cultivated a politics of belonging based on nationality (Chowdhory 2018: 43–71). The "cartographic anxiety" (Krishna 1996) of the nation-state, which has always been based on securing the integrity and contiguity of its territorial possessions when it comes to tussles over borders and territories, has simultaneously complicated the issues of identity, belonging, and membership for its population. Spatiotemporal differences inherent to South Asian society have caused it to digress from a linear pattern of development and modernity, thereby precluding any notion of integrated citizenship. As a result, citizenship in postcolonial states of South Asia is not a policy instrument but relegated instead to being a source and marker of social identity.

In India and Bangladesh, both partitioned on the basis of religious differences, the prerogatives for citizenship were complicated by nationalism being tethered to religious identity. Apart from instituting a physical border that demarcated the territories of nascent states, partition created boundaries that reflected the colonial production of space geared to convenient governance of its subjects, leaving the South Asian states to grapple with a "messy template" (Cons and Sanyal 2013: 6) of aggravated religious, ethnic, and linguistic tensions in their nation-building process.

Indo-Bangladesh enclaves (or *chitmahals*) are at once a cartographic irregularity and an anomaly to sovereign authority. Located along the uneven and porous boundary between India and Bangladesh, they accounted for the bulk of existing enclaves globally until the Land Boundary Agreement of 2015 called for their residents' return to their respective countries. van Schendel (2002: 119) explains that prior to colonial rule, *chits* were fragments of land separated from parent estates of the Mughal Empire held by the predominant landlords of Bengal, whether by challenging or allying with the Mughal state. The lands occupied by the Mughal state and rulers of Cooch (a city in West Bengal), eventually went on to become enclaves and can be traced back to the peace treaties of 1713 (Whyte 2002; Shewly 2015). Because they were never fully incorporated by the Mughal state, the enclaves exhibited quasi-feudatory behavior and remained relatively aloof from the seat of Mughal rule in Dhaka. Worries about preserving their place on the map (Krishna 1996) and protecting their sovereign existence were less prevalent then. The people living in these enclave communities did not experience restrictions to their routine mobility or face ambiguity or precarity in terms of identity and survival, as during postcolonial years. This aspect of precolonial communities needs to be analyzed to understand how, with time, the enclave dwellers' notion of belonging shaped and reshaped their identification with (and differentiation from) the surrounding host population. The people in precolonial traditional communities nurtured multiple allegiances—such as religion, caste, or occupation—without preferential hierarchy to determine belonging; individual or collective identities were considered contextually, and it was not "disreputable or unreasonable" to belong to these "varying layers of community" at the same time (Kaviraj 2010: 12). Modern communities, however, were highly influenced by the colonial system, which converted subjects into quantifiable units belonging to specific and definitive communities.

After the British consolidation of Bengal through the Battle of Plassey in 1757 and Battle of Buxar in 1764, discontinuous territoriality and semi-feudatory status of enclaves continued until the nineteenth century (Cons 2016). Despite acknowledging difficulties in administering the enclave region, the British made no significant effort to facilitate the exchange of the Cooch Behar enclaves and Mughal enclaves (Whyte 2002: 75) because residents preferred the status-quo. Prior to 1949, enclaves formed as the geographical border between India and Eastern Pakistan, with 130 of them physically located in East Pakistan and 51 in India (Van Schendel 2002). After the partition of 1947, the princely state of Cooch Behar was confronted with the restricted choice of joining either India or Pakistan. The partition process based on religious nationalism created two newly independent states, but also generated a huge movement of the population across the newly created borders, including many refugees. Upon Cooch Behar's accession to India, its Indian enclaves acquired the status of Indian territory by transferring their jurisdiction to the most proximal Indian district (Van Schendel 2002).

The modern state-building process in India and Pakistan institutionalized the documentation and verification system for the population traveling across the borders by using a passport and visa regime from 1952 onward, thus hindering the movement of enclave dwellers and complicating their right to passage beyond their respective enclaves. The formation of Bangladesh in 1971 diminished to a considerable extent this vaunted tethering of territoriality with religious identity. Indian enclaves that were located in East Pakistan until 1971, and since then in Bangladesh, attained the status of "international enclaves" along with the similar Bangladeshi enclaves encircled by Indian territory. The consolidation of territories and the sovereign control and management of populations within the territories by the postcolonial states led to a situation in which enclaves were perceived as cartographic aberrations curtailing the mobility of enclave dwellers. Despite bilateral negotiations between the two countries, the enclave issue remained unresolved until 2015 because of discord on a federal level among the central and state elements of Indian government (Appadorai 1981), as well as differences in the religious affinity of both countries (Whyte 2002; Shewly 2015). The dynamics of the bilateral relations between India and Pakistan, and later

India and Bangladesh, has affected not only the legal status of the enclaves but also the marginal existence of their respective residents, rendering them as "immobile subjects" in the absence of the state.

To the new citizens, being mobile meant navigating resettlement camps and waiting for their complete integration as full citizens. As indicated in Table 8.1, fieldwork took place in three camps located in Dinhata, Mekhligunj, and Haldibari. The 47 families living in the Mekhligunj camp comprised 196 members, whereas Dinhata camp had 58 families with 247 members. The Haldibari camp included 96 families with 481 members. Each household was given 30 kilograms of rice, pulses (beans, lentils, peas), oil, salt, and kerosene. Every individual received an Aadhaar card, a unique identification number that was part of the exchange set forth in the LBA. Initially, the ration for each family remained the same in all resettlement camps: basic amenities and housing in structures made of tin, which heated up quickly in summer.

But over time rations dried up and families faced hardships. They were Indian citizens but had difficulty accessing their rights. Many who voluntarily moved to camps when given a choice of citizenship after the LBA of 2015 was ratified still spoke of anxiety and the agony leaving their family members who chose to stay back. Enclave residents with relatively large possessions of land (in comparison to others) had chosen Bangladeshi citizenship and stayed where they were rather than relocate and chose Indian citizenship. Primary concerns emerged from divided families whose members had to go back and forth between the Indian chit and Bangladesh. Some inhabitants had left land behind along with monetary assets in various banks. Other complications emerged among large numbers of enclave members belonging to joint families.

In Mekhligunj (see Figure 8.3), the new citizens expressed happiness to be part of the "matribhumi" (homeland). This settlement has 47 families, and each has a tale to tell.

Our field work indicated that similarly precarious situations existed even after citizens had relocated to transit camps. A third visit to the field in 2017 allowed for long conversations with residents in the three camps we visited in 2016. One such conversation revealed that the "new" citizens' integration into the Indian state remained incomplete. N. Sarkar (NS), M. Sarkar (MS), and A. Sarkar (AS) in the Dinhata camp despairingly agreed, "desher nagarik to holam, kintu a ki rokom je buja jai na" (we have become citizens of this

Table 8.1 The new citizenship

Profile of enclave citizens in North Bengal	Total number of families	Total number of persons
Dinhata	58	247
Meklikunj	47	196
Haldibari	96	481

Source: Data collected from the field visits—2016–2017

Figure 8.3 New citizen settlements in Mekligunj.
Copyright: Nasreen Chowdhory.

country but the nature of it is unknown). As MS suggested, "desh amader tai sob mane niyachi" (we belong in this country hence we have accepted our fate). These "new" citizens were disappointed that after reaching India, and having limited or no access to land they left behind, their desire to be "full" citizens of the country remained unfulfilled. When asked what it meant to be a complete citizen, their responses varied: access to the same resources as their neighbors across the camp, or benefits for their children similar to those of all other students in school, bicycles from a West Bengal government program, and welfare for school and college-bound children, especially girls. Yet Muneer Rahman from Krishimala camp in Dinhata lamented that the camp was becoming a site of violence, requiring vigilance along with protection from the government. He also complained of corruption and trafficking, although other members refuted his claims. Their inaccessibility to social benefits and their continued place at the periphery made them feel like lesser citizens.

The contours of belonging within a geographically defined space are devised by the state, and those who are legally entitled to belong are provided with rights and privileges accruing from citizenship. As the supreme legal instrument of the country, the Constitution of India does not define "citizen." But Article 5-11 stipulates who can be a citizen of the country and confers power on the Parliament to frame the qualifications to acquire and terminate citizenship in India. The Citizenship Act of 1955 bestows citizenship on people residing in the specific territory the state acquires. Prior to

LBA 2015, this decoupling of territory from belonging precipitated the sub-optimal status of "half citizens" for the enclave dwellers. Choosing citizenship required them to abandon their homes, land, and other minor assets they possessed in the enclave territory and relocate to Indian territory. By reaffirming their citizenship and accepting displacement, they expected to gain complete mobility, rights, and protection like any other full citizen of the country. How much of their situation had really improved since they became "new" citizens?

Residents of the Haldibari camp mentioned that none of the children they gave birth to after relocating received birth certificates. The inadequate healthcare infrastructure and medical support currently available were additional sources of worry: "shasto doftor ache, kintu somoi somoi babhosta bhalo na" (there is a health unit but time to time, help is missing). Those with divided families lamented that when in the enclaves, rules had been rather relaxed, which helped them to go back and forth between enclaves with a travel card. But in their present scenario they needed a passport, impossible under the circumstances. Some wanted to return to withdraw money left behind in banks in the former enclaves. Many family members spoke of needing resources, hoping perhaps to sustain their future by collecting what was left behind. One aspect of solidarity among new citizens came from reminiscing about family members they had left behind.

U Goni (UG) from the Dinhata camp shared the stories about leaving three brothers and his aged parents back in Bangladesh. He hoped that they would get another opportunity to join him in India. "Amra onek ke fela asche … kintu bura amma ke rekhe hoyeche" (we had to leave behind family members and [leaving our] mother behind is rather difficult).

> Amara desh bolte India ke jani. Anok asa niya nijar desh a alam … apnara didi r to sob janen. Ekhane bhalo lage, kintu purono chita amader kicchu sompod roye giyacha. Amar ma abonk bhai purono chita ase. Tai mo n kharap hoi. Bhebe chilam je nagorik hole travel korte parbo … kintu passport nai… anok somosha. Jomi ache tobe bhalo dam pai na.

(India is a country and we are happy to be here. We have however left something very precious behind—my relatives—and would like to travel, but getting a passport is not easy. Also land that is left behind needs to be sold, but again we cannot travel).

But in the same camp, Shri Manohar Chandra Burman thought that those left behind in Bangladesh were better settled and more privileged. He said, "Amara aste cheyachilam kintu akhane asa mone hoche bhul holo" (though we wanted to join India, now I regret it). He voiced concerns that his parents and brother enjoyed relatively better amenities than those who had decided to join India. He was disappointed by employment opportunities. Abu Razzak mentioned that in 2017, the situation was slightly better: they were given land where they could relocate, but it was far outside the limits of the city where it was difficult to find new sources of livelihood. And lacking access to land left behind precluded their chance to profit from selling it.

Rather than relocate to Bangladesh, some residents chose to stay behind in former Bangladesh enclaves that LBA 2015 now mandate as Indian territory. Discarding their Bangladeshi citizenship and opting for new Indian citizenship cemented their belonging to their ancestral lands; challenging their previously immobile and marginal status did not involve displacement from the land they inhabit. We call both categories of enclave dwellers—those who either chose or reaffirmed their Indian citizenship—as "new" citizens. But the state addressed those who relocated from Indian enclaves in Bangladesh to transit camps for rehabilitation as "returnees." A Press Information Bureau circular illustrates this distinctive treatment: India's Minister of State for Home Affairs stipulated that adequate initiatives were necessary to address the "development and integration [of] Bangladesh enclaves in India" and the "issues of rehabilitation of 'returnees'" from Indian enclaves located in Bangladesh (Press Information Bureau 2017).

We challenge the suitability of the term "returnees" because it focuses on two questions: Where were these people returning from? Where were they returning to? For a segment of people who were already inhabiting Indian enclaves with Indian citizenship (however nominal it was), and who retained it by relocating from enclaves, the term "returnee" only highlights the nation-state's preoccupation with territory rather than with people who encompass the territory. Despite using their agency to relocate and thereby challenge their status of immobility in enclaves, the state bracketed them as an "alienated other" returning to Indian territory. The anxiety they harbored would be higher than those from the Bangladeshi enclaves in India, considering the land, immovable assets, and families they had to leave behind. To call these relocated "new" citizens "returnees" is nothing short of irony. In that sense, their anxiety can be partly attributed to the fact that by being "new" citizens they were admitted to the hierarchy of national belonging, only to realize that they were placed in its lowermost echelons. In order to access their entitlements, they will have to constantly make a claim of belonging in comparison to the other privileged sections of this hierarchy.

Belonging and the choice of citizenship

The Land Boundary Agreement of 2015 is an effective resolution of the postcolonial nation-states of India and Bangladesh, consolidating their territorial boundaries. Integrating the population, who are former residents of enclaves, to the newly consolidated territorial boundaries of these nation-states involved a corresponding symbolic management of their citizenship. Through their choice of citizenship, the enclave dwellers were reaffirming the "hyphenation" that exists between a citizen of a state and the sovereign territory of the state. The status of citizen, unhyphenated from the geographical space of state's territory, relegated them to a precarious existence where they were systematically denied the rights and privileges of being a citizen. The ambivalent nature of their citizenship as enclave residents made them immobile subjects and precluded them from substantively realizing their constitutionally guaranteed protection.

This "new citizenship" necessitates a closer look from the perspective of enclave dwellers at various everyday aspects of citizenship that are being shaped by identity and belonging. Karim, who inhabited a former Indian enclave, moved to Mekhligunj resettlement camp when he chose Indian citizenship after the ratification of LBA 2015. But his choice entailed a difficult decision to separate from his 70-year-old mother, his sisters, and more than 60 relatives who stayed in Bangladesh. Hossain similarly moved from another enclave to Dinhata camp in Krishimala. He was popular among the people for his religious wisdom, considerable knowledge of Ayurveda medicine, and extrovert personality. He too had to leave behind his parents and whatever land he possessed there. For both Karim and Hossain, new citizenship meant aligning their nationality with territory at the cost of disrupting kinship structures and abandoning the land where they had lived since their birth. Both became anxious about family members left behind in the Bangladeshi chit. Communication with loved ones amounted to sending and receiving notes. Hossein mentioned that he longed to be with his aging father who was unwell, but with the enclave exchange process travel was more difficult. Yet they still hoped to become full-fledged citizens of the Indian state, with equal access to all rights and privileges associated with citizenship.

Interestingly, both Burman and U Goni were unhappy with their precarious existence as half citizens. It comforts them that to enjoy the same mobility as the country's other citizens, they would not have to stop at border posts or acquire numerous permits and fill out documents. Regardless of the specific religious, cultural, or linguistic identity they chose to associate with, the sense of belonging they developed to the Indian nation should be understood as an enabling factor in this choice of citizenship.

As an interdisciplinary concept, "belonging" has been contextualized as "spaces of belonging" (Mills 2006; Nelson 2007), "places of belonging" (Nelson and Hiemstra 2008), "landscapes of belonging" (Trudeau 2006), or "sites of belonging" (Dyck 2005; Tolia-Kelly 2006). The discourse on belonging differentiates two positions. The first, from the purview of the nation-state where the norms and contours of belonging are instituted from above, necessitates a top-down approach. The second, conversely a bottom-up conceptualization, signifies the ways in which individuals perceive their belonging in terms of their association with a particular environment, community, or even the nation. Here, belonging becomes inherently rooted in the way individuals subjectively perceive their own connections, attachments, and relationships to a specific surrounding. The first conceptualization results in a hierarchical understanding of belonging; the second privileges an understanding of belonging that is multilayered and overlapping.

For those in a nation-state, belonging is constantly mediated by membership within the territory of the nation. The characteristic traits of a nation are presupposed to be determined by its collective nature. Those belonging to the nation are legally included by the state through provision of citizenship, while others not opting for citizenship are either excluded or left to languish in the peripheries. In addressing which individuals qualify to belong in the various territorial spaces of specific nation-states, some scholarly postulations focus on the affiliation connecting citizenship and the notion of belonging (see Hopkins 2007; Clark 2008). Luke Desforges and colleagues (Desforges et al. 2005) opine that traditionally, citizenship has been a symbol of belonging to the place of a state and that the substantive realization of citizenship was always enacted in this setting. The symbolic management of the nation-state will often demarcate spaces of belonging that complement its narrative as a nation, simultaneously creating a segment of people who are excluded, as they are not qualified to occupy these spaces (Anderson and Taylor 2005). But the state's territorialized notion of belonging is at times contested by the everyday lives of individuals that get entwined with identity, recognition, and citizenship. These routine interactions are instrumental in shaping an individual's belonging and have been progressively hypothesized as one of the primary domains in which citizenship is expressed, realized, or even refused (Mee and Wright 2009).

Accordingly, when the state-centric notion of belonging is tethered to the traits of a homogeneous nation, it simultaneously creates inclusion and exclusion of people (Anthias and Yuval-Davis 1992). Within this domain of inclusions and exclusions, "hierarchy of belonging" manifests (Hickman et al. 2005; Anthias 2008; Phoenix 2011). Whereas belonging as espoused by the nation-state will occupy the predominant position in this hierarchy, the intersectionality of race, ethnicity, religion, and sexuality causes individuals to be placed differentially in this pyramid. This hierarchical understanding of belonging as construed by the nation-states is perpetuated and affirmed through citizenship. It demarcates people as "us" and "others"—"us" meaning those who would qualify more to belong to the nation-state as opposed to the "other." The differential belonging construed by the state also showcases its power to include a few and exclude others and, in the process, makes borders and boundaries the site of active demarcations. The state incessantly exercises its power in producing and reproducing this hierarchy of belonging, not only by guarding and securing its borders but also by shifting and altering the same borders. Hence, the contours of inclusion and exclusion are constantly altered when the state steps in to undertake "symbolic management," so as to preserve the contrived ethos of the nation it represents. Here, the hierarchy of belonging becomes political and acts at two levels: first is when the nation-state dictates the norms of belonging to suit the narrative of the nationhood it espouses and the second is when individuals and groups make varying claims of belonging to align with or contest the hierarchy of national belonging.

Scholars have analyzed the hierarchy of belonging in terms of its fortification by various aspects, including the legal edifice of state (Bhambra 2016) and the institutional structure of state (Aliverti 2018). The literature has also assessed the exclusion of vulnerable segments like minorities or migrants from this hierarchy (Phoenix 2011; Back et al. 2012), and the various scales at which such a hierarchy of belonging operates (Clarke 2020). Audra Simpson (2014: 23) asserts that the "settler colonialism structures justice and injustice in particular ways, not through the conferral of recognition of the enslaved but by the conferral of disappearance in subject." Enclave dwellers occupy a unique position because their de-territorialized identity caused them to be placed outside the hierarchy of national belonging, which in turn was overtly territorial. Their ambiguous membership status, epitomized by "suspended citizenship" in an absentee state, reflects their exclusion from this hierarchy. Yet, each of these individuals nurtures a belonging to various entities like land, family, enclave community, or even host or home country. Nira Yuval-Davis (2006: 199) suggests that individuals can develop an attachment to multiple objects and simultaneously can "belong" in different ways to different entities. Accordingly, she defines belonging as "an act of self-identification or identification by others, in a stable, contested or transient way," thus making it a dynamic process of "naturalized construction of a particular hegemonic form of power relations" (Yuval-Davis 2006: 199).

Belonging can be studied along three interconnected levels—the first being social locations, the second being individual identifications and emotional attachments, and the third being ethical and political value systems. The positionality of individuals with respect to categories such as gender, class, nation, kinship, or profession constitutes their belonging with respect to socioeconomic locations, which in turn have implications on the power relations in the society. Despite identifying or associating with a particular marker of identity, belonging according to social location is constructed along multiple other markers that individuals possess and affirm but do not uniformly prioritize. That is to say, depending on the particular context in which an individual is situated, certain markers will be more prominent and predominant than others. Individuals could prioritize gender over ethnicity, religion, or class when their social context is mobilized along gender lines. Hence, it can be argued that the sense of belonging internalized by the Indian citizens of former enclaves was as layered and overlapping, yet mutually exclusive as their subjective identity.

Citizenship can also be comprehended from the purview of the "historical process" that causes individuals to develop "shared characteristics" from their common cultural credentials (Shapiro 2000: 79). A sense of belonging rooted in the precolonial and colonial history of the enclaves cuts across the territorial borders instituted by these postcolonial states. In choosing citizenship, this aspect of belonging eased the enclave dwellers' agony and anxiety at being displaced from the only place they had ever known and then re-placed in the territory of a different state. Prioritizing common aspects of belonging nurtured in former members of the Indian nation over belonging to the land caused them to reaffirm and reiterate their Indian citizenship and made them "new" citizens. And this context was necessitated by the symbolic management of the nation-state precipitating a similar symbolic management of its citizens. Effectively, they had to use their agency to claim their belonging in the state and thereby align themselves with the hierarchy of national belonging. Only such a "new citizenship" would facilitate the access to rights and privileges reserved to those who belong in the nation-state.

Conclusion

The anxiety of enclave dwellers in an ambiguous territory is perpetual in nature. The assertion of territorial sovereignty by the postcolonial nation-states of India and Bangladesh caused enclave dwellers to cross the foreign territory of the encircling state for basic needs, livelihood, and daily sustenance. This effectively constrained their mobility and caused enclave citizens to live the lives of "half citizens" anxiously hovering between legality and illegality. The severed geographical positioning of enclaves with respect to the home state reinforced that anxiety. But when India and Bangladesh undertook a spatially oriented "symbolic management" of their respective nation-states by consolidating their territories through the exchange of enclaves, they gave the choice of citizenship (and the anxieties caused by it) to the enclave dwellers. Because citizenship legally validates an individual's belonging to the state, enclave dwellers had to reclaim their belonging by reaffirming their citizenship and thereby becoming "new" citizens if they aligned within its spatial coordinates. While all Bangladeshi enclave dwellers in India stayed back in the Indian territory and became new citizens of India, the former Indian enclave dwellers in Bangladesh had to relocate from their ancestral lands to transition camps before their resettlement and reintegration to Indian sovereign territory.

Notions of belonging for these new citizens manifested in multiple ways. For those who stayed back in enclaves and became new citizens of India, belonging to the only land they had ever known was instrumental in choosing to reaffirm and reiterate their Indian citizenship. This effectively placed them in the territorially determined, state-centric hierarchy of national belonging. But displacement from their land, relatives, and immovable assets—now in the territory of another sovereign state—caused them additional anxiety; navigating everyday life within the transition camps, where their anticipation of being mobile, as full citizens, caused even more anxiety. Constraints that preclude them from accessing formal structures of belonging instituted by the state still prevail for the "new" citizens in the camps, although in renewed ways. As they transition, whether as "half citizen" enclave dwellers or "new" citizens, anxiety remains an indelible factor in their hopes and efforts to realize full citizenship.

References

Aliverti, Ana. (2018). "Law, Nation and Race: Exploring Law's Cultural Power in Delimiting Belonging in English Courtrooms." *Social & Legal Studies* 28: 281–302.

Anderson, Kay, and Affrica Taylor (2005). "Exclusionary Politics and the Question of National Belonging." *Ethnicities* 5(4): 460–85.

Anthias, Floya (2008). "Thinking through the Lens of Trans-locational Positionality: An Intersectionality Frame for Understanding Identity and Belonging." *Translocations* 4(1): 5–20.

Anthias, Floya, and Nira Yuval-Davis (1992). *Racialized Boundaries: Race, Nation, Gender, Colour and Class and the Anti-racist Struggle*. London & New York: Routledge.

Appadorai, Arjun (1981). *The Domestic Roots of India's Foreign Policy 1947–1972*. Oxford: Oxford University Press.

Back, Les, Shamser Sinha, and Charlynne Bryan (2012). "New Hierarchies of Belonging." *European Journal of Cultural Studies* 15(2): 139–54.

Banerjee, R. (1966). "An Account of the Enclaves - Origins and Development." in *Census 1961, West Bengal, Distfict Census Handbook, Cooch Behar*. Calcutta: Super Intendent Government Printing, West Bengal.

Bhambra, Gurminder (2016). "Viewpoint: Brexit, Class and British 'National' Identity." *Discover Society* DS34. Accessed July 3, 2021, https://archive.discoversociety.org/2016/07/05/viewpoint-brexit-class-and-british-national-identity/

Chowdhory, Nasreen (2018). *Refugees, Citizenship and Belonging in South Asia Contested Terrains*. Singapore: Springer Nature.

Clark, Juliet (2008). "Australian Public Opinion on Citizenship and Transnational Ties in Asia." *Australian Geographer* 39(1): 9–20.

Clarke, Amy (2020). "Hierarchies, Scale, and Privilege in the Reproduction of National Belonging." *Transactions of the Institute of British Geographers* 45: 95–108.

Cons, Jason (2013). "Narrating Boundaries: Framing and Contesting Suffering, Community, and Belonging in Enclaves along the India–Bangladesh Border." *Political Geography* 35: 37–46. Accessed July 3, 2021, https://www.researchgate.net/publication/259138725_Narrating_boundaries_Framing_and_contesting_suffering_community_and_belonging_in_enclaves_along_the_India-Bangladesh_border

Cons, Jason (2016). *Sensitive Space: Fragmented Territory at the India-Bangladesh Border*. Seattle & London: University of Washington Press.

Cons, Jason, and Romola Sanyal (2013). "Geographies at the Margins: Borders in South Asia—An Introduction." *Political Geography* 35: 5–13.

Desforges, Luke, Rhys Jones, and Mike Woods (2005). "New Geographies of Citizenship." *Citizenship Studies* 9(5): 439–51.

Dunn, Elizabeth Cullen, and Jason Cons (2013). "Aleatory Sovereignty and the Rule of Sensitive Spaces." *Antipode* 46(1): 92–109.

Dyck, Isabel (2005). "Feminist Geography, the 'Everyday,' and Local-Global Relations: Hidden Spaces of Place-Making." *Canadian Geographer/Le Géographe Canadien* 49(3): 233–43.

Hickman, Mary J., Sarah Morgan, Bronwen Walter, and Joseph Bradley (2005). "The Limitations of Whiteness and the Boundaries of Englishness: Second-generation Irish Identifications and Positionings in Multi-ethnic Britain." *Ethnicities* 5(2): 160–82.

Hopkins, Peter E. (2007). "Young People, Masculinities, Religion and Race: New Social Geographies." *Progress in Human Geography* 31(2): 163–77.

Jones, Reece (2009). "Sovereignty and Statelessness in the Border Enclaves of India and Bangladesh." *Political Geography* 28(6): 373–81.

Kaviraj, Sudipta (2010). *The Imaginary Institution of India: Politics and Ideas*. New York: Columbia University Press.

Krishna, Shankaran (1996). "Cartographic Anxiety: Mapping the Body Politic," 193–214. In H. R. Alker and M. J. Shapiro (eds.), *Challenging Boundaries: Global Flows, Territorial Identities.* Minneapolis: University of Minnesota Press.

Mee, Kathleen, and Sarah Wright (2009). "Geographies of Belonging." *Environment and Planning A: Economy and Space* 41(4): 772–79.

Mills, Amy (2006). "Boundaries of the Nation in the Space of the Urban: Landscape and Social Memory in Istanbul." *Cultural Geographies* 13(3): 367–94.

Nelson, Lise (2007). "Farmworker Housing and Spaces of Belonging in Woodburn, Oregon." *Geographical Review* 97(4): 520–41.

Nelson, Lise, and Nancy Hiemstra (2008). "Latino Immigrants and the Renegotiation of Place and Belonging in Small-town America." *Social & Cultural Geography* 9(3): 319–42.

Ong, Aihwa (1999). *Flexible Citizenship: The Cultural Logics of Transnationality.* Durham, NC: Duke University Press.

Ong, Aihwa (2003). *Buddha Is Hiding: Refugees, Citizenship, the New America.* Berkeley, CA: Sage Publications.

Ong, Aihwa (2006). "Mutations in Citizenship." *Theory, Culture & Society* 23(2–3): 499–505.

Phoenix, Aisha (2011). "Somali Young Women and Hierarchies of Belonging." *Young* 19(3): 313–31.

Press Information Bureau (2017). "India Bangladesh Land Boundary Agreement (LBA)." Accessed July 4, 2021, https://pib.gov.in/newsite/PrintRelease.aspx?relid=159233

Shamir, Ronen (2005). "Without Borders? Notes on Globalization as a Mobility Regime." *Sociological Theory* 23(2): 197–217.

Shapiro, Michael J. (2000). "National Times and Other Times: Re-thinking Citizenship." *Cultural Studies* 14(1): 79–98.

Sheller, Mimi, and Tore Sager (2008). "Mobility, Freedom and Public Space," 25–38. In S. Bergmann (ed.), *The Ethics of Mobilities: Rethinking Place, Exclusion, Freedom and Environment.* Farnham, UK: Ashgate.

Sheller, Mimi, and John Urry (2006). "The New Mobilities Paradigm." *Environment and Planning A: Economy and Space* 38(2): 207–26.

Shewly, Hosna J. (2012). *Life, the Law, and Politics of Abandonment: Everyday Geographies of the Enclaves in India and Bangladesh.* Dissertation, University of Durham.

Shewly, Hosna J. (2013). "Abandoned Spaces and Bare Life in the Enclaves of the India–Bangladesh Border." *Political Geography* 32: 23–31. Accessed July 4, 2021, https://www.sciencedirect.com/science/article/abs/pii/S0962629812001187

Shewly, Hosna J. (2015). "Citizenship, Abandonment and Resistance in the India–Bangladesh Borderland." *Geoforum* 67(2015): 14–23. Accessed July 4, 2021, https://www.sciencedirect.com/science/article/abs/pii/S0016718515002912

Simpson, Audra (2014). *Mohawk Interruptus: Life across the Borders of Settler States.* Durham, NC: Duke University Press.

Tolia-Kelly, Divya P. (2006). "Mobility/Stability: British Asian Cultures of 'Landscape and Englishness'." *Environment and Planning A: Economy and Space* 38(2): 341–58.

Trudeau, Daniel (2006). "Politics of Belonging in the Construction of Landscapes: Place-Making, Boundary-drawing and Exclusion." *Cultural Geographies* 13(3): 421–43.

Turner, Bryan S. (2007). "The Enclave Society: Towards a Sociology of Immobility." *European Journal of Social Theory* 10(2): 287–304.

Van Schendel, Willem (2002). "Stateless in South Asia: The Making of the India-Bangladesh Enclaves." *The Journal of Asian Studies* 61(1): 115–47.

Whyte, Brendan R. (2002). *Waiting for the Esquimo: An Historical and Documentary Study of the Cooch Behar Enclaves of India and Bangladesh*. Melbourne, Australia: School of Anthropology, Geography and Environmental Studies, University of Melbourne.

Yuval-Davis, Nira (2006). "Belonging and the Politics of Belonging." *Patterns of Prejudice* 40(3): 197–214. Accessed July 4, 2021, https://doi.org/10.1080/00313220600769331

9 Climate and non-climatic stressors, internal migration, and belonging in Ghana

Hanson Nyantakyi-Frimpong and
Dinko Hanaan Dinko

Introduction

Climate change is one of the current factors contributing to distress migration, as shoreline erosion, coastal flooding, and extreme drought displace millions of people from their homelands (Kaczan and Orgill-Meyer 2020; Mueller et al. 2020; Paprocki 2020; Van der Geest 2011). In 2018, the Internal Displacement Monitoring Centre (IDMC) reported that 17.2 million people were displaced as a result of sudden extreme climatic events such as rainstorms, cyclones, and hurricanes (IDMC 2019). Available estimates show that climate migrants or climate refugees, as they are sometimes called, would number about 200 million by the year 2050 (Brown 2008). Some scholars have cautioned against using terms like "climate refugees" and "climate migrants," arguing that climate change alone does not cause migration (Warner 2010; Foresight 2011; Renaud et al. 2011). Instead, climatic circumstances exist as one of several factors driving an individual's or a community's decision to migrate. The term "climate refugee" is especially popular in climate-migration literature, but it is problematic as well because *refugee* is a legal category limited to people fleeing persecution. Since the 1990s, there has been an upsurge in research to better understand the experiences of these migrants.

Discussions on climate-induced human mobility have largely focused on cross-border international migration (e.g., see Feng et al. 2010; Benveniste et al. 2020). How climate change influences in-country migration is relatively underexplored in the climate change-migration literature (Chen and Mueller 2018; Wilson 2020). Drawing upon experiences from Ghana, West Africa, we address this research gap and more specifically explore the ways in which climate change and other political–economic factors influence internal migration among the Dagaabas, one of the main ethnic groups in northwestern Ghana (Lentz 2000).

We base the chapter on 65 in-depth interviews conducted as part of a broader study on the human dimensions of climate change. Details about the study's research methods have been published elsewhere (e.g., see Nyantakyi-Frimpong and Bezner-Kerr 2015a and 2015b; Nyantakyi-Frimpong and Bezner-Kerr 2017). Fieldwork took place in Ghana's Upper West Region,

DOI: 10.4324/9781003170686-11

located within the southern fringe of the West African Sahel, an area where annual rainfall volumes have been in long-term decline (Luginaah et al. 2009; Niang et al. 2014). Migrant families from the Upper West Region were traced and interviewed in the Brong-Ahafo Region, located in the southern part of the country (see Figure 9.1).

In this chapter, we explain the long-term historical contexts that have shaped north–south internal migration in Ghana. Part of this history focuses on British colonial rule in the Gold Coast from 1874 to 1957, which left the northern parts of Ghana highly marginalized. Neoliberal economic reforms in the 1980s further entrenched this colonial marginalization, setting in motion internal migration mainly from the north to the south. We further show how in more recent times, climate change and land expropriation have amplified internal migration. We pay particular attention to the migration process from the north, initial integration in the southern parts of the country, and the everyday lived experiences of migrant farmers. Central to our analysis is the issue of migrant identities. More specifically, we show how ethnic identity is used as a justification for differential treatment and the subtle economic exploitation of migrant farmers from northern Ghana. Confronted with economic exploitation and hostility, these northern migrants often question their sense of belonging in the southern parts of the country. In telling this story of north–south migration in Ghana, we let the migrants

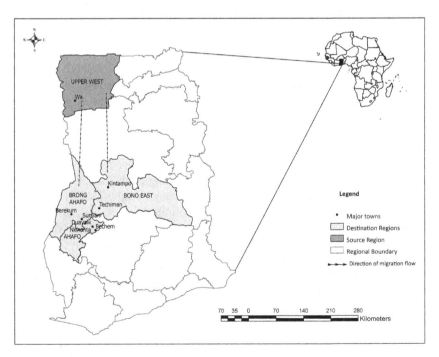

Figure 9.1 Ghana's internal migration source and destination regions (prepared by Dinko Hanaan Dinko).

themselves speak by quoting directly from interviews. Given that most discussions of migrants' identity and belonging are based on cross-border contexts (e.g., Madsen and Van Naerssen 2003), this chapter adds to existing work by providing experiences of internal migration among a population that might be assumed to have a shared culture and national identity.

North–south migration in Ghana: The role of colonial marginalization and contemporary socioecological stressors

Northern Ghana is in many ways distinct from the rest of the country. The region is characterized by general underdevelopment, severe poverty, food insecurity, and malnutrition (Ghana Statistical Service 2014; Atuoye et al. 2019). Available data from the Ghana Statistical Service shows that in the northern parts of the country, poverty rates are two to three times higher than the national average (Ghana Statistical Service 2013). A number of historical factors have shaped this uneven geography of development between northern and southern Ghana. Among these factors, the most noteworthy is British colonial rule in the Gold Coast (now Ghana) from 1874 to 1957.

Although the British established the Gold Coast colony in the coastal areas of the country around 1874, the northern territories were not brought under colonial rule until 1902 (Lund 2003). Initial colonial policies focused on the production of crops that offered the greatest potential for export to Britain, including rubber, cocoa, and coffee (Seini 2002). Little attention was paid to the production of crops that were unattractive in European markets. The colonial administration's emphasis on rubber, cocoa, and coffee production led to the extensive development of infrastructure in coastal southern Ghana, where the forest ecology supported the production of export-oriented crops (Plange 1979). Consequently, the northern parts of the country remained marginalized (Songsore 2011). The north became integrated with the Gold Coast colony only because it provided labor for south-based cocoa and coffee plantations (Lentz 2000). This in turn reduced the labor available for food production in the north, which also intensified food insecurity (Luginaah et al. 2009; Nyantakyi-Frimpong and Bezner-Kerr 2015b).

The colonial government also initiated a policy of forced-labor recruitment from the north to increase the workforce in mines and railway construction projects in the south (Lentz 2000). The railway construction projects were meant to augment the transportation of export products (e.g., cocoa, coffee, rubber) from the forest regions to the coast for export to Europe. In addition to forced-labor recruitment, people from northern Ghana had to travel south to earn enough to pay colonial hut taxes imposed by the British administration (Songsore 2011). All these processes established patterns of north–south migration that have persisted to the contemporary period (Luginaah et al. 2009; Rademacher-Schulz et al. 2014; Kuuire et al. 2016; Nyantakyi-Frimpong and Bezner-Kerr 2017).

After independence from Britain in 1957, the immediate postcolonial government continued to neglect the north, reinforcing the spatial disparity in development under colonialism (Songsore 2011). For example, immediate postcolonial policies focused on the extraction of natural resources and the development of industrial estates mainly in southern cities. With low natural resource endowment in the north, little government revenue was spent on developing this part of the country, resulting in limited infrastructure and access to other social services (Songsore 2011). The cumulative effects of these processes led to intensified migration from the north to the south (Luginaah et al. 2009; Kuuire et al. 2016).

In 1983, the Ghanaian government implemented economic reform policies called Structural Adjustment Programs (SAPs). These SAPs further intensified north–south migration. More than any other sector in the Ghanaian economy, agriculture saw one of the most ferocious restructuring (Hutchful 2002). Noteworthy among structural changes included the removal of subsidies on fertilizers, seeds, and pesticides for agriculture. The government further retrenched agricultural extension staff and dismantled marketing boards that serviced smallholder input requirements (Pearce 1992; Hutchful 2002). Other policy measures included increasing support for large landholders and the abandonment of smallholder, subsistence agriculture (Pearce 1992).

These SAPs unleashed profound social and economic transformations in the Ghanaian countryside, marking a great watershed in the viability of smallholder farming (Pearce 1992). Many farmers were squeezed out of agriculture, as their purchasing power became dramatically eroded. Additionally, input and output markets became volatile, constricted, and competitive. For example, due to trade liberalization policies within SAPs, local products such as rice, maize, beef, and poultry faced stiff competition from highly subsidized and cheap imports from Europe, Asia, and North America (Hutchful 2002). The effects of SAPs were geographically uneven across the country. Northern Ghana experienced the most severe impacts because of general underdevelopment and a heavy reliance on rainfed agriculture (Konadu-Agyemang 2000; Songsore 2011). Thus, SAPs intensified the already uneven regional development in Ghana. With persistent poverty and reduced agricultural productive capacity, many small farmers were driven to cities in the south, where they worked as day laborers for minimal wages (Abdul-Korah 2011; Kuuire et al. 2016).

These historical factors have continuously reinforced north–south migration in Ghana, but recent severe climatic changes are also exacerbating the process. During the course of fieldwork conducted for this chapter, climatic changes were said to be affecting smallholder agriculture and limiting livelihood opportunities in northern Ghana (Nyantakyi-Frimpong and Bezner-Kerr 2015a). Trying to adapt, farmers then leave northern Ghana for the country's forest regions in the south (see Figure 9.1), where the favorable climate supports agriculture. One farmer summed up a recurring concern that many study participants articulated:

> We have always been migrating from the north to the south. Our great-grandfathers migrated to the south. Our elderly fathers also migrated

to the south. Today, we are also migrating. Our grandfathers migrated to look for work in the south because of the history of this place. Some were forced to work in the mines. Today, we are also migrating because of the same history, plus new problems such as severe changes in rainfall. All our crops are failing because of longer droughts. That is why we migrate to engage in farming in [southern Ghana].

Although climate change was said to be intensifying north–south migration, land expropriation was also highlighted as a major factor. The majority of farmers reported having lost their lands to a process called land grabbing or Large Scale Land Acquisitions (LSLAs) (Vermeulen and Cotula 2010; Nyantakyi-Frimpong and Bezner-Kerr 2017). These LSLAs are also tied to the long-term impacts of structural adjustment policies in Ghana. In an effort to promote foreign direct investment, the Ghanaian government has created an attractive investment climate for large-scale agriculture, mining, and biofuel production by foreign corporations, often in partnership with domestic elites. Different sources estimate that between 2004 and 2010, more than one million hectares of land were allocated to foreign-based corporations for investments in gold mining and commercial agriculture (Nyantakyi-Frimpong and Bezner-Kerr 2017; Hausermann et al. 2018). Most of these lands have been appropriated in northern Ghana, causing landlessness and intensifying out-migration. In an earlier survey conducted as part of this study, landless households reported a higher average number of migratory members (six persons), as compared to near-landless households (three persons), or land-rich households (one person) (Nyantakyi-Frimpong and Bezner-Kerr 2017). Lacking access to land, these households were unable to survive in northern Ghana without relying almost exclusively on hiring out their labor power in the south. In the next section, we describe how farmers initiate the migration process, and we focus on the everyday lived experiences of migrants in southern Ghana.

The migration process, initial integration, and fulfilling aspirations

During in-depth interviews, many farmers described the north–south migration process, initial integration, and how they were living the life envisioned in southern Ghana. Some described the migration experience as "a very simple process." These respondents were mostly young and unmarried, or recently married couples without children. As one interview participant noted:

For my wife and myself, it was a very simple process because we had no children when we decided to migrate to southern Ghana in 2011. We sold my motorbike to gather some funds and then we set the date to finally move here to the south. We came with only two small bags, stayed with my uncle in one of the market towns for six months, and then we were on our own. After that time, we rented a single bedroom.

For male respondents who had multiple wives and children, however, the migration process was described as quite complicated. Some had to arrange for family members to take care of their children. Others had to decide which of the multiple wives should join the migration journey, or stay behind and look after land and other properties. One male migrant farmer narrated his experience:

> I have three wives and six children, so it was a hard decision. ... I migrated because the rains were getting very unreliable and then I also lost about 90 percent of my land to a mining project initiated by the government. When I decided to migrate to look for other opportunities, it was a very hard decision. You cannot move with six children and three wives at a time. So, it was a process of constant arguments with the women to decide who should go and who should stay behind. In the end, I came with [name removed], the youngest wife, who had no child at the time. Every six months or so, I still have to visit my remaining two wives and children back in the north.

Another participant added:

> The journey itself to southern Ghana is not difficult because it is just two or three hours away by bus. What is very difficult is finding a job after arrival, and making arrangements for the uptake of the family left behind. If you want to go into farming after arriving in the south, renting a land is also difficult, since people usually do not come with a lot of money. Normally, most of us northern migrants arrive in the south with nothing.

The choice of migrants' destination was an issue also raised frequently by the interviewees. Ghana's Brong-Ahafo Region, with a dense forest ecology within the southern part of the country, was the main area where these migrants often settled. Although a few settled in urban areas of the Brong-Ahafo Region, the majority settled in rural parts of the region, where there are numerous opportunities for agriculture. In the interviews, many of the migrant farmers explained that they decided to settle in the Brong-Ahafo Region mainly to live near people of the same ethnic group, because the region already has a large number of northern migrants. As one respondent revealed:

> Brong-Ahafo was the first choice when I decided to migrate. Because there is already a big Dagaaba community here. There are even farming settlements named after communities in the Upper West. The Dagaaba community is big here so I knew I would feel at home.

Other existing case studies have revealed similar findings, showing why northern Ghanaians choose the Brong-Ahafo Region as a migrant destination (e.g., Abdul-Korah 2007; Baada et al. 2019).

The interview findings further illustrated that upon arrival, most of the farmers were not quite living the life they envisioned in southern Ghana. For example, those who settled in urban areas were experiencing a form of hidden homelessness. Some were sleeping in uncompleted residential buildings, while others lived in bus terminals with poor, unsafe conditions. The following interview narrative reflects some of these living arrangements:

> Since I arrived from the Upper West, I have been trying to save money to rent a place of my own. Currently, I am staying in an uncompleted building with five other friends. The building is roofed, but it has nothing more—no windows, electricity [and] water. ... We are staying in this uncompleted building to protect the property for the landlord. We fixed our own doors and windows in one of the rooms, where all six of us are currently living. ... Most guys have come to stay here while figuring out what to do next. Some boys sleep in the bus terminal when they first arrive. Where I stay, at least you are with others and you feel somehow safe. I am working as a farm laborer and a porter, but I cannot save much because cost of living here is very high. I did not imagine that life would be very hard even here in the south.

Many of these migrant farmers were emotionally and financially stressed because they needed to accept poorly paid work. Throughout the interviews, many revealed working in insecure, exploitative, and typically low-wage employments in southern Ghana. Some of them reported working as casual day laborers (locally called *by-day labor*). Others worked as tenants and sharecrop farmers on a piece-rate or fixed-contract basis. With the latter, the land is often obtained through rent-in-kind sharecropping, with rents fixed at one-third of the harvest if the tenant advances the production cost, and two-thirds otherwise (Luginaah et al. 2009; Nyantakyi-Frimpong and Bezner-Kerr 2017). Yet as we show in the next section, these sharecropping arrangements could be very exploitative. Aside from disadvantageous work relations, isolation and attendant feelings of loneliness were also key concerns for some of the migrants we interviewed, particularly the young and unmarried.

Economic exploitation and the politics of belonging

Migrants who worked as farm laborers were discriminated against and exploited simply because of their identity as northerners. In her anthropological research, Enid Schildkrout (1979) has documented the disparaging manner in which, since the precolonial period, Asante ethnic groups in southern Ghana have related to ethnic groups from the north. As Schildkrout (1979: 188) has noted, the

> Asante had basically two types of relations with northerners in the precolonial period: they knew them as slaves and as traders. Although the

political and economic context of these relations has changed in the twentieth century, one can still discern continuity.

She goes on to show that the "Asante held notions of superiority vis-à-vis northerners" (Schildkrout 1979: 191), a comment also repeatedly emphasized by northern migrants interviewed in this study. Historical research by Kwesi Prah (1975) further shows how this precolonial north–south relation was further entrenched through British colonial policies of keeping the north as a labor reserve (see also Lentz 2000; Songsore 2011).

Several interview accounts reflected the notion that southern Ghanaians are superior and northern migrants are cheap laborers who could easily be exploited. For example, the conventional acceptable rate for farm labor work was approximately $2 per day at the time of fieldwork. But northern migrant laborers were being paid $0.75 per day. This amount was often reduced by half ($0.38) during off-peak farming seasons. A farm owner from southern Ghana made the following comment in an interview: "But Northerners have always been cheap laborers since our great-grandfathers' time. How can you expect a farm laborer from the north to be paid the same wage as a person from the south?" Indeed, this participant's account reinforces the view that southerners are superior and should be treated differently than groups from the north. Another older study participant also made the following comment, more explicitly reflecting the colonial practice of recruiting northerners as cheap laborers to be used in southern-based plantations:

I am a businessperson with a profit mindset. That is why I prefer the northern ethnic group of migrant laborers. They are hardworking and cheap. They are desperate, so they do not complain. No offenses to this group, but their great-great-grandfathers were all cheap laborers during the colonial era.

Farm laborers were not the only ones economically exploited: migrants who engaged in sharecropping were victimized as well, and left to feel that they didn't belong in southern Ghana. As indicated earlier, rent-in-kind sharecropping is a major activity that migrant farmers undertake upon arrival in southern Ghana. For migrant farmers engaged in sharecropping, the study findings revealed that they were not only confined to unproductive contracts but also were barely earning enough to maintain themselves. For example, commenting on sharecropping relations between southern landowners and northern tenant farmers, one study participant said:

When the farm owners see you as a migrant from the north, then they manipulate the sharecropping contract. My landlord told me that for the first three years, there was nothing to be shared. The sharecropping contract started in year 4. This is something he could never do to a farmworker from the south.

Furthermore, some of the migrant farmworkers reported that their share-cropping contracts were terminated as a result of extended illness, farm injury, or poor harvests. Ironically, these same sharecropping contractual relations were never applied to tenant farmworkers who come from southern Ghana.

Thus, for northern migrant farmworkers, these exploitative relations raised serious questions about their belongingness. Most of them said these relations often remind them that although they are Ghanaian, they do not seem to belong in the southern parts of the country. A female study participant articulated this point most profoundly in the following excerpt from an interview about sharecropping contracts:

> The landowner was a woman, and I thought she is going to be nice because I am also a woman. She looked in my face and said there was nothing to be shared until year 3 of the sharecropping contract. Then, I got a farm injury in year 2. In year 3, she said the harvest was small, so she terminated the contract. And this was at the point we should have started sharing the farm produce. I was exploited in my own country, just 50 miles away from my hometown.

Indeed, many of these migrant farmers felt that exploitative work relations should typically occur in cross-border international migration, as reported in many case studies (e.g., Holmes 2013; Horton 2016). These migrants never felt they could face economic exploitation within their own home country. Numerous scholars have highlighted findings similar to the ones here but focus on conflict-prone contexts, where internally displaced people face various forms of exclusion and exploitation (e.g., Brun et al. 2017).

Many of the migrants in this study who lack a sense of belonging shared their visions of hope and excitement at the prospects of going back home and creating a new life in the north. But these visions were always qualified with comments like "land being available," as most had migrated because the government allocated farmland for mining and biofuel production (Nyantakyi-Frimpong and Bezner-Kerr 2017; Alhassan et al. 2018). Participants who had plans to return remained strongly connected to family members in the north through money transfers and gifts, as well as frequent telephone calls. We emphasize, however, that not everyone wanted to go back home to the north. Findings from the interviews showed that roughly a quarter of the respondents thought that going back home would be perceived as a total failure. One study participant described this failure by saying:

> You have come here for a better life because of climate stress in the north, so you cannot go empty-handed. Going back would be a defeat. It would be a failure. After living here in the south for three years, I have settled in and plan to stay and call this place my home. If you see this place as your home, then you can get focused.

These findings are similar to Abdul-Korah's (2007) research among Dagaaba migrants, in which a large percentage of participants showed that despite the harsh challenges, they plan to stay and call southern Ghana home.

Conclusion

In this chapter, we have sought specifically to examine north–south internal migration among the Dagaaba ethnic group in Ghana by showing the environmental and political–economic circumstances that lead to migration and the choice of destination. As we have demonstrated, the initiation of migration is a fairly simple process for young people, the unmarried, and recently married couples. For people who are older or in polygamous marriages, however, the migration process could be complicated, with difficult negotiations around who should migrate and who should stay behind. We have further shown the struggles and difficulties that characterize the everyday lives of these Dagaaba migrants in southern Ghana. Noteworthy among these difficulties include hidden homelessness, emotional and financial stress caused by the need to accept low-paying jobs, and economic exploitation. The issue of economic exploitation is deeply rooted in a view that northern migrants are cheap laborers who could easily be exploited, a perception with a history in British colonial policies of recruiting forced labor from northern Ghana. Narratives from the migrants revealed that the majority do not feel welcome or experience a sense of belonging in their own home country of Ghana.

Overall, these findings have important implications for the scholarly literature on migration, identity, and belonging. They demonstrate that challenges linked to belonging are not only inherent in international border-crossing, as the migration literature largely demonstrates, but also are evident in internal migration contexts. Thus, the evidence we present in this chapter supports recent calls for a more careful examination of the varied geographies at which belonging is articulated and experienced, with more research not only within transnational spaces but also within national spaces (e.g., Fiddian-Qasmiyeh 2020). Bringing more of these internal migration experiences to the fore, and unpacking how they differ from transnational migration contexts, is a task that awaits scholars interested in human mobility and place-belongingness.

Acknowledgments

Many thanks to Tamar Mayer and Trinh Tran for critical feedback on initial drafts of this chapter. Early stages of the fieldwork were funded by the International Development Research Centre (IDRC), Canada [Grant # 106690-99906075-013]; the Centre for International Governance Innovation (CIGI), Canada; and the Land Deal Politics Initiative (LDPI), International Institute of Social Studies, The Netherlands. More recent fieldwork activities and data analyses have been funded by the University of Denver's Faculty Research Fund [Grant # 86906-145015]; the Department of Geography and the Environment, University of Denver; and a Public Good Fund from the

University of Denver's Center for Community Engagement to advance Scholarship and Learning (CCESL) [Grant # 86847].

References

Abdul-Korah, Gariba (2007). "Where Is Not home?": Dagaaba Migrants in the Brong Ahafo Region, 1980 to the Present. *African Affairs* 106(422): 71–94.

Abdul-Korah, Gariba (2011). "Now If You Have Only Sons You Are Dead": Migration, Gender, and Family Economy in Twentieth-century Northwestern Ghana. *Journal of Asian and African Studies* 46(4): 390–403.

Alhassan, Suhiyini, Mohammed Shaibu, and John Kuwornu (2018). Is Land-grabbing an Opportunity or a Menace to Development in Developing Countries? Evidence from Ghana. *Local Environment* 23(12): 1121–40.

Atuoye, Kilian Nasung, Roger Antabe, Yujiro Sano, Isaac Luginaah, and Jason Bayne (2019). Household Income Diversification and Food Insecurity in the Upper West Region of Ghana. *Social Indicators Research* 144(2): 899–920.

Baada, Jemima Nomunume, Bipasha Baruah, and Isaac Luginaah (2019). "What We Were Running From is What We're Facing Again": Examining the Paradox of Migration as a Livelihood Improvement Strategy among Migrant Women Farmers in the Brong-Ahafo Region of Ghana. *Migration and Development* 8(3): 448–71.

Benveniste, Hélène, Michael Oppenheimer, and Marc Fleurbaey (2020). Effect of Border Policy on Exposure and Vulnerability to Climate Change. *Proceedings of the National Academy of Sciences* 117(43): 26692–702.

Brown, Oli (2008). *Migration and Climate Change.* No. 31. United Nations Publications.

Brun, Catharine, Anita Fàbos, and Oroub El-Abed (2017). Displaced Citizens and Abject Living: The Categorical Discomfort with Subjects out of Place. *Norsk Geografisk Tidsskrift–Norwegian Journal of Geography* 71(4): 220–32.

Chen, Joyce, and Valerie Mueller (2018). Coastal Climate Change, Soil Salinity and Human Migration in Bangladesh. *Nature Climate Change* 8(11): 981–85.

Feng, Shuaizhang, Alan B. Krueger, and Michael Oppenheimer (2010). Linkages among Climate Change, Crop Yields and Mexico–US Cross-border Migration. *Proceedings of the National Academy of Sciences* 107(32): 14257–62.

Fiddian-Qasmiyeh, Elena (2020). *Refuge in a Moving World: Tracing Refugee and Migrant Journeys across Disciplines.* London: University College London Press.

Foresight: Migration and Global Environmental Change: Future Challenges and Opportunities (2011). *Final Project Report.* London: The Government Office for Science.

Ghana Statistical Service (2013). *The 2010 Population and Housing Census: Regional Analytical Report for the Upper West Region.* Accra, Ghana: Ghana Statistical Service.

Ghana Statistical Service (2014). *Ghana Living Standards Survey Round 6 (GLSS6): Poverty Profile in Ghana (2005–2013).* Accra, Ghana: Ghana Statistical Service.

Hausermann, Heidi, David Ferring, Bernadette Atosona, Graciela Mentz, Richard Amankwah, Augustus Chang, et al. (2018). Land-grabbing, Land-use Transformation and Social Differentiation: Deconstructing "Small-scale" in Ghana's Recent Gold Rush. *World Development* 108: 103–14.

Holmes, Seth (2013). *Fresh Fruit, Broken Bodies: Migrant Farmworkers in the United States. California Series in Public Anthropology.* Vol. 27. Berkeley: University of California Press.

Horton, Sarah Bronwen (2016). *They Leave Their Kidneys in the Fields: Illness, Injury, and Illegality among US Farmworkers*. Berkeley: University of California Press.

Hutchful, Eboe (2002). *Ghana's Adjustment Experience: The Paradox of Reform*. Geneva: Unrisd.

Internal Displacement Monitoring Centre (2019). Global Report on Internal Displacement 2019, ICMC. https://www.internal-displacement.org/global-report/grid2019/

Kaczan, David, and Jennifer Orgill-Meyer (2020). The Impact of Climate Change on Migration: A Synthesis of Recent Empirical Insights. *Climatic Change* 158(3): 281–300.

Konadu-Agyemang, Kwadwo (2000). The Best of Times and the Worst of Times: Structural Adjustment Programs and Uneven Development in Africa: The Case of Ghana. *The Professional Geographer* 52(3): 469–83.

Kuuire, Vincent, Paul Mkandawire, Isaac Luginaah, and Godwin Arku (2016). Abandoning Land in Search of Farms: Challenges of Subsistence Migrant Farming in Ghana. *Agriculture and Human Values* 33(2): 475–88.

Lentz, Carola (2000). Colonial Constructions and African Initiatives: The History of Ethnicity in Northwestern Ghana. *Ethnos* 65(1): 107–36.

Luginaah, Isaac, Tony Weis, Sylvester Galaa, Mathew Nkrumah, Rachel Benzer-Kerr, and Daniel Bagah (2009). Environment, Migration and Food Security in the Upper West Region of Ghana. In Isaac Luginaah and Ernest Yanful eds., *Environment and Health in Sub–Saharan Africa: Managing an Emerging Crisis*, 25–38. Dordrecht: Springer.

Lund, Christian (2003). "Bawku Is Still Volatile": Ethno-Political Conflict and State Recognition in Northern Ghana. *Journal of Modern African Studies* 41(4): 587–610.

Madsen, Kenneth, and Ton Van Naerssen (2003). Migration, Identity, and Belonging. *Journal of Borderlands Studies* 18(1): 61–75.

Mueller, Valerie, Glenn Sheriff, Xiaoya Dou, and Clark Gray (2020). Temporary Migration and Climate Variation in Eastern Africa. *World Development* 126: 104704.

Niang, I., O. C. Ruppel, M. A. Abdrabo, A. Essel, C. Lennard, J. Padgham, et al. (2014). Africa. Chapter 2. In *The Fifth Assessment Report of the Intergovernmental Panel on Climate Change*, 1199–265. Cambridge: Cambridge University Press.

Nyantakyi-Frimpong, Hanson, and Rachel Bezner-Kerr (2015a). A Political Ecology of High-input Agriculture in Northern Ghana. *African Geographical Review* 34(1): 13–35.

Nyantakyi-Frimpong, Hanson, and Rachel Bezner-Kerr (2015b). The Relative Importance of Climate Change in the Context of Multiple Stressors in Semi-arid Ghana. *Global Environmental Change* 32: 40–56.

Nyantakyi-Frimpong, Hanson, and Rachel Bezner-Kerr (2017). Land Grabbing, Social Differentiation, Intensified Migration and Food Security on Northern Ghana. *The Journal of Peasant Studies* 44(2): 421–44.

Paprocki, Kasia (2020). The Climate Change of Your Desires: Climate Migration and Imaginaries of Urban and Rural Climate Futures. *Environment and Planning D: Society and Space* 38(2): 248–66.

Pearce, Richard (1992). Ghana. In Alex Duncan and John Howell eds., *Structural Adjustment and African Farmers*, 14–47. London: Overseas Development Institute.

Plange, Nii (1979). Underdevelopment in Northern Ghana: Natural Causes or Colonial Capitalism? *Review of African Political Economy* 6(15–16): 4–14.

Prah, Kwesi (1975). The Northern Minorities in the Gold Coast and Ghana. *Race & Class* 16(3): 305–12.

Rademacher-Schulz, Christina, Benjamin Schraven, and Edward Salifu Mahama (2014). Time Matters: Shifting Seasonal Migration in Northern Ghana in Response to Rainfall Variability and Food Insecurity. *Climate and Development* 6(1): 46–52.

Renaud, Fabrice, Olivia Dun, Koko Warner, and Janos Bogardi (2011). A Decision Framework for Environmentally Induced Migration. *International Migration* 49: 5–29.

Schildkrout, Enid (1979). The Ideology of Regionalism in Ghana. In William Shack and Elliott Skinner eds., *Strangers in African Societies*, 183–207. Berkeley: University of California Press.

Seini, Wayo (2002). *Agricultural Growth and Competitiveness Under Policy Reforms in Ghana: ISSER Technical Publication No. 61*. Legon: University of Ghana.

Songsore, Jacob (2011). *Regional Development in Ghana: The Theory and the Reality* (new edition). Accra, Ghana: Woeli Publishing Services.

Van der Geest, Kees (2011). North-South Migration in Ghana: What Role for the Environment? *International Migration* 49: 69–94.

Vermeulen, Sonja, and Lorenzo Cotula (2010). Over the Heads of Local People: Consultation, Consent, and Recompense in Large-scale Land Deals for Biofuels Projects in Africa. *The Journal of Peasant Studies* 37(4): 899–916.

Warner, Koko (2010). Global Environmental Change and Migration: Governance Challenges. *Global Environmental Change* 20(3): 402–13.

Wilson, Tamar Diana (2020). Climate Change, Neoliberalism, and Migration: Mexican Sons of Peasants on the Beach. *Latin American Perspectives* 47(6): 20–35.

10 Henancun in Beijing, a parallel society in the making

Jia Feng and Guo Chen

Introduction

In the summer of 2009, we followed the literature that documented the four different migrant enclaves in Beijing to look for Henancun[1]—a place where mostly Henan migrants work in the informal recycling sector. We gave the address to a taxi driver who took us to the front gate of the Olympic Village. Once there, we immediately knew that Henancun had moved or had been moved again. We learned almost nothing about Henancun from local Beijing residents, so we instead had to follow one recycler who worked in a gated community in Changping District (to the north of the Olympic Village). Finally, we arrived at the biggest migrant-run recycling center in Beijing—the Dongxiaokou Henancun (Figure 10.1).

This recycling center is located on the edge of the developed city near the Dongxiaokou village on the border between the Changping and Haidian districts. While Dongxiaokou itself awaits redevelopment, the undesirable empty rural land is perfect for hosting a recycling market to generate extra revenue for the village, which also provides affordable housing for the rural migrants. With about 75% of the rural migrants from Henan Province in Dongxiaokou, this Henancun has become the largest recycling center in Beijing since the one previously located at Datun Road and Wali Village was demolished and redeveloped to host the 2008 Beijing Summer Olympic Games. The Dongxiaokou Henancun resembles previous recycling centers documented by Ma and Xiang (1998) and Gu and Liu (2008), whereby the almost identical group of migrants continued the same form of livelihood and businesses in a more remote location in Beijing. The continuity and transient nature of Henancun caught our attention as other similar migrant enclaves—Zhejiangcun and Xinjiangcun—have faded away amid Beijing's urban development processes in the past 30 years (Ma and Xiang 1998; Wang, Zhou, and Fan 2002; Xiang 2005; Jie and Taubmann 2008).

In this chapter, we focus on the Henan recycling migrants and their experiences in Beijing amid the large-scale rural-to-urban migration that has occurred since the 1980s in China. In particular, we intend to understand how rural migrants cope with the unfriendly, often exclusionary urban environment by trying hard to construct a home space and parallel society

DOI: 10.4324/9781003170686-12

Figure 10.1 Dongxiaokou recycling market overview.
Copyright: Jia Feng.

between Beijing and their rural hometown. In addition, facing a unique situation of being connected and yet doubly excluded from both their hometowns and their migration destination, we examine how migrants redefine their existence by creating a functioning and self-justifying temporary parallel society in Beijing.

Rural-to-urban migration in China

Since China adopted a "socialist market economy" in the late 1970s, various economic and social policies have encouraged the freed rural labor to seek opportunities in the urban sector. Because of the long-existing *hukou*[2] policy, most rural migrants retained their rural registration status in the cities, by which they are designated as the "floating population." In 2015, 247 million (18.0%) out of the 1.375 billion Chinese people are living outside their *hukou*-registered locations (NHCPRC 2018; NBSC 2020). In Beijing, there are 7.6 million non-*hukou* residents in 2018, which makes up about 35.5% of the total city population (BMBS 2019). Due to *hukou*'s internal passport function since the 1980s (Fan 2002), the Chinese rural-to-urban migration resembles international immigration, in that the unskilled migrant labor remained excluded in the cities despite their contribution to the economy (Roberts 1997). Meanwhile, the Chinese state condoned this situation as part of the shift from focusing on "equality" under a planned economy to prioritizing "efficiency" in the socialist market economy (Wang 2008) in the post-Mao era, where urban development benefits from the "temporary" and informal migrant labor regime (Fan 2008).

Myriad studies have shown that the rural migrants in cities are living and working in inferior economic, political, and social conditions (Chan 1996; Solinger 1999; Fan 2002; Shen 2002; Wang 2004). Scholars have examined this dichotomized or segregated economic and social situation in cities through different angles. The neoclassical view takes the human capital and economic disparity between regions as a motivation to explain the rural-to-urban migration (Fan 1996), whereas the dual-labor-market perspective argues that the institutional barrier—*hukou*—channels rural migrants into a separate and informal job market in cities (Solinger 1999; Wang and Zuo 1999; Ding et al. 2001; Carrillo 2004). Social capital also plays an essential role for migrants to use various kinship or geographical ties to form a chain and channelized migration to the city (Ma and Xiang 1998; Massey et al. 1999; Xiang 2005; Jie and Taubmann 2008).

Furthermore, on the premise that migration is followed by integration and/ or assimilation to the receiving society, studies have adopted dualistic views regarding the origin and destination of migration (Hu et al. 2011; Boccagni 2014; Marta et al. 2020). In the Chinese rural-to-urban studies, plenty of findings have either supported or challenged the view that migrants desire to stay in cities, as researchers argue that the majority of rural migrants expect to return to their hometowns (Solinger 1995; Hare 1999; Fan 2008). In a recent project, Nalini Mohabir et al. (2017) point out that migrants have a preference to remain in megacities as opposed to small ones or their rural areas. And Cindy Fan (2008, 2016) also examines a permanent migrant paradigm, which suggests that migrants are not pursuing a permanent settlement in the city but rather that temporary migration allows migrants to enjoy the benefits from both worlds. Zhang Jijiao has also reviewed the concept of "belonging" and notes that under the two-tiered *hukou* system, the discrepancy of migrants' "being" and "longing" constructed a complicated notion of home and belonging for migrants (Zhang 2013). Dror Kochan uses a translocal approach and argues that the "ancestral home (*laojia*[3]), the city home, and the material home" complexity reflects the migrants' home-making practices with flexibility and mobility under the precarious urban context (Kochan 2016: 30). The structural constraints for migrants to make sense of identity and belongingness need to be accounted.

Facing stigmatization and discrimination in the cities, Henan migrants in Beijing have been trying to build their own community since the 1990s. In Beijing, urban villages provided affordable housing for migrants from the same hometowns (Zheng et al. 2009), and four different ethnic enclaves— Zhejiangcun, Henancun, Fujiancun, Xinjiangcun—emerged in the city (Ma and Xiang 1998; Jeong 2000; Xiang 2005; Gu and Liu 2008; Jie and Taubmann 2008). But all of the above-mentioned migrant enclaves have disappeared except for Henancun. In this chapter, we look at migrants' everyday experiences and try to understand how and why Henan migrants construct their "home" and community in Beijing.

We have conducted multiyear observations between 2009 and 2015 with a seven-month period of fieldwork between 2012 and 2013. The fieldwork took

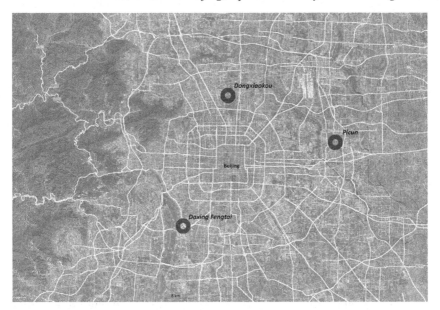

Figure 10.2 Fieldwork locations in Beijing.
Copyright: Jia Feng. Based on Google Map.

place at three locations in the northern (Dongxiaokou), eastern (Picun) and southwestern (DXFT) parts of Beijing (Figure 10.2). The results are based on 304 valid questionnaire responses from the three locations collected with the assistance of 10 local college students in early January 2013. With a proportionally stratified sampling strategy, the practical plan was based on a systematic sampling with replacement plan (SRS) to survey every other recycling yard and every other residential house in each alley in the three recycling enclaves. We conducted over 60 interviews between 2009 and 2013 with various recycling enclave stakeholders, including community recycling actors, local community residents, NGO members, village government leaders, and district government officials. In 2015, we revisited several of the stakeholders on a separate project, which granted us an opportunity to reevaluate our initial interpretation of the situation.

Never belonged

To get to Henancun, we needed to provide detailed instructions to guide the taxi driver deep into the villages to the north of Beijing. Passing the heavy traffic on the fourth and fifth ring roads, the high-rise commercial centers, renovated *danwei*-style gated communities, and newly constructed main traffic roads with young trees on the sides, the taxi driver became skeptical about where we were heading, when all of a sudden Henan restaurants and truckloads of plastic packages, scrap metal, and mixed paper signaled that we had arrived.

Mr. Zhou was one of the first to accept our interview request. As a yard recycler, his yard was in the middle of a private recycling company that hosted more than 120 family-run recycling yards focusing on a variety of recyclable materials. Mr. Zhou was among the early wave of migrants from Henan Province who arrived in Beijing in the late 1980s. From our conversation, we cannot detect the Henan accent that is rather common in the recycling yards. With a clean-shaven face, metal-frame eyeglasses, and a carefully tucked-in shirt, he looked like any other businessmen in Beijing. Sitting on the fragile plastic stool in his yard house, he laughed about my question concerning his interactions with the city outside the recycling center. "I almost never go there anymore. Actually, I haven't been outside for quite some time." Like his peer recyclers, he refers to the city he has migrated to and lives in as the "outside," while referring to the company and the yard as "inside." When asked why he didn't take the subway or buses to visit "outside" places, he further explained:

> When I was on a bus, I didn't even know where to put my hands or where to look. Look at the Beijing people or those well-off ones, when they get on the subway or bus, they have watches on their clean hands. When they pull on those grab handles, you can tell that they are doing good. If you look at us, with hands in our sleeves, we are not doing good. It's not about how many times you wash your hands or how much you clip your nails. Hands are different.

Since Mr. Zhou arrived at Beijing in the late 1980s, he has worked different labor-intensive jobs. Because most of his fellow hometown relatives and friends joined recycling in the city, he started as a roving bicycle recycler in the 1990s. When asked about the most significant experience he had in Beijing that had changed him, he remained silent for a while, hesitated, and after taking a deep smoke, he said one word—*shourong*.

The *shourong* policy, also known as the *shourong qiansong* (meaning, respectively: custody and repatriation) policy, was in effect between 1982 and 2003. In 1982, the State Council of China instituted the *shourong* policy to help, educate, and organize urban vagrants in the massive rural-to-urban migration. Gradually, rural migrants with no fixed address, no livelihood, and no residence permit came to be known as "Three-No People (三无人员)," who existed under the radar of this policy in the 1990s. When Three-No People were identified, they would be locked up (*shourong*) and sent back to their hometowns (*qiansong*). Not too long after it began, the policy shifted to target rural migrants in cities, mostly by extracting fines and extorting bribes. Under the controversial *shourong* policy, many migrant workers had to buy their way out of such threats. Our survey suggests that more than one-third of the rural migrants who were in Beijing before 2003 were locked up at least once under the *shourong* policy.

Almost all interviewees in their 40s or older had the exact reaction to the question on *shourong*: shaking their heads in retrospect, waving their hands, and repeating in a very serious tone: "that was a dark history nobody wants to remember." Then they would silently take a deep breath and look away.

As an itinerant tricycle recycler—that is, one who roams the streets on a bike or trike, looking for recyclable materials—Mr. Qian was a victim of *shourong* multiple times. His experience as a "veteran" in his field is typical among the responses in the study; the first time it happened to him was in Wali (the later Olympic Village) in 2000.

> I was in Wali in 2000 when I recycled around Changqiao in the city. When you were in the street recycling, they[4] didn't care what you said, they just took you and sent you to your hometown, Henan for me. If you had money, they'd let you go. It was brutal. At that time, I was riding my tricycle, but it didn't matter whether you were walking or riding. They could easily tell that you were not local, and they'd take you without explanation, get you to the police station to register, then send you away.
>
> In 2000, we needed to pay 300 *yuan*. If you had money, give them 300, [they'd let you go], then 200, then 100. If you didn't have any money, they'd take you all the way to Henan and let you go. So, if you paid, they would let you go from the train at the cargo stations. I was let go somewhere close to Baoding or Shijiazhuang, still more than 1,000 km from Henan. Then they just didn't care anymore, no matter you chose to walk [back to Beijing] or whatever. If you knew someone whom they knew of, you could go too. Otherwise, you had to be sent out of Beijing. … At that time [after I was locked up], my wife could not get hold of me and worried a lot. What she knew was that I simply disappeared and nobody knew where I was. When I was released, I was able to buy a [train] ticket and came back [to Beijing]. But, many times, they didn't send you immediately because they needed to gather enough people to fill the train. Sometimes, I was locked in Qiliqu (七里渠) for two or three days. But luckily, since there were lots of Henan migrants on the streets, it was easy to fill the train to Henan [bitter smile]. Only our trains went very frequently because we had so many in Beijing.

Although most veteran migrants have experienced *shourong* the way Mr. Qian has, the experience for others involved violence, as some migrants were beaten up in custody. These experiences significantly affected the way migrants lived and worked in the city toward the end of the 1990s. Mr. Zhou, introduced earlier in the chapter, explained that after being caught several times, no one dared to run the streets anymore. Facing the threat of *shourong*, migrants started to seek protection from the recycling companies and live inside the gated and walled recycling yards.

> When locked up by *shourong,* we would contact the people [the recycling company managers] who knew someone in the defense team or the police to get us out. But nobody had cell phones in the 1990s, so if you were caught alone, you wouldn't be able to come out since you couldn't contact the people who could help. During those years, nobody would go

out alone. When we had to go on the street, we had to have at least two people going together. When we two went out, we had to go separately, as I needed to walk about 50 meters ahead of you. If I got caught, you could go back to find help and get me out. You also needed to pay someone to get other people out too. This catalyzed the emergence of a peculiar job to fish people out from the defense team, *shourong* or the police station. They were mostly local people who knew or had connections with someone from the inside.

When the *shourong* policy was over in 2003, migrants no longer lived the same lives as before. Many continued to rent a space inside the migrant recycling companies and chose to specialize in one or several items for recycling purposes. Many migrants who have been in Beijing since before 2003 have expressed their concerns walking or taking public transportation in the city even after the end of *shourong*. The walled and gated recycling companies, as a safe haven, have become the prototype of Henancun, which later nurtured the development of a parallel society between migrant hometowns and their destination.

Throughout our interviews, we discovered that while some migrants have been in Beijing for over 20 years, they never feel like they belong or are welcome in the city. Discrimination based overtly or covertly on their appearance and very existence has significantly squeezed migrants' business and livelihood space into a rather segregated and excluded area.

An identity divided

Just as the international immigrants carry their hometown legacy in a new country, rural-to-urban migrants also hold an identity from both worlds. During the interviews with the migrants in Henancun, it is easy to identify the fusion of a formal business environment and a rural hometown atmosphere. The suit-and-tie clad businessmen, the luxury cars, and the truckload of packed and squared recyclable materials all narrate a story of success in the urban commodity supply chain. Meanwhile, the hanging cured meat, fish and sausages, groups of live chickens, ducks and yard dogs, the tire marks truck leave on the muddy road after rain, the scattered one-floor brick houses, people holding a bowl of meal in the yards, and the smell of burning wood and charcoal all tell a story about the quasi-country livelihood that migrants experience every day. From many perspectives, Henancun is more than a migrant space, as the construction of its identity is an ongoing and never-ending subject among the rural migrants.

A formal businessman in recycling

Among the migrant recyclers we interviewed, many adhere to a strict dress code every day. Although this situation is uncommon among other migrant groups in Beijing, several recyclers indicated that the dress code is meant to ensure a formal business attitude. In Dongxiaokou, many recycling yards provide shade-cloth-covered parking spaces for luxury cars, which symbolize their qualification in the bidding war for a recycling contract in the city. "To win a bidding competition among migrant recyclers," Mr. Zhang, a metal recycler in his 40s, told us,

> you need to dress well and drive a nice car to impress people and show your success in the business. So, they know you are qualified to have their business. ... Did you see my Lexus outside? The only time I drive it is to sign contracts or talk about business.

The side-by-side display of the luxury car and haphazard pile of recyclable materials mirrors the migrant story of constructing a mixed local-migrant identity in the city. While the shiny local identity resides in the formal business meetings on the "outside," the dirty and hardworking parts of the migration are usually kept invisible, inside the recycling yards.

A celery controversy

When large-scale rural-to-urban migration started in the 1980s, many rural migrants left their hometowns because of poverty, arguing that "migration would at least save food for others at home." The rural-to-urban migration continued for over 20 years, but things have not changed much in migrants' rural hometowns. When asked about his hometown in Henan province, Mr. Zhou said,

> In Beijing, I can eat celeries that are 1 yuan per 1 *jin* (½ kilo); at my old home [hometown], celeries are sold for 1 *mao*[5] [per 1 *jin*], but I can't afford it. There is just no income at my hometown.

This controversy is very common for the rural migrants in the city. While the migrants have increased their wealth with the city's rapid economic development, the hometown economy lagged far behind. Although migrants were excluded and segregated in the city, they do not have the financial freedom to move back to their hometown either. This celery controversy shines a light on the economic and financial reality: the rural migrants are locked in a place between an excluding urban society and an unreturnable rural hometown. While the international immigrants usually have "a foot in two worlds," the rural migrants in Henancun experience some of the extreme versions of divided identity that do not associate with either end of their migration.

Connect to disconnect

While migrants are working in the recycling business, they are required to keep an institutional tie with their rural home. One of the most important ties is through the complex educational system defined by the *hukou* policy.

The *hukou* registration provided at birth makes migrants' children adopt their own rural registration from their parents and they have to take college entrance exams in the province their *hukou* status specifies. Although Henancun has several migrant-run elementary and junior high schools, and Beijing has relaxed the restriction banning migrant children from the local public schools, the education quality is believed to be much better under the public-school system in migrants' hometowns. One common response we received has illustrated a very interesting perspective on strategies migrants use when it comes to their families: "If the children are doing good in the schools here, we will send them back to the hometown. If they are not doing well [at school], we will keep them by our side." From a very practical perspective, whether migrants send their children back to their hometowns is not about the education they receive but rather depends on the chance that the child will succeed in the college entrance exam to be able to leave the hometown through the formal channel of education.

This situation leads many migrant families to use a split-family strategy, which is a highly constrained choice to effectively access needed institutional resources for their children's education. This strategy is usually adopted later when a child enters junior high school, years before they start to prepare and eventually take the college entrance exam. For migrant families with children attending primary schools, children from 75% of the families go to hometown primary schools. For junior high schools, the percentage increases to 86%. According to our survey, while almost all of the migrant families with children adopt a split-family strategy of sorts at some point, about 25% of the families have one parent returning to their hometown to take care of the child for a relatively long duration, and the remaining 75% leave the children to stay with a relative while both parents work in Beijing. The situation reflects a paradox that migrants have been facing.

In order to permanently escape the discriminated migrant status in the city, many are forced to maintain their rural identity, which is required for accessing the crucial educational resources in rural areas that would help bid a successful future and an urban hukou for their children in the city. This cyclical process reinforces a feeling of being trapped as "floating migrants".

In addition, Henan migrants' actions and plans for housing reflect a different kind of attitude toward their hometowns. From experiences living in Henancun, most migrants understand that it is not a permanent setting. The hometown still represents a sense of affordable and attainable "permanency" that never existed in Beijing. Most migrants choose to invest in the housing market in their hometowns. Among the more than 300 responses to the question, "Have you bought, built, or planned to buy or build housing in Beijing, your hometown, or other rural places?" none indicated that they either would build/buy or plan to build/buy houses in Beijing. Once financially capable, they often prepare to seek permanent security in their hometown facing the unforeseeable future in the city.

Living between two identities, migrants have been navigating through their options to secure a future for their families. If the education-based strategy is a forced choice for migrants to stay connected to the hometown they have escaped, the housing-based arrangement is just an opposite option for migrants to connect to their hometown as an added security for the potential challenge during their stay in Beijing. Shaped by the economic, financial, educational, social, and institutional constraints and opportunities at both ends of migration, a divided identity becomes the last reserved option for migrants to secure a future.

Belonging under construction

While Henancun stays under the radar because of the exclusion and discrimination in the city, the rural migrants strive to transform the urban space in order to achieve a sense of belonging in Beijing. While many international immigrants or rural-to-urban migrants strive to establish their rights in the city, Henan migrants have long acknowledged the temporality and transient nature of both Henancun and their stay. Accordingly, they have devoted their limited resources to transforming Henancun into a migrant space that is both comfortable and functional when it comes to sustaining their livelihood and businesses in Beijing. When asked how they feel living in Henancun, many migrants imply that

> we have everything we need, the groceries, street market, breakfast place, and hometown flavors. There are also buses going directly to Gushi [the county where most Henan migrants come from]. … We don't go outside because we have everything we need.

Constructing a migrant community in Henancun

To migrant recyclers, Henancun is not only a safe haven for avoiding the *shourong* policy but also provides social services for migrants to have a sustainable life inside. In the largest Henancun around Dongxiaokou, many migrant businesses start to enrich the livelihood by providing hometown-flavor restaurants, small grocery stores, supermarkets, public bathing facilities, hotels, long-distance bus services, car repair, kindergarten, elementary schools, and junior high schools. Most, if not all, of the services are owned and operated by the migrants sharing the same hometowns. During the years of our fieldwork, an afternoon street market usually started at 4 p.m. every day, selling everything from vegetables and fruit, live fish, and hometown deli to clothes and toys. A self-sufficient community provides migrants the luxury to live like they are "at home" again.

One evening, we were invited to dinner by a migrant friend, Mr. Sun, who has become a used-market vendor after spending more than 10 years in the recycling business. The recycling and reuse sectors are never separated, and now Mr. Sun and his wife sell used and refurbished furniture in a large supermarket setting. They live in a one-story rental house in Henancun that is hidden behind street-facing businesses that specialize in Gushi cuisine, quick breakfasts, or selling grocery. After parking their tricycle against the wall in the pathway, Mr. Sun told us that there was an agreed rule that no recyclable or reusable commodities are allowed along the pathway so that everyone can enjoy the public space as a community. After checking the contents of his refrigerator as his wife began to cook, Mr. Sun told us that he was going to buy a special dish from his hometown restaurant. He led us through the alley, shouting "hi" to almost everyone whose door was left open, or when the sound of a TV could be heard from inside. Then he opened a back door without any sign and walked right into a restaurant, where he was obviously a regular. Speaking in a hometown dialect, he ordered some specialty goose dish and started searching for his favorite channels on the restaurant TV. He also randomly chatted with the other customers in the restaurant. He spoke directly, again in his hometown dialect, to discuss the price of recyclable materials and what the future of business would look like after the 2008 world financial crisis. After about 10 minutes, the dish was brought out in a plastic bag inside a large white ceramic bowl. Mr. Sun took the bowl, left the money on the counter, and told the owner, "I'll bring the bowl back after dinner." Then he led us back to his house.

As we started eating dinner, several of his friends knocked and pushed open the door to check "why it smelled so good." Seeing us in the house, they hesitated to join the dinner, but it was easy to tell that they are hometown friends in a very close relationship and similar businesses. Although Henancun stays relatively excluded and segregated from the city, the migrant space is active and alive. Not following a typical city manner, the community is built to resemble a way of life familiar to the migrant families. The open-gate environment in the alley, where people communicate by shouting through windows and walk right in without knocking on the door, represents a close-knit community. The hometown-flavor restaurants that dot the Henancun street fronts not only provide services to cure the homesick but also become focal points for migrants to share stories and build relationships.

On the outside, Henancun is distinguished from the city by various store signs that advertise Henan cuisine as well by the recycling shops and truck-loads of recyclable materials. But behind those signs, the migrants are building a community woven together through their hometown identities and shared experiences in Beijing.

Working and living with a migrant schedule

Meanwhile, this unique migrant community also runs on its own schedule. As the city is about to rest with people returning home in the evening, Henan migrants start their workdays. As dusk gradually calms down the busy city traffic, migrant recyclers from all the communities in the city start riding their tricycles and driving their trucks back to Henancun to sell. All the recycling yards in the various companies start to welcome the incoming vehicles and direct them to unload the mixed, half-sorted, or well-sorted materials into the yards. Because of Beijing's road space rationing system after the 2008 Summer Olympics, many trucks can only drive out of the central city after 10 p.m., and many migrant yards have to remain open even after midnight.

Mr. Gao from a paper-recycling yard told us, "I have to be ready all the time because if you are not there when they knock on your door even after midnight, you lose their business forever." For Mr. Gao's paper-recycling yard, a day's work never ends based on time but only stops when no one knocks on his door anymore. Usually, morning starts around 6:30 a.m. in the yard when six hired workers sit in front of three huge piles of recycled paper weighing between 8 and 15 tons in the yard and start to sort them into newspapers, print paper, books, paperboard, and corrugated paperboard. The day's job is to sort the materials, load at least two 11-meter trucks with different paper products, ship and sell them to either the packaging or paper companies, and clear the yard to prepare for the next full yard load of recyclable paper to come in after dusk. Mr. Gao laughed and told us: "The work never ends. We never have breaks during weekends because Saturdays and Sundays are usually the time for people to sell their materials. We also never visited anywhere in the city since we just don't have the time." Ironically, although Henancun in Dongxiaokou has developed to be a stand-alone and self-sustainable migrant space, the schedule and function of this social space are about business only.

Nurturing an exclusive migrant space

In addition, this space and the associated recycling business are designated almost exclusively for rural migrants. Throughout our questionnaire, 303 of 304 respondents are rural migrants, as one person has changed his *hukou* status to Beijing after serving in the army. Of the 303 migrants, 74% are from Henan Province, while 14% are from Hebei Province. Among the Henan migrants, 95% are from the city of Xinyang (97% of whom are from the Gushi county of Xinyang). In interviews with several recycling company managers and house owners, they all indicate that the recycling experience is one important factor of success, and the people with an "accent" or "look" from "those" places have more promise in achieving success and staying for a long time. Gradually, Henancun as a recycling center has become a nearly exclusive migrant space through chain and channelized migration, which gradually led people to join the recycling sector in Beijing through blood relationships, location relationships, and work relationships (*Gushi* Labor Report 2009).

Meanwhile, this chain and channelized migration does not stop after migrants arrive at the city. While the migration remains transient and temporary, the recycling business becomes the only feature to retain some form of permanency for the rural migrants in Beijing. Throughout the years in Beijing, Henan migrants have achieved their social connections and capital accumulation step by step. Mr. Zhou upgraded from paper-recycling to plastic recycling before focusing on foam to recycle at the time of the interview. For each upgrade, he said,

> I just gave the business and [business] connections to my relatives and friends from my hometown. I can't just abandon the business because of the hard work I put in it. After I gave away my previous business, I upgraded to a better item to recycle but it [the new item] usually requires more investment and specialized knowledge, which can only be gained by watching the market and learning the policy changes all the time.

It is not uncommon for rural migrants to secure a profitable job opportunity from their relatives and friends before they migrate. This form of chain and channelized migration helps strengthen the migrant identity of Henancun in the city.

Building a better "home"

Given Henancun's transient nature, migrants are usually reluctant to invest to better their living conditions. But recyclers who specialize in reusing and repairing appliances and furniture have contributed their skills to gradually make Henancun an important hub by providing local residents and migrants alike with various used furniture and appliance options. Interestingly, these efforts differ from the recycling services they provide to the city. This value-based trade of the used/refurbished/repaired products is less of a profit-driven business and more of a community-supporting activity for all migrants who share similar experiences in the city. While Henancun benefits greatly from this specialty service, the meaning of community for all rural migrants in the city is strengthened.

Interviewed in front of his rental house, Mr. Qian indicated that

> all the appliances in my house here are ones I bought either on the street or from other recyclers. ... I have everything I need, and they are all cheap and usable. ... I bought my air conditioner [AC] for 400 yuan, and if I move again, I can just sell it for that price. If it stops working, I can disassemble it, and sell it piece by piece for different types of materials.

Inside his house, Mr. Qian has about everything in a typical household, a refrigerator, a washing machine, a TV, a floor fan, a computer, and an AC. During an interview with Mr. Qian in September, he explained that

> the AC is not working properly because of the unstable electricity voltage. Also, electricity is expensive as one kwh costs about 1 yuan, and this AC uses more than one kwh per hour. As it's not so hot anymore, I'm going to take it down and sell it since it's of no use in the wintertime.

Because the recyclers are aware of Henancun's short lifespan, they do not bother decorating or upgrading their living environment. Meanwhile, the recycling business provides a near-perfect solution for migrants to get by with affordable appliances and furniture for those few years in one place, as they only pay a basic cost of the product instead of price reflected by the market or brand value.

With the recognition of social exclusion and discrimination, Henan migrants are actively constructing their community with economics in mind. On the one hand, Henancun offers temporary space to accommodate migrants' needs in the city, and on the other hand, migrants must take measures to turn the migrant space, where temporality and precarity prevail, into a shared community. Centering on their businesses in Beijing, the migrant space takes on a hometown (*laojia*) tone, given the workers' flexible schedules, the regional cuisine and habits, and the sense of shared community. To migrants, the space signifies more than a temporary stopover to make money. Rather, their temporary stay has been permanent enough so that they strive to improve the quality of their lives. The permanency is so spatially precarious, however, that they live a base-line existence that is literally represented by paying only the material value of everything they need to sustain their stay.

Conclusion

Beijing was home to over 21 million residents in 2013, among which 8 million (37.8%) are rural-to-urban migrants who have a temporary residential status because of their rural *hukou* status. In the city, about 200,000 are working in the recycling business, representing about 1% of the total city population. Since the mid-1980s, migrant recyclers have gradually carved out a space of their own because of the discriminatory and stigmatizing urban policies. Just as Henri Lefebvre (1992) showed how the urban social space mimics a flaky *mille-feuille* pastry, the emergence and permanent temporality of Henancun reflected migrants' political, social, and economic interpretation of their precarious migration in the midst of China's rapid economic development. Among all the migrant spaces in Beijing, Henancun remains one of only a very few to adapt to the urban system by building a parallel society between the city and the rural. We argue in this chapter that in addition to the institutional explanations of migration, migrants as active agents adapted themselves to the urban system and used various coping strategies to sustain their temporary stay and manage the transiency of their migrant space in the city.

Based on the interview and observation records in our study, we argue that rural migrants have been facing discrimination overtly and covertly in the city. Only when the discriminatory *shourong* policy started to threaten their existence did Henan migrants begin to build a segregated space to protect themselves in the city. Being excluded from their migration destination, rural migrants also face the dilemma of a divided identity between the city and their hometowns. While migrants have built their livelihood in the city, their rural *hukou* and identity keep reminding them where biased institutional policies determine where they belong (Chen 2012). The literal meaning of a divided identity becomes evident for many migrants who are forced to use a split-family strategy in hope to escape their hometown through the institutional channel of education. The debate of migration belonging, or identity, can no longer be determined by answers to a multiple-choice question; rather, the debate must include an understanding that the divided sense of belonging is embedded in migrants' everyday lives all the time.

Meanwhile, facing precarious migration conditions, the rural migrants become active agents who try hard to ameliorate their difficulties by constructing a migrant-centered community. They have not only followed their agriculture-like flexible working schedule but also constructed a migrant center by providing value-based used/refurbished/repaired furniture and appliances to improve their living experience in the city. The value-based trade adds community cohesion by drawing migrants together to cope with the transient and temporary experience of city life. Remaining excluded in the city, the migrant community serves as an important parallel society that exists between migrants' rural hometowns and their migration destination.

Drawing on a statement about guest worker programs that circulated among researchers and policymakers—"there is nothing more permanent than temporary migration"—we argue that Henancun, a de facto parallel society in Beijing, has been shaped by the city and reshaped by the migrants to construct a shared community by all migrants living outside either world of their long-sought-for belonging.

Notes

1 Henan is a rural and inland province in China that is not immediately adjacent to Beijing. The suffix "cun" means village, which is the smallest unit in the Chinese administrative divisions. Henancun indicates a residential settlement of Henan migrants in Beijing.
2 *Hukou*: a household registration system, which entitles one person's registration status to be rural or urban based on the family's registration status. This policy is related to people's access to medical and educational resources. Rural migrants in the cities usually cannot enjoy the social welfare available to the locally registered urban residents.
3 *Laojia*: the literal meaning is "old home." It usually refers to a migrant's hometown.
4 The word "they" here is referring to the local police and the public order joint defense team (治安联防队).
5 1 *mao* is 0.1 yuan; 1 yuan (RMB) is about 15 cents in USD.

References

BMBS (2019). "2018 年全市常住人口发展变化情况 [The Development and Change of Local Residents of Beijing in 2018]." Accessed January 24, 2021, from http://tjj. beijing.gov.cn/zt/rkjd/sdjd/201907/t20190709_143423.html#:~:text=2018%E5%B9 %B4%EF%BC%8C%E5%85%A8%E5%B8%82%E5%B8%B8%E4%BD%8F%E5 %A4%96%E6%9D%A5,%E9%99%8D%E5%B9%85%E8%B6%8B%E5%8A%BF %E8%BF%9B%E4%B8%80%E6%AD%A5%E5%8A%A0%E5%A4%A 7%E3%80%82

Boccagni, Paolo (2014). "What's in a (Migrant) House? Changing Domestic Spaces, the Negotiation of Belonging and Home-making in Ecuadorian Migration." *Housing, Theory and Society* 31(3): 277–93.

Carrillo, Beatriz (2004). "Rural-Urban Migration in China: Temporary Migrants in Search of Permanent Settlement." *Portal Journal of Multidisciplinary International Studies* 1(2): 1–26.

Chan, Kam Wing (1996). "Post-Mao China: A Two-class Urban Society in the Making." *International Journal of Urban and Regional Research* 2(1): 134–50.

Chen, Guo (2012). "Structural Evaluation of Institutional Bias in China's Urban Housing: The Case of Guangzhou." *Environment and Planning A* 44(12): 2867–82.

Ding, Jinhong, Leng Xiliang, Xiukun Song, B. Hammer, and Yuehu Xu (2001). "中国对非正规就业概念的移植与发展 [Interpretation and Development of Informal Job Opportunities in China]." *Chinese Journal of Population Science. 中国人口科学* 6: 8–15.

Fan, Cindy (1996). "Economic Opportunities and Internal Migration: A Case Study of Guangdong Province, China." *The Professional Geographer* 48(1): 28–45.

Fan, Cindy (2002). "The Elite, the Natives, and the Outsiders: Migration and Labor Market Segmentation in Urban China." *Annals of the Association of American Geographers* 92(1): 103–24.

Fan, Cindy (2008). "Migration, the State, and the Household", 1–210. In *China on the Move: Migration, the State, and the Household*. London: Routledge.

Fan, Cindy (2016). "Household-splitting of Rural Migrants in Beijing, China." *Trialog: Journal for Planning and Building in a Global Context* 1–2(116/117): 17–21.

Gu, Chaolin, and Haiyong Liu (2008). "Social Polarization and Segregation in Beijing", 198–211. In *The New Chinese City*. Hoboken, NJ: John Wiley & Sons.

Gushi Labor Report (2009). *Gushi* County Government, *Xinyang* City, *Henan*, China.

Hare, Denise (1999). "'Push' versus 'Pull' Factors in Migration Outflows and Returns: Determinants of Migration Status and Spell Duration among China's Rural Population." *The Journal of Development Studies* 35(3): 45–72.

Hu, Feng, Zhaoyuan Xu, and Yuyu Chen (2011). "Circular Migration, or Permanent Stay? Evidence from China's Rural–Urban Migration." *China Economic Review* 22(1): 64–74.

Jeong, Jong-Ho (2000). "Renegotiating with the State: the Challenge of Floating Population and the Emergence of New Urban Space in Contemporary China." PhD diss. Yale University, New Haven, CT.

Jie, Fan, and Wolfgang Taubmann (2008). "Migrant Enclaves in Large Chinese Cities," 181–197. In *The New Chinese City*. Hoboken, NJ: John Wiley & Sons.

Kochan, Dror (2016). "Home Is Where I Lay Down My Hat? The Complexities and Functions of Home for Internal Migrants in Contemporary China." *Geoforum* 71: 21–32.

Lefebvre, Henri (1992). *The Production of Space*. Malden, MA: Wiley-Blackwell.

Ma, Laurence, and Biao Xiang (1998). "Native Place, Migration and the Emergence of Peasant Enclaves in Beijing." *The China Quarterly* 155: 546–81.

Marta, Joan, Akhmad Fauzi, Bambang Juanda, and Ernan Rustiadi (2020). "Understanding Migration Motives and Its Impact on Household Welfare: Evidence from Rural–Urban Migration in Indonesia." *Regional Studies, Regional Science* 7(1): 118–32.

Massey, Douglas, Joaquin Arango, Graeme Hugo, Ali Kouaouci, and Adela Pellegrino (1999). *Worlds in Motion: Understanding International Migration at the*

End of the Millennium: Understanding International Migration at the End of the Millennium. London: Clarendon Press.

Mohabir, Nalini, Yanpeng Jiang, and Renfeng Ma (2017). "Chinese Floating Migrants: Rural-Urban Migrant Labourers' Intentions to Stay or Return." *Habitat International* 60: 101–10.

NBSC (2020). "国家数据 [National Data by National Bureau of Statistics in China]."

NHCPRC (2018). "国家卫生健康委员会:2018　中国流动人口发展报告内容概要 [National Health Commission: Briefing of the Report of Chinese Floating Population Development in 2018]." Accessed January 24, 2021 from, http://www.199it.com/archives/813002.html

Roberts, Kenneth (1997). "China's "Tidal Wave" of Migrant Labor: What Can We Learn from Mexican Undocumented Migration to the United States?" *The International Migration Review* 31(2): 249–93.

Shen, Jianfa (2002). "A Study of the Temporary Population in Chinese Cities." *Habitat International* 26(3): 363–77.

Solinger, Dorothy (1995). "The Floating Population in the Cities: Chances for Assimilation?" In D. Davis, ed., *Urban Spaces in Contemporary China: The Potential for Autonomy and Community in Post-map China*. Cambridge: Cambridge University Press and Woodrow Wilson Center Press.

Solinger, Dorothy (1999). *Contesting Citizenship in Urban China*. Berkeley: University of California Press.

Wang, Shaoguang (2008). "Double Movement in China." *Economic and Political Weekly* 43(52): 7–8.

Wang, Yaping (2004). *Urban Poverty, Housing and Social Change in China*, 1st ed. London: Routledge.

Wang, Wenfei, Shangyi Zhou, and Cindy Fan (2002). "Growth and Decline of Muslim Hui Enclaves in Beijing." *Eurasian Geography and Economics* 43(2): 104–22.

Wang, Feng, and Xuejin Zuo (1999). "Inside China's Cities: Institutional Barriers and Opportunities for Urban Migrants." *The American Economic Review* 89(2): 276–80.

Xiang, Biao (2005). *Transcending Boundaries. Zhejiangcun: the Story of a Migrant Village in Beijing*. Leiden: E. F. Brill.

Zhang, Jijiao (2013). "Shifting Two-Tiered Boundaries of Belonging: A Study of the Hukou System and Rural-Urban Migration in China," 136–63. In L. Pries, ed., *Shifting Boundaries of Belonging and New Migration Dynamics in Europe and China*. London: Palgrave Macmillan.

Zheng, Siqi, Fenjie Long, Cindy Fan, and Yizhen Gu (2009). "Urban Villages in China: A 2008 Survey of Migrant Settlements in Beijing." *Eurasian Geography and Economics* 50(4): 425–46.

Part III

Re-creating home away from home

11 Uprooted: living between two worlds—German postwar refugee

Narratives on displacement and exile

Andreas Kossert and Tamar Mayer

Introduction

> They were expelled from their home, but they never arrived in our home.
> They settled here with us, they found refuge in our town, but actually car-
> ried on living in their lost home. Constantly, they spoke about what they
> have lost and nobody in town was willing to listen to them.
>
> (Hein 2005: 35)[1]

Throughout history, refugees have never been welcomed. To this day, refugees
continue to provoke; they are seen by some as disturbing nuisances, illegal
intruders, and a threat to sedentary worlds. Many refugee groups in various
parts of the world know what it means to be forcibly displaced and then to
arrive in a foreign and often hostile environment. More than 14 million German
refugees experienced this uprootedness after 1945. These ethnic German refugees—whose situations are reflected in the above quote by the novelist Christoph
Hein, himself a refugee from Silesia—received an unwelcome, even hostile
reception from their fellow Germans upon their arrival to a "cold home" in
Germany. But for years, hardly anyone in Germany told their stories.

In 2014, the British Museum in London organized a successful exhibition
titled, *Germany: Memories of a Nation*, presenting just 100 objects to cover
2,000 years of German history. It was a carefully selected range of key visuals
explaining major developments in German culture and society, including the
first printed book as well as the Volkswagen Beetle. High-profile media cover-
age in Germany praised the rather small London exhibit as sensational, and
it focused especially on one particular item: a simply crafted wooden handcart (see Figure 11.1), which German refugees from Pomerania used while
fleeing the Russians in 1945 (MacGregor 2014). Many refugees, trying to
escape at the last minute, carried their few personal belongings with them on
the handcart. German society, it seemed, was surprised by this key visual—a
provisional means of transportation that symbolized the uprootedness and
homelessness of refugees. The British curators chose this object in an effort
to better understand Germany and its postwar society. The successful exhibit
crystallized how an external perspective that served as a virtual mirror was
necessary for a deeper understanding of German forced migration and its

DOI: 10.4324/9781003170686-14

Figure 11.1 Handcart from Pomerania, used by refugees fleeing west.

Copyright: Deutsches Historisches Museum/A. Psille.

impact on postwar Germany. In other words, the London project not only exposed this past to the world and to Germany but enabled the German public to reflect on the magnitude of German uprootedness after 1945.

Even though the numbers of uprooted Germans were astounding, the uprooted themselves were never fully acknowledged by German society: it took an exhibit in London to raise awareness of the fate of Germans after the redrawing of Europe's post-1945 borders. The work of an outsider (the exhibit's curator) helped expose this chapter in German history and gave voice to the pains associated with this enormous migration influx. Until the British exhibit in 2014, this story was little known, not only outside but also inside Germany. Within Germany, the influx of refugees was often seen as nothing more than "Germans" arriving back in "Germany." But as this chapter demonstrates, that story was so much more complicated than a simple return "home." Ethnic Germans, now refugees, who shared language and culture with other Germans, were rejected by German society upon their arrival and endured harassment and hostility. These experiences remain yet one more part of German history that cannot be mourned (Mitscherlich and Mitscherlich 2007).

Regardless of the painful past and until very recently, Germans, including many politicians, celebrated what they perceived as the "successful integration" of the German refugees. Over decades, a master narrative had developed, which told the following congratulatory story: a war-torn society managed to successfully integrate millions of homeless refugees despite hardship and sacrifice of the new arrivals and those absorbing them. Politicians as well as many refugee representatives praised the "full integration," calling postwar Germany the new home for every German, including those refugees who just arrived. Yet for decades this master narrative, advanced by the political establishment, illuminated only a narrow understanding of the forced migrants' experiences. It highlighted a materialistic understanding of integration while ignoring the complexity and pains of the arrival process and the ensuing spectacular cultural clashes with other Germans. Given the pains experienced by these German refugees after 1945, it is important to understand what "integration" for refugees means and how this process complicates their national and cultural identity. We use the German case as a way to explore refugee experiences and their political, social, cultural, and material challenges. We contextualize the German refugee experience into a broader European and global perspective, trying to provide answers to universal patterns of refugee narratives on displacement and exile as their lives split between two worlds. All refugees are regularly challenged by questions of belonging, which means constantly shifting between new and old identities.

German refugees after 1945

As a result of World War II and Stalin's politics, the map of Europe changed fundamentally as Germany lost a quarter of its territory, including Silesia, East and West Prussia, large parts of Pomerania, and Brandenburg (see Figure 11.2). All ethnic Germans living in these territories were forced to leave their homes based on a decision made by the Allies, victors over Nazi Germany, which reshaped the borders in Central Europe and shifted Germany's borders westward. Even though Germans called these places home, they were considered the enemy, and therefore the Allied Forces decided to ethnically "unmix" the former German regions and expel the remaining German native population to make a place for the incoming (mostly Polish) settlers.

Ethnically German refugees also came from other regions, which for centuries they called home and were never part of modern Germany. These included Bohemia, Transylvania (or Latvia), which after 1945 became parts of Czechoslovakia, Hungary, Yugoslavia, Romania, Poland, and the Soviet Union. As a consequence of the war, millions were either forced to flee the advancing Red Army or were expelled and deported to the western zones of Germany following the Potsdam Conference, where the future of defeated Germany was negotiated and where the new postwar order was set. At this conference, which took place at the former Imperial Palace Cecilienhof in Potsdam during the summer of 1945, the victorious Allies agreed to transfer millions of people in an "orderly and humane" manner. But this peaceful transfer turned out to be nothing more than wishful thinking: several hundred thousand refugees and expellees were killed during the flight and displaced by either the advancing Soviet army or later at the hand of Polish, Czech, or Yugoslavian militia; others died from disease and starvation along the way and upon their arrival (Douglas 2013).

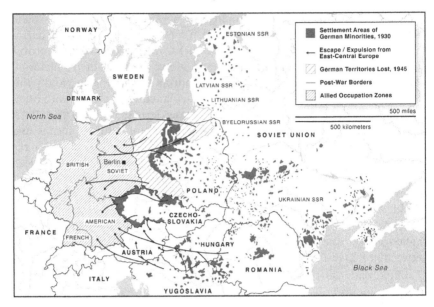

Figure 11.2 German refugees and expellees, 1945–1949.

Based on "Deutsche Flüchtlinge und Vertriebene 1945–1949" (from Kossert 2009).
Data & Sources: Natural Earth Data; "Atlanta-Karte der Besatzungs-Zonen" (Atlanta-Service 1946).
Cartographer: Gregory T. Woolston.

Altogether as many as 14 million Germans arrived in the Allied control zones of Germany after 1945. On top of this, to date, an additional estimated 4.3 million displaced ethnic Germans or those of German descent "returned" to Germany from Central and Eastern Europe, especially from the former Soviet Union. In total, close to 20 million German refugees needed to be absorbed. This means that today every third living ethnic German is either a refugee or the descendent of one. We would expect that the infusion of so many people into the society would have a lasting cultural impact on the absorbing society, but this was not the case here. The cultural impact of ethnic German displacement hardly existed. For example, one could not find restaurants serving regional cuisines, such as Bohemian, Silesian, or Danube Suebian. Apart from some folkloristic niches founded by refugee organizations themselves, such as the traditional Landsmannschaften trying to keep the cultural legacy of their home regions alive, there were no serious attempts to preserve their local cultures, customs, or unique German dialects.

Nevertheless, the impact of this influx on German society was visible and everlasting in two areas in particular: the religious composition of Germany and the housing stock. The once-homogenous cities and villages were forced to absorb ethnic Germans of different religious affiliations, and this led at times to deep divisions along religious sectarian lines. After 1945, Silesian Catholics resided in the entirely Protestant Frisian Islands on the North Sea coast; Lutheran East Pomeranians found themselves in the deeply Catholic Bavarian countryside. The homeless Germans from the east, who settled in Germany, arrived with no hope of returning to their homes there, thus challenging local societies and their century-old traditions and habits (Beck 2002). For example, the Catholic parish of Bergen, which covers the entire island of Rügen on the Baltic coast and was always predominantly Lutheran, counted 1,254 members in 1934. In October 1946, there were nearly 10,000 members, a rise of 786% (Holz 2003: 398). Another example comes from western Germany, the Rhineland, which was always mostly Catholic. Here it was the other way around: Protestant refugees arrived in vast numbers and helped shape a much more religiously diverse society. The Protestant Church in the Rhineland almost doubled, increasing from 2.2 to 4.1 million members within 15 years between 1949 and 1964 (Goltz 1966: 8).

The physical appearance of many German cities, old villages, and town centers had changed as well, as hundreds of thousands of housing units, all in the same architectural style, were built during the 1950s and 1960s to absorb the millions of expellees. This response to the housing shortage dictated the architectural chapter of postwar refugee history. The very existence of such a shortage symbolizes the massive changes that German society faced after 1945, when millions of homeless refugees tried to find new homes. The newly built houses provided a safe haven for those who spent many years in provisional housing, which included refugee camps or, often, cohabitation in the houses of local residents. Initially, right after the war, 14 million refugees needed to find shelter. By Allied decree, local house owners were ordered to accommodate refugees, whether they liked it or not, and very often were forced to share their house or apartment for years with newcomers who were total strangers to them. After the establishment of the West German state in 1949, the government initiated a program that provided basic government support and interest-free loans, the so-called Lastenausgleich, to help refugees reestablish their lives. It was never a full compensation for their material losses or their emotional suffering, but at least the government funds provided some help for them to start a new life.

But material support and new housing on the village peripheries could neither heal nor compensate for the loss and dislocation that mass displacement had caused. Refugees had regularly experienced angst of expulsion, traumas of forced labor and deportation, and of physical and sexual violence. For decades, many Germans were not aware of these pains, and those who were often turned a blind eye. Not only were these refugees forced to leave their homes, but upon arriving in postwar Germany they were met with great resentment from their host co-national Germans. Postwar refugees found themselves between a rock and a hard place. It did not matter that they spoke German, or that they too supported the National Socialist Party during the 1930s and into the early 1940s. They were still rejected and in many ways never fully absorbed, and they could never feel fully at home.

That is why after 1945, the German world was divided between natives and refugees. The very presence of millions of expellees reminded many West German natives of a mutually lost war that most people wanted to forget. Only a few years earlier, most Germans hailed the Führer. But now, facing defeat, there were hardly any signs of solidarity. On the contrary, many native

Germans rediscovered their regional identities. They wanted to be Badeners, Suebians, Lower Saxons, or Holsteiners. Sometimes they blamed the expellees for Hitler's rise to power. Some thought that these new arrivals, who had been forced to leave their homes, must have been expelled, and were thus now ostracized, because they were dedicated to the ruthless Nazi cause; such thinking often turned into the general accusation that they got what they deserved. Refugee narratives often recall how they felt when exposed to such verbal abuse and discrimination. At the same time, old resentments fueled new versions of Nazi propaganda, as people from "the East" were looked down upon and denounced as "Polacks" or "Gypsies." The racial ideology so present in the Nazi party did not stop at the end of the war. For millions of refugees, the racism that they experienced came as a shock, which they could not have imagined after having survived an often-dramatic escape or even expulsion for being German.

The refugees depended heavily on other people's mercy in their hostile environment. In the northwest of Germany, in the Emsland region, there was a saying: "The three great plagues are wild boars, the potato beetle [or Colorado beetle] and the refugees" (Meißner, quoted in Eiynck 1997: 495). To outside observers entering Germany, the country seemed to have suffered the likes of a biblical plague, but it didn't influence the course of history. Allied troops had to use military force in order to make room for refugees in natives' homes. Sometimes residents were only willing to accommodate refugees after British and American soldiers threatened them with guns. Most refugees headed to the rural areas because the big cities were mostly destroyed. Having experienced uprootedness and destruction, they joined rural communities that hardly suffered any material damage in the war.

Entering these communities and encountering hostile reactions was the first experience many of the refugees had faced after their displacement. The stark contrast between locals and refugees was witnessed by all, including US Army officers who noted that "in Bavaria or perhaps the whole of Germany there is no difference between a Nazi and Anti-Nazi, Black and Red, Catholic or Protestant. The only difference is between natives and refugees" (Erker 1990: 384). In a Bavarian village in March 1947, a handwritten poster circulated with the message: "Refugees out of our village! Treat them with a whip and not with providing accommodation—the human trash from Sudeten! Long live our Bavaria!" (Erker 1990: 387). The Bavarian Catholic Church was even afraid that Bohemian Catholics might come together in poverty and despair to foster a "religious bolshevism" within the rural and conservative Bavarian parishes (Erker 1990: 387, 397).

In the Soviet Occupied Zone, which in October 1949 became East Germany (German Democratic Republic, GDR), the refugees' arrival was equally challenging. But unlike the Western Occupied Zones, the Communist government—the so-called first German anti-Fascist state—adopted a different approach of how to deal with the incoming 4.3 million refugees and expellees in their own territory. Initially, they experienced the same hardship and homelessness, as well as an unwelcoming reception by the East German local population. But quickly, the East German government took a more radical step. Being totally dependent on the Soviet political and military power, it feared nothing more than a separate refugee group identity that would challenge the Soviet and East German narrative of the Soviets having liberated Germany from the Nazi regime.

As a result, millions of ethnic German refugees from Eastern Europe, now under Soviet control in East Germany, experienced their exodus quite differently. Already in the late 1940s, the East German government eliminated the term "refugee" and replaced it with the term "resettlers." All refugees disappeared from official statistics as if they never existed. And in the early 1950s, even the term "resettlers" was gone. Finally, refugees officially were called "new citizens." The East German state tried to eliminate any traces of a separate German refugee identity. They forced the refugees to be silent about their loss and experiences. In public, they were forbidden to even mention what happened to them. Often, private refugee reunions were screened by the GDR secret police; many ended up in prison just for remembering their homelands. The leading communist party, the SED, introduced a "strict assimilation concept" (Amos 2009: 9, 18), enforcing the taboo on publicizing the collective experiences of millions of their citizens. And to make things worse, they received no substantial material compensation or support. Finally, for many refugees in the GDR, hiding their biographical details caused additional hardship. Only after reunification in 1990 were the ethnic German refugees and former GDR citizens able to reconcile their experiences by being part of a new public discourse. For them it was the first time, after the more than 40 years since losing their homelands, they were able to speak publicly about their experiences.

Living between two worlds: narratives on displacement and exile

The arrival of the refugees from the east into the Occupied Zones of Germany was accompanied by a major cultural clash. Not only did they lose everything that was once familiar—their sense of security, parental homes, the specific dialect, and flavors and smells of their family cuisine—but they came to rural areas where they were *othered*. They settled initially in areas with no visible physical war damage, but the impact of that war defined who they would become. They were troubled by questions about belonging, identity, and roots, to which the receiving community had no access. There was a great dissonance between being "repatriated" to Germany and never feeling at home there among fellow Germans.

When we listen to individual refugee stories, it becomes clear that "arriving" in the purest sense cannot be forced by official statements and cannot be heralded by claims of successful integration. Many refugees lived for decades in a transitional world between their homes of origins and their present locations. When old refugees in Germany were talking about "home," they meant their old places in Bohemia, Transylvania, Silesia, or East Prussia. Even today, after 75 years, some of the refugee generations still struggle with how to define "home." Feeling alien or not at home is very often a fate that refugees could not control. The Polish author and recent Noble Prize–winner Olga Tokarczuk, born in 1962, is a daughter of a Polish refugee family in former German Silesia who has written about her experiences, yet her words are just as relevant for German postwar refugees and their families.

> Even though I do not belong to that generation, which survived all this, the history of my family could be a good addition to it. In private memories, in family narratives, the drama returned with the stubbornness of a nightmare—lost family ties, lost family members, burnt documents, an undefined nostalgia for old birthplaces, the fascination for material things, which seemed to be more real in times of chaos then human beings and their memories; the feeling of homelessness in this world, … and the feeling to have suffered from injustice (Tokarczuk 2004: 8–9).

Olga Tokarczuk not only reflects a universal experience but also an often-ignored and hardly recognized essential part of German postwar history. The traumas of most refugees in the world were very often the result of the violence and humiliation they endured. They kept these stories behind closed doors and did not share them with the general public. The weight of the past was too heavy for many to carry. Very often, older refugees could not live with these memories, etched in their souls and bodies, and collapsed both physically and mentally. Many died because of homesickness, yet this condition has never been included in any medical handbook as a reason for death. The novelist Christa Wolf, herself a refugee, described the impact of homesickness soon after World War II in her novel *Kindheitsmuster*:

> For the elderly—for those who have talked for years about death as a way? Yes! to provoke the younger folk—it was time to be silent—because now what followed was their death, they felt it immediately, they aged in weeks as much as they would have normally in years, … but for various reasons, could it be called typhus, hunger? Or simply homesickness, which was a perfect reason to die.
>
> (Wolf 2002: 412)

Experiences of humiliation and an unwelcoming reception, as well as mourning over lost homelands, were hidden from view. The historian Hans-Ulrich Wehler has called it the "privatization of mourning." But stories are retold from one generation to the next and *are* kept alive. Sometimes, they are told only with the death of parents or grandparents, and it is then that refugee descendants meet with the traumas in their family history. Writing about his mother who was expelled from the Sudetenland, the journalist Martin Tschechne wrote:

> It wasn't much my mother left behind, when she died six years ago … some clothes, letters, some jewelry—and a worn out copy of a letter from spring 1945. All German inhabitants were told, first in Czech, then in German, to gather in the market square with one piece of luggage. They

were ordered to make sure the homes they left behind were clean, and to put new sheets on the beds for the new inhabitants. My mother was expelled. She never got over the loss of her home. It was, I am not exaggerating, a pain, she never lost for a single day for the rest of her life and I am quite surprised, how little she told me and my brother about it. But we haven't asked her either. Because we were not interested? Because it might have been too bitter? Or were we trying to protect her feelings?

(Tschechne 2004)

The 14 million Germans who arrived in what was left of Germany after World Wat II were not given alternative destinations or any options of going elsewhere. A refugee from Silesia summarized their long-term dilemma: "The expulsion destroyed entire families, many families collapsed mentally too. We long ago overcame the material loss, but we never really came to terms with the actual loss of our home, where we were born" (Leuchtenberger 1997: 418).

This collective experience has shaped Germany much more than is visible at first glance. The refugee legacy survives in many families and could ideally resonate and yield empathy for present-day refugees. Rupert Neudeck, the founder of the rescue campaign of the ship *Cap Anamur*, which saved Vietnamese boat people in the 1980s, was a German refugee himself and remembers his flight from Danzig at the end of World War II. "The old images lasted and shaped my future life—along with the important message: most peoples' biographies are actually linked to migration or to being refugees. Even those, who are sure that their families are rooted where they now live—should not be too sure or complacent. It could always happen that they or their descendants will be forced to leave. Because there is a refugee in all of us" (Neudeck 2016: 7f, 21).

Is there a successful integration? Reflections on refugees and sedentary societies

Although the story of the German refugees in the post-World War II era may be unique in its own way, the contours of rejection felt by the refugees seem to be similar to that of other refugees, even at present. Therefore, the German experience can provide answers to more universal questions faced by all refugees; leaving home, mostly for good, is a dramatic turning point in peoples' lives and very often beyond imagination. Outsiders can hardly perceive what it means to leave *everything* behind—except, of course, one's life. Flight and fleeing are not an adventure. What was "forgotten" at home can never be retrieved. In the moment, even refugees themselves are rarely aware that theirs is a journey from which there is no return. What does it mean for a farmer to leave his cattle behind? What does it mean for old people to bid farewell for good to their house, their neighbors, and family? Flight and displacement interrupt an unwritten contract with all ancestral traditions. In the blink of an eye, material things gathered over time become worthless. Wills, investment, property, and saving accounts all become meaningless as people are forced to leave. Refugees must not only abandon their material belongings but also their ancestors. Graveyards and family graves are left behind, overgrown and neglected as time goes on. And finally, what does it mean for parents or grandparents who are left behind because they cannot embark on the demanding and uncertain trip—when children and grandchildren flee for good, and in many cases without further contact?

Uprooted: This word metaphorically unfolds an enormous dynamism as it reflects various concepts and conflicts. Uprooted refugees are confronted with sedentary societies whose inhabitants still hold the privilege of having a home. Those inhabitants are the ones who decide the refugees' fate. Refugees depend on their decisions, which determine whether they are welcomed, accepted, or rejected. After being forcibly displaced and arriving in their host societies, refugees are often confronted with hostile reactions that challenge their identity anew. The writer Reinaldo Arenas experienced his exile in the United States as a constant conflict between *here* and *there*. In an interview for the *New Yorker* in 1983, he was asked if New York was a good place for him to write. He answered that his exile from Cuba was only a physical one:

> Every person who lives outside his context is always a bit of a ghost, because I am here, but at the same time I remember a person who walked those streets, who is there, and that same person is me. So sometimes I really don't know if I am here or there.
>
> (quoted in Slater 2013)

For Reinaldo Arenas, the inner ambivalence caused by his uprootedness remains, and he has no means to overcome it. Like him, many refugees live in permanent exile, unable to take root again in new places. He shares his experience with many other refugees, who even after decades are not able to find a new home. Their confusion often results in the physical disorientation described by a female German refugee from Sudetenland after her arrival in Munich in 1945:

> I felt tremendously homesick and longed to be back in Troppau. Even many years later, I still got lost in the streets of Munich because I thought if I turned around at the next corner, I would be in Troppau. Even today, it happens to me sometimes.
>
> (Wagnerová 1990: 41)

As mentioned earlier, the German master narrative praised the so-called successful integration of millions of postwar refugees. By taking a closer look at the German case, a counter-narrative can be developed that questions the myth of successful integration. Refugee worlds mean turmoil, disarray, and waning certainties. The German case study provides an example that demonstrates a stark contrast between the master narrative and individual refugee stories.

Those who could always be certain about their home never needed to ask questions about belonging, identity, or roots—and those who lost their home were constantly forced to raise such questions. The word *home* itself carries significant weight, and the word's politically and emotionally charged German translation (*Heimat*) is very often abused, exploited, and caricatured. But in English there is no exact equivalent to the German word *Heimat*. As Jean Améry once said, in a statement that rings true for all refugees: Home very often only gains importance after you become uprooted (Améry 1997: 81). This is why refugees so often idealize their lost homelands. For them, home becomes an ultimate projection. Initially, refugees long to return but after realizing that this is no longer possible, their lost home becomes an idealized space.

As Reinaldo Arenas has emphasized, a refugee does not live in his chosen home but exists as a ghost between two worlds—in both his place of origin and in exile at the same time. Arenas never overcame his homesickness, which sometimes grew even stronger than his pragmatic understanding of having found refuge.

> I have realized that an exile has no place anywhere, because there is no place, because the place where we started to dream, where we discovered the natural world around us, read our first book, loved for the first time, is always the world of our dreams. In exile one is nothing but a ghost, the shadow of someone who never achieves full reality. I ceased to exist when I went in to exile, I started to run away from myself.
>
> (Arenas 2000: 293)

His case exemplifies the fact that *arriving* does not necessarily mean being at home. "I feel like a tree uprooted. You can plant it somewhere else, but it will never be the same," said Ariel Sabar's father after he was expelled from Kurdish Iraq in 1951 (Sabar 2009: 284). Tsering Wangmo Dhompa was born in North India to a Tibetan mother and witnessed her mother's suffering in exile: "All of her exiled life she waited to return home," Dhompa recalls, remembering her mother's words. "She spoke of exile as something that would be expunged over time. When *this* is over, we can go home. She waited from year to year. She carried a hope that if we waited long enough, *this* would end" (Dhompa 2013: 1). Another refugee, André Aciman, had to leave the Egyptian city Alexandria as a child. "What makes exile the pernicious thing it is not really the state of being away, as much as the impossibility of ever *not* being away—not just being absent, but never being able to redeem this absence" (Aciman 1999a: 10). It is the transitional nature of being a refugee. Being uprooted means a turbulent and insecure existence in exile and in between. "It reminds me of the thing I fear most," Aciman continues,

> that my feet are never quite solidly on the ground, but also that the soil under me is equally weak, that the graft didn't take. In the disappearance of small things, I read the tokens of my own dislocation, of my own

transiency. An exile reads change the way he reads time, memory, love, fear, beauty: in the key of loss.

(Aciman 1999b: 22)

Reviewing the German postwar case can help to better understand the challenges refugees are still facing today. Moreover, the term "integration" needs to be critically questioned—starting with its very meaning. In the context of refugees' lives, *integration* is all too easily applied in material terms, as if its "success" comes from attaining a job, shelter, or a house in the new environment. Instead, as we have seen, the counter-narrative of individual stories often reveals *disintegration* or *exile* as refugees sacrifice a huge part of their existence. They suffer a tremendous material and immaterial loss, including their culture and identity of origin. Again, after their arrival their rescued identities are challenged by foreign and often hostile societies. Refugees very often feel uprooted. It can change over time, yes, and integration is possible, but there is no guarantee. Often, refugees remain in a mental exile until the end of their lives, invisible to outsiders. They feel as though they have yet to arrive and do not belong, while the surrounding world easily mistakes their existence as perfect integration. For that reason, it remains crucial to listen to refugees' narratives. The handcart used by German refugees in 1945 symbolizes not just another historic case study of uprootedness and homelessness of refugees but also reminds us of the ongoing challenge refugees face in the twenty-first century.

Conclusion

As this chapter shows, the German postwar example provides a perfect case study that illustrates the extent to which refugees live between two worlds: the one they lost and the one in which they were forced to settle. As it was shown in the case of German refugees, finding their space in a hostile environment (even if only partially hostile) is a distressing and often a painful process; it can take up a lifetime, or even beyond by having an impact on the second or third generation. Integration is neither written in stone nor a self-fulfilling prophecy. Even displaced ethnic Germans faced tremendous difficulties in postwar Germany. This is why the term "successful integration" presents its own limitations and should never be taken for granted, especially when we fully acknowledge and do justice to the ramifications for mental health. For that reason, it remains fundamentally important to listen to refugees' narratives. The handcart used by German refugees in 1945— described at the beginning of the chapter as the key visual feature in the exhibition at the British Museum, and as a revelation even for the German public—therefore symbolizes not just another historic case study of uprootedness and homelessness of refugees, but reminds us of an everlasting challenge of how we look at forced displacement and the people affected by it also in the twenty-first century.

Note

1 All translations from German to English are by Andreas Kossert.

References

Aciman, André (1999a). "Permanent Transients," 9–14. In André Aciman ed., *Letters of Transit. Reflections on Exile, Identity, Language and Loss.* New York: The New Press.

Aciman, André (1999b). "Shadow Cities," 15–34. In André Aciman ed., *Letters of Transit. Reflections on Exile, Identity, Language and Loss.* New York: The New Press.

Améry, Jean (1997). "Wieviel Heimat braucht der Mensch?" 74–101. In Jean Améry ed., *Jenseits von Schuld und Sühne. Bewältigungsversuche eines Überwältigten,* 3rd ed. Stuttgart: Klett-Cotta.

Amos, Heike (2009). *Die Vertriebenenpolitik der SED 1949 bis 1990.* München: Oldenbourg.

Arenas, Reinaldo (2000). *Before Night Falls.* Dolores M. Koch, trans. New York: Penguin Books.

Beck, Wolfhart (2002). *Westfälische Protestanten auf dem Weg in die Moderne. Die evangelischen Gemeinden des Kirchenkreises Lübbecke zwischen Kaiserreich und Bundesrepublik.* Paderborn: Ferdinand Schöningh.

Dhompa, Tsering Wangmo (2013). *A Home in Tibet.* New Delhi: Penguin Books.

Douglas, Raymond (2013). *Orderly and Humane: The Expulsion of the Germans after the Second World War.* New Haven, CT: Yale University Press.

Erker Paul (1990). "Revolution des Dorfes? Ländliche Bevölkerung zwischen Flüchtlingszustrom und landwirtschaftlichem Strukturwandel," 367–425. In Martin Broszat, Klaus-Dietmar Henke and Hans Woller eds., *Von Stalingrad zur Währungsreform. Zur Sozialgeschichte des Umbruchs in Deutschland.* München: Oldenbourg.

Goltz, Fritz (1966). *Veränderungen in der evangelischen Kirche im Rheinland durch die Vertriebenen und Flüchtlinge.* Neuß: Rheinischer Heimatbund. Verlag Gesellschaft für Buchdruckerei.

Hein, Christoph (2005). *Landnahme.* Frankfurt/Main: Suhrkamp.

Holz, Martin (2003). *Evakuierte, Flüchtlinge und Vertriebene auf der Insel Rügen 1943–1961.* Köln/Weimar/Wien: Böhlau.

Kossert, Andread (2009). *Kalte Heimat. Die Geschichte der deutschen Vertriebenen nach 1945.* Munich: Pantheon.

Leuchtenberger, Johannes (1997). "Interview," 418. In Andreas Eiynck eds., *Alte Heimat—Neue Heimat. Flüchtlinge und Vertriebene im Raum Lingen nach 1945.* Lingen: Emslandmuseum.

MacGregor, Neil (2014). *Germany: Memories of a Nation.* London: Penguin Books.

Meißner, Manfred (1997). "Interview," 495. In Andreas Eiynck ed., *Alte Heimat—Neue Heimat. Flüchtlinge und Vertriebene im Raum Lingen nach 1945.* Lingen: Emslandmuseum.

Mitscherlich, Alexander and Margarete Mitscherlich (2007). *Die Unfähigkeit zu trauern. Grundlagen kollektiven Verhaltens,* 27th ed. München: Piper.

Neudeck, Rupert (2016). *In uns allen steckt ein Flüchtling. Ein Vermächtnis.* München: C.H. Beck.

Sabar, Ariel (2009). *My Father's Paradise. A Son's Search for His Family's Past.* Chapel Hill, NC: Algonquin Books.

Slater, Ann Tashi (2013). "The Literature of Uprootedness: An Interview with Reinaldo Arenas." *New Yorker*, May 12. Accessed January 21, 2021, http://www. newyorker.com/books/page-turner/the-literature-of-uprootedness-an-interview-with-reinaldo-arenas

Tokarczuk, Olga (2004). "Eine Freske menschlicher Schicksale. Vorwort." In Helga Hirsch ed., *Schweres Gepäck. Flucht und Vertreibung als Lebensthema*. Hamburg: Edition Körber.

Tschechne, Martin (2004). "Leid, das nicht vergeht. Helga Hirsch spricht mit den Nachkommen der Vertriebenen—ihr Buch dient der Wahrheit und der Versöhnung." In *Die Zeit*. Literatur-Beilage (November 2004).

Wagnerová, Alena (1990). *1945 waren sie Kinder. Flucht und Vertreibung im Leben einer Generation*. Köln: Kiepenheuer & Witsch.

Wolf, Christa (2002). *Kindheitsmuster*. München: Sammlung Luchterhand.

12 Palestine in exile

Blurring the boundaries and re-creating the homeland

Anne Irfan

Introduction

Modern Palestinian history is defined by displacement. In 1948, more than three-quarters of the Palestinian Arab population were forced into exile, becoming dispersed across the Middle East and the wider world. Numerous further displacements followed: the most prolific one in 1967, and the most recent one following the outbreak of the Syrian conflict in 2011. The duration of the Palestinian exile means that their case epitomizes many of the extremities of protracted displacement: it is the longest-running refugee crisis in modern history; it has entailed multiple displacements; and until recently, the Palestinians were the single largest recognized refugee population in the world[1] (ISI, ASKV, and ENS 2019; PCBS 2019; UNHCR n.d.; UNRWA n.d.a). Yet while these characteristics could make the Palestinian case into the archetypal story of forced migration, more often their experiences of displacement have been exceptionalized within the field of migration and refugee studies (Hammami and Tamari 1997; Kagan 2009; Albanese and Takkenberg 2020).

Palestinian refugees' "exceptionalism" appears affirmed by the existence of a distinctive regime for them, via the United Nations Relief and Works Agency (UNRWA). While all other refugees worldwide receive services from the UN High Commissioner for Refugees (UNHCR), Palestinian refugees in the Middle East are served instead by UNRWA. This makes them currently the only group in the world to have their own designated UN agency, and consequently, the only people excluded from UNHCR's mandate (Akram 2002; Kagan 2009).

Yet despite the distinctiveness of the UNRWA setup, the Palestinians' so-called exceptionalism is often overstated. Although the longevity of Palestinian exile is undoubtedly unusual, protracted displacement per se is not uncommon. Defined by the UN as a situation whereby at least 25,000 refugees of the same nationality have been in exile for at least five consecutive years, protracted displacement accounted for as much as 78% of the global refugee population in 2018. While Palestinian refugees have lived in exile for more than 70 years, other cases of displacement have also lasted decades: there has been a sizeable Burundian refugee population in Tanzania for

DOI: 10.4324/9781003170686-15

nearly 50 years, for example, while elsewhere populations of Vietnamese, Afghan, and Iraqi refugees have all lived in continuous exile for more than 40 years (UNHCR 2018). The reality, then, is that the Palestinian refugee case is unusual without being entirely exceptional.

With this in mind, this chapter examines how Palestinian refugee history can inform broader understandings of forced migration. It is premised on the contention that, far from being exceptional, Palestinian displacement can be highly instructive as a case study for rethinking paradigms around migration in general and forced migration in particular. Specifically, Palestinian refugee experiences challenge two paradigmatic binaries in the field. First, the underlying fact of Palestinian statelessness means that their various migrations have been engendered by structural conditions that cannot be reduced to the "push/pull" factors of particular events, or to the reductive dynamics of countries "sending" or "receiving" migrants. Over the last 70 years, the factors driving the direction of continual Palestinian movement have been so entwined, and in some cases mutually dependent, that it is near impossible to categorize them separately. Instead, continual Palestinian migration in the modern era should be examined as a long-term structural phenomenon.

Second, and perhaps most compelling, Palestinian refugees themselves have challenged the perceived binary between homeland and exile by working to "re-create" Palestine outside its historical borders, most notably in its refugee camps. As a result, Palestinian identity and consciousness of the homeland has persisted and remains strong even among those generations born in exile—thus complicating the assumption that homeland and exile are always mutually exclusive. By examining these two challenges to conventional binaries, this chapter invokes Palestinian refugee history in order to reconceptualize broader understandings of forced migration.

The origins of Palestinian exile

The Palestinians' long history of displacement began in the late 1940s, with events known in Arabic as the Nakba (catastrophe). Through large-scale violence and expulsions, the Palestinian people were dispossessed, displaced, and dispersed, as the new state of Israel was established in 1948 on 78% of Palestine. The successful resettlement of Jewish refugees and migrants as Israeli citizens was, thus, tied from the outset to the dispossession of the Palestinian people. The Israeli state's establishment not only left them stateless but also resulted in the exile of the majority; during the period between 1947 and 1949, around 750,000[2] Palestinians fled their homes or were violently expelled by Zionist militias (UN ESM 1949; Pappe 2006; Khalidi 2020). Many Palestinians who became refugees at the time later recalled feeling that they had been literally "replaced" by incoming Jewish migrants from Europe. Salman Abu Sitta, a Palestinian refugee from Beersheba, described his confusion as a teenager in 1948: "Who were these people? They were not Arab Jews. People said they were a motley assortment of Jews imported from across the sea"[3] (Abu Sitta 2016: 61).

Most of the Palestinians who left did so in the belief that their flight would be temporary. Accordingly, they did not prepare for a long exile or even a long journey. Salah Khalaf, who was 14 when his family fled Jaffa, recounts in his memoir:

> Confident of a speedy return, [my parents] left all their furniture and possessions behind, taking with them only the bare necessities. I can still see my father, clutching our apartment keys in his hand, telling us reassuringly that it wouldn't be long before we could move back.
>
> (Abu Iyad and Rouleau 1981: 12)

Shafiq Al Hout, whose family also fled Jaffa in 1948, similarly recalls standing on the boat leaving Palestine and thinking: "No doubt we would be going back. Two or three weeks at the most and we would be back" (Al Hout 2011: 12). Such recollections are commonplace. Jean Said Makdisi, whose family lost their home in Jerusalem in 1948, has observed how long it took for many Palestinian refugees "to realize that there was to be no return. Usually it was years before they came to this understanding" (Said Makdisi 2005: 34).

Among those forced to flee, large numbers sought sanctuary in the two parts of Palestine that were not absorbed by Israel in 1948: the West Bank, including East Jerusalem, and the Gaza Strip. Others fled Palestine altogether, with the majority going to the neighboring states of Jordan, Lebanon, and Syria. Each of these states received around 100,000 Palestinian refugees from 1947 to 1949, while another 13,000 Palestinians fled to Egypt (Y. Sayigh, 1997; Khalidi 2007; R. Sayigh 2007; El-Abed 2009: 1; Al Hardan 2016; Gabiam 2016: 19). Fewer still sought refuge elsewhere in the Middle East and North Africa, while a small number with the means and connections were able to build new lives in Europe and North America.

Less than a quarter of the Palestinian population were able to avoid exile and remain in what became Israel, although many of them were nevertheless internally displaced (Y. Sayigh 1997: 37–38). While this group would eventually become Palestinian citizens of Israel, they shared the fate of their refugee brethren in terms of dispossession. They were placed under Israeli martial law from 1948 to 1966, and as was the case for those in exile, their Palestinian passports became defunct (Pappe 2011). In other words, their national identity as Palestinians was stripped of any formal recognition in the nation-state order. The Palestinian people, thus, entered a condition of statelessness. They have remained in that condition ever since.

Structural statelessness and continual displacement

Palestinian displacement is best understood not as an event but as a structure. As a stateless people surviving in the era of the nation-state, the Palestinians have lived in a condition of chronic instability ever since the Nakba. Lacking the protection of their own state, they have been exposed to the whims of the various governments hosting them and are accordingly always vulnerable to further displacements. The ramifications of their state-lessness in this regard have been compounded by the political instability of the Middle East, where the vast majority of Palestinian refugees have lived since the Nakba.

Palestinian displacement has, therefore, functioned as a continual process, repeated many times in the decades after the original forced migration of the Nakba. Further Israeli-induced displacements of Palestinians occurred in the Gaza Strip in 1956 (Filiu 2014: 96–106); and on a larger scale from both Gaza and the West Bank during the 1967 War, when more than 300,000 Palestinians were forced into exile, around half for the second time (Forsythe 1983; Rempel 2006; Filiu 2014: 127). Fifteen years later, the Israeli invasion of Lebanon and siege of Beirut created further displacements, as thousands of Palestinian homes were destroyed (Khalidi 2020: 143–62). These events repeated themselves, albeit on a smaller scale, in the 2006 Israel-Hezbollah War, when military aerial bombardments led to the internal displacement of around 16,000 Palestinian refugees (UNRWA 2006).

Yet it was not only Israeli military actions that engendered repeated Palestinian displacements after the Nakba. In fact, since 1948, the majority of Palestinian forced migration has taken place within the Arab world (Albanese and Takkenberg 2020; see Table 12.1). Often, this has been in the context of war or internecine conflict, such as during the Lebanese Civil War (1975–90), the two Iraq wars (1990–91 and 2003–11), and the ongoing con-flict in Syria (2011–present).[4] In all these conflicts, the Palestinians were far from the only group displaced; the 2006 Israel-Hezbollah War displaced around a million people within Lebanon, the vast majority of whom were Lebanese citizens (Human Rights Watch 2007). Meanwhile, the recent Syrian conflict has created a Syrian refugee population of 5.6 million around the world, not including those internally displaced (UNHCR n.d.), while more than 2 million Iraqi citizens became refugees during the First Gulf War of

Table 12.1 Major Palestinian cross-border displacements (internal displacements not included)

Year(s)	Country	Reasons for flight	Numbers*
1947–49	Palestine	Expulsions & war	750,000
1967	West Bank & Gaza Strip	War & Israeli occupation	300,000
1970–71	Jordan	Expulsion, PLO war with Jordan	100,000
1991	Kuwait	Expulsion	400,000
1995	Libya	Expulsion	30,000
2003	Iraq	War	Data unavailable
2011–present	Syria	Civil war, attacks by Daesh/ISIS	120,000

* Numbers are approximate.

Sources: UN ESM 1949; Forsythe 1983; Viorst 1984; Lamb 1995; Shiblak 1995; Pappe 2006; Rempel 2006; Al-Nakib 2014; Filiu 2014; Khalidi 2010; UNRWA 2019a; Khalidi 2020.

1990–91 (Galbraith 2003; UNHCR 2003). Nevertheless, Palestinians have been particularly vulnerable to these instances of forced migration in two regards: as noncitizens of the country of origin (e.g., Syria or Iraq); and as a stateless people (Fiddian-Qasmiyeh 2015; Almustafa 2018).

Given their forced migration, Palestinian refugees have sometimes been unable to access essential aid in the immediate wake of displacement. International responses to the Syrian crisis, for example, are designed to reach *Syrians*, meaning that Palestinian refugees from Syria—who do not have Syrian citizenship—can end up falling through the gaps (Abu Moghli et al. 2015). Their suffering can be particularly acute if they find themselves in countries where UNRWA does not operate, such as Turkey or Egypt.[5] This exclusion has been a central part of many Palestinians' experiences of further displacement. Mohamad Jabeti, a Palestinian forced to flee Damascus for Germany in 2015, told journalists, "I feel like a second-class person. ... I don't have the same rights as everyone else [even though] I am in the same situation as other Syrians" (Bolongaro 2016).

Moreover, the Palestinians' statelessness has made many governments reluctant to accept them as twice-over refugees, for fear that without their own country they are more likely to settle permanently in the host state. Thus, Jordan, which already hosts a significant number of Palestinian refugees from 1948, closed its borders to Palestinian refugees from Iraq in the wake of the 2003 US invasion, while accepting Iraqi citizens driven into exile by the same war. Syria followed suit in 2006, going on to deport hundreds of Palestinian refugees from Iraq back across the border (Cohen 2008). More recently, first Lebanon and then Jordan have sealed their borders to Palestinian refugees from Syria, while continuing to accept Syrian citizens fleeing the same conflict (Meier 2016; UNHCR 2017).

Alongside these conflict-induced displacements, Palestinians have also been the target of numerous expulsions since 1948. Strikingly, most have been carried out at the behest of Arab leaders, who continue to rhetorically support the Palestinian cause while mistreating and even deporting the people. Three such cases are particularly prominent. From 1970 to 1971, the Jordanian government went to war with the Palestine Liberation Organization (PLO), on account of the latter's perceived challenge to its sovereignty. Jordanian victory resulted in the expulsion of as many as 100,000 Palestinians from the country, mostly PLO cadres and their families (Viorst 1984: 86). Twenty years later, the Kuwaiti government expelled the country's established Palestinian community of just under 400,000 people, in retaliation for the PLO's support for Saddam Hussein during the Iraqi invasion (Al-Nakib 2014: 23). Salman Abu Sitta, who became a twice-over refugee when forced to flee Kuwait at the time, recounts telling friends on his arrival in Jordan: "I am a Palestinian. It is my umpteenth displacement." (Abu Sitta 2016: 254). Then from 1993 to 1995, Muammar Qadhafi showed his opposition to the Oslo Accords, which the PLO had signed with the Government of Israel, by expelling around 30,000 Palestinians from Libya (Lamb 1995; Shiblak 1995; Khalidi 2010).

All this meant that displacement was an iterative process for many Palestinians, with their statelessness leaving them devoid of either legal state-based protection or a country to which they could return. Instead, they were often forced to move continually between the various Arab host states. In the early 1970s, for example, the majority of Palestinian expellees from Jordan went to Lebanon, largely because of the latter's weak central state (PLO official Shafiq Al Hout [2011: 106] described it as "open, like a garden without a fence"). This community would go on to face the threat of further displacement shortly afterward, when the Lebanese civil war began in 1975. Twenty years later, many Palestinian expellees from Kuwait sought refuge in Jordan, where around 250,000 of them held citizenship (Ibrahim 1991). Lebanon and Jordan have, therefore, hosted not only Palestinian refugees from 1948 but also large numbers of further-displaced Palestinians in more recent years.

This trend has been exacerbated in the context of the recent Syrian crisis. While both countries have closed their borders to Palestinian refugees from Syria—Jordan in 2013 and Lebanon in 2014—they nevertheless host significant numbers who entered before these entry bans were put into effect. As of 2019, there were around 17,000 Palestinian refugees from Syria in Jordan and 30,000 in Lebanon (UNRWA 2019b). In view of the earlier expulsion of Palestinians from Jordan, and the numerous displacements of Palestinians within and from Lebanon, this situation could be described as something of a revolving door. With no structural power, the collective Palestinian refugee population face being continually "transferred" between and across the Levantine states.

To return to this chapter's main contention, Palestinian refugee history thus gives the lie to conceptualizing displacement as a necessarily singular event. In the Palestinian case, it makes far more sense to frame displacement as a continual process, engendered by the structural conditions of long-term statelessness. This is what has made Palestinian refugees so vulnerable to further forced migrations since 1948, and it has rendered them particularly exposed in situations of conflict and war. Moreover, Palestinian refugees' experiences challenge the paradigm whereby migration is determined by particular push-and-pull factors. If people's continual movement is being driven instead by long-term structural conditions—specifically dispossession and statelessness—then the push/pull factors of each event can only tell part of the story.

This is not the only challenge that the case of Palestinian displacement poses to paradigms around migration. Perhaps most importantly, Palestinian refugees themselves have disrupted the perceived binary between home and exile by acting to "re-create" their lost homeland of Palestine while in exile. Although they are not the only group to have done so, the longevity and repeated crises of their exile mean that Palestinian refugees' voices and agency in this regard are particularly significant. Oral histories, commemoration practices, and organized forms of national politics have all been used by

Palestinian refugees to help sustain and buttress their common identity in exile (Khalili 2006; Sayigh 2007). In these ways, they have asserted the tenacity of their national identity against conditions that could have compelled their capitulation on this front. The remainder of this chapter accordingly focuses on such activism in those spaces that were the biggest driving force in this regard: the Palestinian refugee camps.

Home and exile in the Palestinian refugee camps

In the immediate aftermath of the Nakba, many of the poorest Palestinian refugees were forced to seek shelter in the camps springing up across the Levant. Starting out as collections of flimsy tents, over the decades the camps evolved into semipermanent structures often resembling shanty towns (Peteet 2005). As of 2020, there are 58 UN-recognized Palestinian refugee camps in the five areas where UNRWA operates (Jordan, Syria, Lebanon, the West Bank, and the Gaza Strip). They are home to less than half of all registered Palestinian refugees today (UNRWA n.d.c). Yet while the majority of Palestinian refugees have never lived in the camps, the historical and political importance of the camps belies this. A disproportionately high number of Palestinians who joined the nationalist guerrilla fighters known as *fida'yyin*[6] originated from the refugee camps, possibly because the poverty and abject conditions therein gave them the least to lose and the most to gain from militant activism. As a result, the camps gained a reputation as hubs of the Palestinian nationalist movement, or "nests of the resistance." Academic and former PLO official Karma Nabulsi conveyed this when she wrote in 2002, "Palestinian strength is in the camps. ... [That] is where, should anyone desire to discover it, one finds the will of the Palestinian people" (Nabulsi 2002).

The camps' spatial function increased their significance. Their demarcation enabled the refugees to carry Palestine with them into exile and keep alive the memory of the homeland. Although the degree of the camps' separateness varied across the host states—those in Lebanon were mostly closed, while those in the West Bank saw free movement with outside areas (Hanafi 2010)—even the most "open" camps were distinguished from their environs by overcrowding, high levels of poverty, and the visible presence of UNRWA installations. This meant that they could function as "incubators" of Palestinian national and political consciousness. Fawaz Turki, who was forced to flee Haifa as a young child during the Nakba, reflected on this when describing his later childhood years in Burj al-Barajneh camp in Lebanon: "As we grew up [in Burj al-Barajneh], we lived Palestine every day. We talked Palestine every day. For we had not, in fact, left it in 1948. We had simply taken it with us" (Turki 1988: 36).

Thus, while the host states may have seen the camps' containment as a means of controlling the refugees, that same containment enabled the Palestinians to retain their communal consciousness and national identity in exile (Sayigh 1977; Sayigh 2007). As Turki (1993) explained, the refugee camps provided not only physical shelter but also protection for a Palestinian national identity that might otherwise be corroded by host state integration.

Critically, this was not simply the inevitable result of the camps' physical setup but was also made possible by the refugees' voices, agency, and activism. Over the seven decades of their exile, Palestinian camp communities have acted to re-create the lost homeland of Palestine in these spaces by way of naming practices, decorative imagery, and territorial organization. From the early aftermath of the Nakba, refugees named camp streets and quarters after the villages they had had to flee (Peteet 2005). In some cases, the practice was extended to camps themselves, such that Jerash camp in Jordan and Wavel camp in Lebanon became known informally as Gaza and al-Jalil (Galilee), respectively. Within the camps, schools and clinics were often named after places in Palestine, with examples including Haifa hospital in Burj al-Barajneh camp and Gaza hospital in Shatila camp, both in Lebanon. To this day, Palestinian students attend Deir Yassin high school in El Buss refugee camp, also in Lebanon, named after a Palestinian village that suffered an infamous massacre in 1948.

Palestinian refugees' success in "re-creating" Palestine through these means proved important in maintaining the consciousness of a national community, despite their displacement and dispersal. Often, this was mapped directly onto the spatial geography of the camps, with neighborhoods organized such that refugees from the same parts of Palestine were housed in the same camp quarters in exile (Davis 2011). Again, naming practices could be critical here. For example, the "Amqa quarter of Ein el-Helweh camp in Lebanon, and the Tarashha quarter of Burj al-Barajneh camp, were named after the respective Palestinian hometowns of their residents" (Peteet 2005: 111–12).

Many refugee women further maintained their connections to their ancestral villages in Palestine through their clothing, as the traditional Palestinian peasant *thoub*[7] indicated the wearer's place of origin through its particular design, colors, and embroidery patterns. The *thoub*'s significance in this regard was such that after 1948 it quickly evolved into a symbol of Palestinian identity, with even women from nonpeasant urban backgrounds wearing it as a political statement (Karmi 2002; Atwan 2007). Such practices allowed Palestinian refugees to maintain connections to their original towns and villages, expressing their regional affiliations as well as national identity. This was particularly important for the younger generations who had been born in exile. Partly as a result, ancestral towns and villages could remain central to these generations' sense of identity despite their geographical distance from them. As Shafiq Al Hout (2011) writes, it meant that even decades after the Nakba, generations born in the camps could cite their families' places of origin.

The activities of many Palestinian refugees in the camps, thus, served to collapse any notion of a binary between "home" and "exile." Instead, the refugees successfully remade the camps into a substitute or midpoint between the two, effective in sustaining memories of the lost homeland and enabling a continuing connection to it among younger generations. Yousif Qasmiyeh, who grew up in Baddawi camp in Lebanon, describes this when he writes that the camps served as "transitional places … with an amalgamation of details: those which were carried from Palestine and those which have grown in the whereabouts of these camps" (Qasmiyeh 2016: 304).

All this leads to some important conclusions. Once again, it shows how the realities of Palestinian displacement disrupt the neat categorizations of conventional binaries around migration. The protracted nature of the Palestinians' exile means that they are often described as existing "in limbo" between their lost homeland, on the one hand, and the unwelcoming host states, on the other. While valid in many ways, this discourse should not be reduced to a simplistic binary between home and exile—not least because such a binary neglects the agency of the refugees themselves. In fact, despite the overwhelming confluence of factors placing them in a position of marginalization and disempowerment, Palestinian refugees have continually and successfully asserted their national identity—not least by transforming their refugee camps into sites of "Palestine in exile."

Conclusion

Even a brief examination of Palestinian refugee history disrupts paradigms of migration that draw binaries between home and exile and between push-and-pull factors. The reality is more nuanced, complex, and messy. Having reoccurred repeatedly since the original dispossession of 1948, Palestinian displacement is best understood as a structure rather than a singular event. The Palestinians' underlying statelessness as a people has rendered them susceptible to continual marginalization and further displacements over the last 70 years. As such, their movement cannot be explained solely by the push-and-pull factors of particular instances. Instead, it is rooted in the dispossession that has rendered them vulnerable in a world of nation-state normativity. At the same time, the structural conditions of Palestinian displacement do not tell the whole story. Palestinian refugees themselves have further challenged binaries, specifically the one often drawn between home and exile, by working to re-create their lost homeland in the spaces of their refugee camps. Their actions in this regard arguably form a mode of resistance to their continuing unwanted exile and statelessness.

Although this chapter's discussion of Palestinian history has placed it outside the conventional binaries of migration and refugee studies, this should not be taken as evidence of Palestinian exceptionalism. Instead, the purpose of this analysis is to highlight the deeper understandings and nuances that can come from situating the Palestinian case within broader studies of forced migration. At its core, Palestinian displacement is an unusual but not exceptional story, its very value emanating from how it complicates simplistic binaries. As such, it can prove instructive far beyond the field of Palestinian history and can also speak to bigger questions around agency, displacement, and belonging.

Notes

1 There are an estimated 13 million Palestinians in the world today, the vast majority of whom are refugees (Palestinian Central Bureau of Statistics 2019). As of 2019, 5.6 million Palestinians were registered with the UN as refugees, along with the same number of Syrian refugees (UNHCR n.d.; UNRWA n.d.c). It is estimated that a further 3 million Palestinians are unregistered refugees (Albanese and Takkenberg 2020).

2 The exact figure is disputed. This is the most widely agreed estimate, commensurate with UN records (UN ESM 1949).

3 Here Abu Sitta distinguishes European Jewish (Ashkenazi) immigrants from long-standing Middle Eastern Jewish communities in Palestine, whom he describes as "Arab Jews."

4 The Syrian war has been the most devastating, resulting in the internal displacement of 280,000 Palestinians (UNRWA n.d.b), and the cross-border flight of more than 120,000 (UNRWA 2019a).

5 UNRWA is mandated to provide relief services to registered Palestinian refugees in Syria, Lebanon, Jordan, the West Bank, and the Gaza Strip (the "five fields"). As a result, Palestinian refugees are excluded from the mandate of UNHCR, which serves all other refugees worldwide. When Palestinian refugees find themselves outside the five fields they should theoretically be able to access UNHCR services, although in reality they often face obstacles in doing so (Abu Moghli et al. 2015).

6 Literally meaning "those who sacrifice themselves," this Arabic term is generally used to denote Palestinian nationalist guerrillas.

7 The *thoub* or *thobe* is a long embroidered dress with long sleeves, sometimes described as a kaftan and traditionally worn by peasants in pre-1948 Palestine.

References

Abu Iyad, and Eric Rouleau (1981). *My Home, My Land: A Narrative of the Palestinian Struggle*. New York: Times Books.

Abu Moghli, Mai, Nael Bitarie, and Nell Gabiam (2015). *Palestinian Refugees from Syria: Stranded on the Margins of Law*. Al Shabaka Policy Brief. Accessed January 22, 2021, from https://al-shabaka.org/briefs/palestinian-refugees-from-syria-stranded-on-the-margins-of-law/

Abu Sitta, Salman (2016). *Mapping My Return: A Palestinian Memoir*. New York: AUC Press.

Akram, Susan (2002). "Palestinian Refugees and Their Legal Status: Rights, Politics, and Implications for a Just Solution." *Journal of Palestine Studies* 31(3): 36–51.

Al Hardan, Anaheed (2016). *Palestinians in Syria: Nakba Memories of Shattered Communities*. New York: Columbia University Press.

Al Hout, Shafiq (2011). *My Life in the PLO: The Inside Story of the Palestinian Struggle*. Hader Al Hout and Laila Othman, trans. London: Pluto Press.

Al-Nakib, Mai (2014). "'The People are Missing': Palestinians in Kuwait." *Deleuze Studies* 8(1): 23–44.

Albanese, Francesca, and Lex Takkenberg (2020). *Palestinian Refugees in International Law*. Oxford: Oxford University Press.

Almustafa, Maissaa (2018). "Relived Vulnerabilities of Palestinian Refugees: Governing through Exclusion." *Social & Legal Studies*, 27(2): 164–79.

Atwan, Abdel Bari (2007). *A Country of Words: A Palestinian Journey from the Refugee Camp to the Front Page*. London: Saqi.

Bolongaro, Kait (2016). "Palestinian Syrians: Twice Refugees." *Al Jazeera English*. Accessed January 22, 2021, from https://www.aljazeera.com/indepth/features/2016/03/palestinian-syrians-refugees-160321055107834.html

Cohen, Roberta (2008). *Iraq's Displaced: Where to Turn?* Brookings Institution. Accessed January 22, 2021, https://www.brookings.edu/wp-content/uploads/2016/06/10_iraq_cohen.pdf

Davis, Rochelle (2011). *Palestinian Village Histories: Geographies of the Displaced*. Stanford, CA: Stanford University Press.

El-Abed, Oroub (2009). *Unprotected: Palestinians in Egypt since 1948*. Washington, DC: Institute for Palestine Studies.

Fiddian-Qasmiyeh, Elena (2015). "On the Threshold of Statelessness: Palestinian Narratives of Loss and Erasure." *Ethnic & Racial Studies* 39(2): 301–21.

Filiu, Jean-Pierre (2014). *Gaza: A History*. London: Hurst & Company.

Forsythe, David (1983). "The Palestine Question: Dealing with a Long-term Refugee Situation." *Annals of the American Academy of Political and Social Science* 467: 89–101.

Gabiam, Nell (2016). *The Politics of Suffering: Syria's Palestinian Refugee Camps*. Bloomington: Indiana University Press.

Galbraith, Peter W. (2003). *Refugees from War in Iraq: What Happened in 1991 and What May Happen in 2003*. Washington, DC: Migration Policy Institute.

Hammami, Rema, and Salim Tamari (1997). "Populist Paradigms: Palestinian Sociology." *Contemporary Sociology* 26(3): 275–79.

Hanafi, Sari (2010). "Palestinian Refugee Camps in Lebanon: Laboratory of Indocile Identity Formation," 45–74. In Muhammad Ali Khalidi, ed., *Manifestations of Identity: The Lived Reality of Palestinian Refugees in Lebanon*. Beirut: Institute for Palestine Studies.

Human Rights Watch (2007). *Why They Died: Civilian Casualties in Lebanon During the 2006 War*. Accessed January 22, 2021, from https://www.hrw.org/report/2007/09/05/why-they-died/civilian-casualties-lebanon-during-2006-war

Ibrahim, Youssef M. (1991). "Jordan a Grim Refuge for Kuwait Palestinians." *New York Times*. Accessed January 22, 2021, from https://www.nytimes.com/1991/10/03/world/jordan-a-grim-refuge-for-kuwait-palestinians.html

ISI, ASKV, and ENS (2019). *From Syria to Europe: Experiences of Stateless Kurds and Palestinian Refugees from Syria Seeking Protection in Europe*. Accessed January 22, 2021, from https://files.institutesi.org/from_Syria_to_Europe.pdf

Kagan, Michael (2009). "The (Relative) Decline of Palestinian Exceptionalism and Its Consequences for Refugee Studies in the Middle East." *Journal of Refugee Studies* 22(4): 417–38.

Karmi, Ghada (2002). *In Search of Fatima: A Palestinian Story*. London: Verso.

Khalidi, Rashid (2007). *The Iron Cage: The Story of the Palestinian Struggle for Statehood*. Oxford: One World.

Khalidi, Rashid (2010). *Palestinian Identity: The Construction of Modern National Consciousness*. New York: Columbia University Press.

Khalidi, Rashid (2020). *The Hundred Years War on Palestine: A History of Settler Colonial Conquest and Resistance*. London: Profile Books.

Khalili, Laleh (2006). *Heroes and Martyrs of Palestine: The Politics of National Commemoration*. Cambridge: Cambridge University Press.

Lamb, David (1995). "Arab Countries Reluctant to Receive Expelled Palestinians." *The Tech* 115(40): 3. Accessed January 22, 2021, from http://tech.mit.edu/V115/PDF/V115-N40.pdf.

Meier, Daniel (2016). "The Blind Spot: Palestinian refugees from Syria in Lebanon." In Maximilian Felsch and Martin Wahlish, eds., *Lebanon and the Arab Uprisings: In the Eye of the Hurricane.* Accessed January 22, 2021, https://halshs.archives-ouvertes.fr/halshs-01947542/document

Nabulsi, Karma (2002, September 17). "Our Strength Is in the Camps." *The Guardian.* Accessed January 22, 2021, https://www.theguardian.com/world/2002/sep/17/comment

Palestinian Central Bureau of Statistics (PCBS) (2019). "On the Occasion of the International Population Day 11/7/2019." Accessed January 22, 2021, from http://www.pcbs.gov.ps/post.aspx?lang=en&ItemID=3503#

Pappe, Ilan (2006). *The Ethnic Cleansing of Palestine.* Oxford: One World.

Pappe, Ilan (2011). *The Forgotten Palestinians: A History of the Palestinians in Israel.* New Haven, CT: Yale University Press.

Peteet, Julie (2005). *Landscape of Hope and Despair: Palestinian Refugee Camps.* Philadelphia: University of Pennsylvania Press.

Qasmiyeh, Yusif (2016). "My Mother's Heels," 303–05. In Y. Suleiman, ed., *Being Palestinian: Personal Reflections on Palestinian Life in the Diaspora.* Edinburgh: Edinburgh University Press.

Rempel, Terry (2006). "Who Are Palestinian Refugees?" *Forced Migration Review* 26: 5–7.

Said Makdisi, Jean (2005). *Teta, Mother and Me: An Arab Woman's Memoir.* London: Saqi.

Sayigh, Rosemary (2007). *The Palestinians: From Peasants to Revolutionaries.* London: Zed Books.

Sayigh, Rosemary (1977). "The Palestinian Identity among Camp Residents." *Journal of Palestine Studies* 6(3): 3–22.

Sayigh, Yezid (1997). *Armed Struggle and the Search for State: The Palestinian National Movement, 1949–1993.* Oxford: Clarendon.

Shiblak, Abbas (1995). "A Time of Hardship and Agony: Palestinian Refugees in Libya." *Palestine-Israel Journal of Politics, Economics and Culture* 2(4). Accessed January 22, 2021, https://pij.org/articles/596

Turki, Fawaz (1988). *Soul in Exile: Lives of a Palestinian Revolutionary.* New York: Monthly Review Press.

Turki, Fawaz (1993). *Exile's Return: The Making of a Palestinian American.* New York: Free Press.

UN ESM (UN Economic Survey Mission for the Middle East) (1949). *First Interim Report.* Accessed January 22, 2021, from https://unispal.un.org/DPA/DPR/unispal.nsf/0/648C3D9CF58AF0888525753C00746F31

UNHCR (n.d.). *Syria Emergency.* Accessed January 22, 2021, from https://www.unhcr.org/uk/syria-emergency.html

UNHCR (2003). *Chronology: 1991 Gulf War Crisis.* Accessed January 22, 2021, from https://www.unhcr.org/uk/subsites/iraqcrisis/3e798c2d4/chronology-1991-gulf-war-crisis.html

UNHCR (2017). *Return and Readmission of Palestinian Refugees from Syria (PRS) to Lebanon and Jordan.* Accessed January 22, 2021, from https://www.refworld.org/pdfid/5ab8cf9d4.pdf

UNHCR (2018). *Global Trends: Forced Displacement in 2018.* Accessed January 22, 2021, from https://www.unhcr.org/globaltrends2018/

UNRWA (n.d.-a). *Frequently Asked Questions* Accessed January 22, 2021, from https://www.unrwa.org/who-we-are/frequently-asked-questions

UNRWA (n.d.-b). *Syria Crisis.* Accessed January 22, 2021, from https://www.unrwa.org/syria-crisis

UNRWA (n.d.-c). *Where We Work.* Accessed January 22, 2021, from https://www.unrwa.org/where-we-work

UNRWA (2006). *UNRWA Flash Appeal: Lebanon Final Report 2006.* Accessed January 22, 2021, from https://reliefweb.int/sites/reliefweb.int/files/resources/ED91BE2F09570A78C1257418003A15AE-Full_Report.pdf

UNRWA (2019a). *Syria Regional Crisis: Emergency Appeal 2019.* Accessed January 22, 2021, from https://www.unrwa.org/sites/default/files/content/resources/2019_syria_ea_final.pdf

UNRWA (2019b). *Palestine Refugees in Syria: A Tale of Devastation and Courage.* Accessed January 22, 2021, from https://www.unrwa.org/newsroom/features/palestine-refugees-syria-tale-devastation-and-courage

Viorst, Milton (1984). *UNRWA and Peace in the Middle East.* Washington DC: Middle East Institute.

13 Displacement, diaspora, and statelessness

Framing the Kurdish case

Naif Bezwan and Janroj Yilmaz Keles

Introduction: dissent and displacement

The Kurds—considered "the largest stateless nation in the contemporary world"—live under the national jurisdictions of Turkey, Iran, Iraq, and Syria (Vali 1998: 82). A recurrent pattern of repression and resistance has come to dominate the relationship between Kurdish national communities and these four states. As a consequence, the Kurdish diaspora that began in the 1980s continued so that currently the Kurds not only constitute one of the largest stateless nations in the world, but also one of the largest diasporic communities. As an emergent sociopolitical formation, the Kurdish diaspora results from multiple and recurrent strategies of state-led demographic engineering and displacement to which the Kurdish communities in each of the four states, to varying degrees and intensity, have been subjected.

Diaspora is an old, evolving, and contested concept. One influential definition in the field has been offered by Rogers Brubaker, who refers to diaspora as "dispersion in space, orientation to a homeland," and "boundary-maintenance, involving the preservation of a distinctive identity vis-à-vis a host society or societies" (Brubaker 2005: 5–7; Alexander 2017: 1557). His main argument is that diaspora should not be considered as a "bounded entity, but rather as an idiom, a stance, a claim" (Brubaker 2005: 12). As invocations of diaspora necessarily involve references to a group, one should think of diaspora not in substantialist terms but primarily "as a category of practice, and only then ask whether, and how, it can fruitfully be used as a category of analysis" (Brubaker 2005: 12). Claire Alexander (2017: 1553), in turn, highlights the role of violence and the importance of connecting the complex engagements between "here" and "there" both conceptually and empirically, "while recognizing that neither places of origin nor arrival remain unchanged through this process." What matters instead is "the intersection of the *then* in the now, the entanglement of past and present and, indeed, future" (Alexander 2017: 1551).

We refer to diaspora both as a concept of practice and as a descriptive-analytical concept. The former points toward contentious politics and the political process at large in relation to diasporas, while the latter calls for scientific inquiry of the problems, policies, narratives, dynamics, and agencies involved in the politics, formation, and transformations of diasporas.

DOI: 10.4324/9781003170686-16

In the remainder of this chapter, we first outline cases of internal and external displacement in the recent past to provide context to the debate that follows. Next, we discuss the formation of the Kurdish diaspora, its collective action, and the impact of exclusionary policies. We then interrogate the relationship between belonging and identity in the diaspora. Given the paramount importance attached to statelessness, we consider the role and consequences of statelessness among Kurdish communities by presenting a framework for understanding this phenomenon. Finally, we conclude by summarizing the main insights.

It is beyond the scope of this chapter to deal with the whole spectrum of state-engineered internal and external displacements of the Kurdish communities; in what follows, we instead outline some of the key events leading up to the forced migration and thus the formation of Kurdish diaspora.

Cases of displacements in recent history

The Kurdish diaspora can be seen as emerging from the cumulative effect of many critical junctures beginning in the late 1970s and continuing over four decades. One pivotal event was the Islamic revolution in Iran and the ensuing Iran–Iraq War (1980–1988). Following a temporary respite from repression for the Kurdish community in Iran, the newly installed theocratic regime began to exert "a new and more systematic use of concentrated violence and savage repression" (Vali 2020: 185). Both the theocratic power grab in 1979 and the brutal war that followed caused the displacement of tens of thousands of people and a massive influx of refugees from Iran and the Kurdistan Region of Iran to Europe and elsewhere. Since its inception, the ethno-theocratic state of Iran has continued to conduct "systematic and steady demographic engineering and displacement to undermine the distinct ethnic profile of Kurdistan" (Mohammadpour and Soleimani 2020: 10).

In the context of this bloody war, massive flows of forced migration and refugees across international borders also took place in the Kurdistan of Iraq, mainly caused by a twin genocidal campaign, namely the Anfal and the Halabja chemical attack (March 16, 1988), both of which were conducted at the end of the Iran–Iraq War in Kurdistan-Iraq. The Anfal campaign was carried out by the Iraqi army under the Baathist Regime of Saddam Hussein in Kurdistan-Iraq in 1988 (van Bruinessen 1998: 5ff; Hardi 2011).[1] Anfal consisted of a series of eight military offensives that annihilated Kurdish rural life between February and September 1988 (Hiltermann 2008). During this genocidal campaign

> over 2,600 villages were destroyed and an estimated number of 100,000 civilians were murdered. This includes people who were shot in the mass graves and died as a result of the shelling and gas attacks, life in the prison camps, and during their flight to Iran and Turkey.
>
> (Hardi 2011: 13)

Another major development that led to Kurdish mass migration was the Gulf War (August 2, 1990–January 17, 1991) that followed the Iraqi invasion of Kuwait. The subsequent Kurdish uprising was brutally crushed, again causing mass migration and flow of refugees across international borders. Added to that was the migration to the Western countries of Germany, Sweden, Denmark, the United Kingdom, and the Netherlands caused by internal conflict, especially between the Patriotic Union of Kurdistan (PUK) and the Kurdistan Democratic Party (KDP) (1994–1997).

Similar patterns of state-generated migration are also strikingly present in the Kurdish region of Turkey. One of the most significant instances of mass migration relates to the coup d'état of September 1980, especially after the escalation of military conflict between Turkey and the Kurdistan Workers' Party (Partiya Karkerên Kurdistan, PKK) in 1984. Following an almost decade-long period under martial law (from April 1979 to July 1987), a new and more extensive state of emergency regime (also known as State of emergency region or by its Turkish acronym OHAL, 1987–2002) in the Kurdish region was established. Under this emergency regime, designed for the Kurdish regions, the forced deportation from some 3,000 villages created thousands of internally displaced people and refugee flows of more than three million Kurds (cf. Baser et al. 2015: 132). On the pretext of destroying the PKK-led insurgence, a decade-long dirty war was imposed over a whole region from the late 1980s to the end of 1990s, accompanied by mass expulsion, systematic torture, massive human rights violations, destruction of livelihoods, depopulation of thousands of villages, as well as the ban of the Kurdish language. During this time the Turkish military engaged in widespread destruction in the Kurdish regions burning over 4,000 villages and forcing as many as two million Kurds to flee from their homes (cf. van Bruinessen 1998: 3).

The most recent case of internal and external displacement took place after the collapse of peace talks in the summer of 2005, which resulted in armed clashes between Turkish security forces and allegedly PKK-affiliated groups. Some 2,000 people were killed during security operations from July 2015 to December 2016 in the Kurdistan region of Turkey.[2] The killings, according to the UN Office of the High Commissioner for Human Rights (OHCHR), "were reportedly invariably followed by mass displacement of the survivors and the destruction of their homes and of local cultural monuments. Over 355,000 of residents of Kurdistan region of Turkey, mainly citizens of Kurdish origin, were displaced" (UN Report 2017: 5).

Finally, the Syrian civil war, the rise of the Islamic State (ISIS), as well as Turkish military interventions have all generated yet more forced displacement and flows of migrants and refugees (Chatty 2017: 223 ff.). The genocidal violence committed by the ISIS against the Yazidi–Kurdish community in the region around Mount Sinjar (Kurdish Shingal) in northwestern Iraq in August 2014 has caused internal and external displacement of almost the entire community.[3] The seizure of Kobanê by the ISIS, and the internationally acclaimed resistance against it, led to widespread expressions of solidarity among Kurdish communities in both the homeland and the diaspora. This has also strengthened a sense of belonging among Kurdish communities in the diaspora (cf. Eccarius-Kelly, 2019: 29).

As suggested above, this momentum for mobilization in the diaspora and in the countries of origin was undermined by multiple Turkish military campaigns, leading to the policies and processes of displacement, destruction, and military invasion. While fundamentally affected by the ongoing bloody civil war in Syria, the Kurdish community in Rojava has been the target of a dirty war conducted by the Turkish army and its proxies. The first major attack was launched in January 2018, aimed at the Kurdish city of Afrin, a strategically and economically important area located in the heart of the Kurdish region in Syria (Bezwan 2018: 63). The latest offensive was conducted on October 9, 2019, when the Turkish air force launched airstrikes on Kurdish towns and settlements across the Syrian border, which immediately followed the sudden, and internationally condemned, withdrawal of US forces from northeastern Syria on October 6, 2019, by the Trump administration. These Turkish military incursions have caused the expulsion of tens of thousands of refugees and thousands of deaths, including civilians.

Having outlined recent cases of internal and external displacement across all regions of Kurdistan, it can be concluded that the Kurdish diaspora has expanded considerably over the past decades in Europe through a combination of "labour migration, refugee migration, family reunion and settlement of the second and third generation" (Keles 2015: 78). Currently, it is estimated that around two million Kurds live in Europe, of which approximately one million reside in Germany alone. There are also sizable Kurdish communities in Austria, Belgium, France, Finland, Greece, the Netherlands, Norway, Sweden, Switzerland, and the United Kingdom. In addition, beyond Europe, Kurds have also established communities in the United States, Canada, Australia, New Zealand, Japan, Lebanon, Jordan, ex-Soviet republics (Armenia, Georgia, Kazakhstan, Azerbaijan), and elsewhere.

To conclude, although a considerable number of Kurds from Turkey migrated to Europe as "guest workers" in the 1960s, the actual formation of the Kurdish diaspora in Europe and beyond has been the outcome of the catastrophic developments, mass violence, and related state policies presented above. Put differently, the emergence of the Kurdish diaspora is thus causally related to the politics of the "host" states under whose administration the Kurdish communities live. This has had a tremendous impact on the ways in which the Kurds define their belonging, identity, and collective action.

Diaspora, collective action, and its discontents

The Kurdish diaspora is considered as one of the most active diasporic communities. A variety of concepts ranging from "long-distance nationalism" (van Bruinessen 1998) to "ethnic separatism and mobilizing ethnic conflict" (Lyon and Uçarer 2001), to the "mechanisms of diaspora mobilization" (Adamson 2013: 65 ff) and "Kurdish transnational spaces" (Keles 2015:

181) has been used to characterize the nature of this diaspora activism. Whereas earlier literature mainly focused on Kurdish diasporic communities in terms of nationalism and "ethnic conflict," recent work takes a social movement approach to Kurdish diasporic activities, with one scholar defining the Kurdish movement as a "transnational indigenous movement" (Demir 2017: 66).

A particular emphasis in the relevant literature is also placed on what can be called diasporic "nation-building" activities through use of various media, including television and social media, as well as the political agency of Kurdish diasporans (cf. Keles 2015). The creation of a large number of institutions and activities undertaken by the Kurdish diaspora has been referred to as "Euro-Kurdistan" and understood "as a dynamic process of Kurdish collective-identity formation in and through Europe" (Ayata 2011: 525).

Key to this process has been the use of technologies such as satellite broadcasting, internet, and desktop publishing, all of which has not only challenged the "state discourses that deny or suppress Kurdish identity" but also helped to reconstruct Kurdish identity, history, and language (Romano 2002: 148). Of significant importance in that context was the launch of Kurdish MED TV from London in 1995 and its successors channels Medya TV, Roj TV, Nuce, and Sterk. These Kurdish satellite TV channels in Europe caused Turkey, Iran, Iraq, and Syria to lose their monopoly over broadcasting, which they had used to violate the right of people in Kurdish-populated regions to get information in the Kurdish language (cf. Hassanpour 1998). Moreover, by broadcasting in Kurdish, and providing news from a Kurdish perspective—as well as discussions about self-determination, conflict resolution, Kurdish culture, and language—Kurdish television channels have disrupted the Turkish, Arabic, and Persian nationalistic discourse (Keles 2015). As a result, these channels have effectively rendered meaningless any restrictions (including outright bans) placed on Kurdish identity and language (cf. Ayata 2011: 525 ff; Zeydanlıoğlu 2012: 125 ff).

But parallel to the rise of the Kurdish diaspora and its political and cultural activities, even though banned and persecuted in the homeland, the persecuting states have intensified their efforts to counteract these activities by using diplomatic, political, and surveillance instruments, and have even resorted to assassination in some cases. For example, Dr. Abdul Rahman Ghassemlou, the then secretary-general of the Democratic Party of Iranian Kurdistan (PDKI), was assassinated by Iranian emissaries on July 13, 1989, in Vienna (Encyclopaedia of Iranica n.d.). Some three years after his murder, on September 17, 1992, his successor, Sadiq Şerefkendî, along with his comrades Fattah Abdoli, Homayoun Ardalan, and their translator Nouri Dehkordi, were also assassinated at the Mykonos Greek restaurant in Berlin, Germany. The Iranian regime was implicated in the assassination of the Kurdish leaders, but as is often the case with state crime involving the Kurds, the real preparators have not been officially identified, let alone brought to justice.

Turkey has put massive pressure on many European countries to ban the cultural and political activities of Kurdish migrant communities and organizations by using its diplomatic clout, espionage, jamming of television broadcasts into Turkey, and various forms of intimidation. Turkish governments received considerable support from governments of EU member states that feared a "spillover effect" from the Turkish–Kurdish conflict. In 1993, Germany and France "enacted a ban on Kurdish political and cultural organization" (Eccarius-Kelly 2002: 91). In field research conducted by this chapter's coauthor, Janroj Yilmaz Keles, the interviewees repeatedly underlined the ongoing intervention and surveillance of the Turkish governments in Europe, considered by our research participants as a deliberate attempt to import a Turkish hegemonic and nationalistic ideology into the European context (Keles 2015).

Persecution and surveillance of the migrant communities, on the one hand, and diplomatic pressure on (or in some cases collaboration with) the country of settlement, on the other hand, have been used to contain the Kurdish diaspora. This has been emphatically put forward by key interviewees during multiple fieldwork conducted by Keles in Sweden, Germany, and the United Kingdom. For example, in an interview for our research project in 2009, Alan, the director of the Kurdish Centre in Berlin, stated that when they first established the center in 1984, the Turkish Foreign Ministry had strongly reacted and contacted the German Foreign Ministry to shut down the center because its name included the word "Kurdish" (Keles 2011). Similarly, the Turkish Embassy in Copenhagen tried to stop the program devoted to teacher training for the Kurdish-language education of migrants, sponsored by the Nordic Cultural Foundation in Denmark, "by pointing out that participants were still Turkish citizens and thus were not entitled to break Turkish law, whatever country they were in, and in Turkish law Kurdish is a forbidden language" (Hassanpour 1992: 135, quoting Skutnabb-Kengas).

To conclude, despite multiple obstructions, large sections of Kurdish immigrants consider diaspora as a transnational space in which the processes of what may best be called "cognitive liberation" (McAdam 1982: 48) along with "long distance" decolonial collective engagement for self-rule in the homeland take place and are enacted on a regular basis. The diasporic political activism and associated decolonial discourse also define the very parameters in terms of which belonging to and political identifications are expressed.

Belonging and identity in the diaspora

As strikingly expressed by Hannah Arendt in her 1943 essay, "Refugees driven from country to country represent the vanguard of their peoples—if they keep their identity" (Arendt 2007: 274). The question that presents itself to all diasporic communities, whether old or new, is how to keep what they see as their identity while being a part of host societies. To provide an answer to this question, we first briefly consider the concept of diaspora before dealing with the Kurdish case in some detail.

Generally, diaspora is described in terms of six interrelated components: dispersal or immigration, location outside a homeland, community, orientation to a homeland, transnationalism, and group identity (Grossman 2019: 1268 ff.). Diaspora can then be understood as a community that seeks to maintain, redefine, and negotiate its identity in ever-changing circumstances, both in countries of immigration and countries of migration (cf. Bauböck and Faist 2010: 9–13 ff). As an old and evolving concept, diaspora is located at the intersection of the society of settlement with the society of origin "understood as a transnational social organisation relating both to the country of origin and the country of exile" (Wahlbeck 2002: 222). It thus takes a variety of forms ranging from "stateless," "state-linked," "historical," "modern," and "incipient" (cf. Sheffer 2003: 249).

The aspects of diaspora and transnationalism presented above also feature in the scholarly literature on the Kurdish diaspora. By drawing on conceptual tools provided by transnationalism and diaspora studies, scholars have highlighted various aspects of Kurdish diasporic communities, their networks and activities between Kurds in the diaspora and in Kurdistan, as well as between Kurds in different countries in the diaspora (cf. Wahlbeck 2002: 222–25; Keles 2016). One common theme emerges here: the Kurdish communities in Europe, while stemming from a multitude of different political orientations, religious beliefs, linguistic groups, social classes, educational backgrounds, and gendered experiences, are capable of collective action and cooperation of matters of common interest (Wahlbeck 2002: 224).

The most significant question underlying the broader Kurdish context relates to the role of the state and statelessness. Studies on Kurdish diaspora show that statelessness is perceived by Kurdish immigrants from all regions of Kurdistan as a foundational experience informing their orientation, self-understanding, and belonging (cf. King et al. 2008; Keles 2015; Eliassi 2016: 1415; Tas 2016; Syrett and Keles 2019).

While fundamentally affecting the Kurdish communities across four states, the phenomenon of statelessness also influences the situation of Kurdish migrants in their country of immigration. For example, they are officially considered as nonexistent because migrants are registered according to the state they come from. The Kurdish migrants, therefore, simply do not appear in much of the statistical data collected on workers or official data like the census, as the census questions relating to nationality provide Kurds with no option other than to define themselves as Turkish, Iraqi, Iranian, or Syrian. The result is that the Kurdish migrants are, even in the diaspora, compelled to take a national identity that most have spent their lives opposing (Keles 2015).

This invisibility, evidenced by the lack of comprehensive statistical data on the Kurdish population in Europe (Keles 2015), has two primary effects. First, it hinders the Kurdish migrants benefitting from resources otherwise available for migrants and, thus, having access to information in their language. Second, it prevents Kurdish migrants from being part of multicultural policies and practices, particularly in Germany where Kurdish protesters have been heavily criminalized and imprisoned because of their involvement in diasporic Kurdish political movements.

Studies show that the issue of statelessness has much wider social and political consequences. For example, in research on Kurdish immigrants in Sweden and the United Kingdom, Barzoo Eliassi has found that statelessness is perceived by many participants not only as a form of political marginalization, but also as source of "vulnerability and exposure to political and physical violence by the sovereign state" (Eliassi 2016: 1410). Statelessness is regarded as "a structural status injury," haunting "the stateless in their everyday life" (cf. Eliassi 2016: 1416). Another study carried out between April 2014 and May 2015, focusing on Kurdish migrants living in the United Kingdom and Germany, shows that the majority of interviewees described themselves as "being stateless despite holding one or more citizenships" (Tas 2016: 50). This state of affairs is referred to as "social statelessness," which goes beyond de jure and de facto statelessness as they both "focused on the legal connections between individuals and the state" (Tas 2016: 49). Janet Klein, in turn, takes a critical approach to the use of statelessness in relation to the Kurds as they are overwhelmingly nationals of the states in question (Klein 2010: 227 ff.).

Given the paramount importance that statelessness occupies in the broader Kurdish context, we next provide a conceptual approach to this problem and its implications.

Understanding the conundrum of statelessness

As indicated above, at its most basic level, statelessness "describes people who are not nationals of any state" (Blitz and Sawyer 2011: 1). But important as it is, lack of citizenship cannot exhaustively define the phenomenon of statelessness (cf. Sawyer 2011: 70). As the overwhelming majority of the Kurds are citizens of the respective states ruling over the Kurdish region,[4] it cannot be the lack of formal citizenship that matters. As we point out in this chapter, formal citizenship neither provides the Kurds with recognition of communal existence and rights, nor does it lead to the enjoyment of citizens' rights and liberties. Put differently, in the Kurdish case legal citizenship does not provide the "right to have rights," as cogently stated by Hannah Arendt in the context of her own statelessness (see also on that dictum see Kohn 2007: xxiv; Staples 2012: 100 ff.; Sawyer 2011: 106). If citizenship, however, does not protect Kurds from being dominated and denied communal existence and rights, it can only take the form of what can best be defined as "exclusion by inclusion," as suggested by Giorgio Agamben's concept of "bare life," *homo sacer.* This ambivalent figure, Agamben states, demarcates a zone of undecidability where human life is included in the juridical order "solely in the form of its exclusion [that is, of its capacity to be killed]" (Agamben 1998: 8). In this way, bare life "remains included in politics in the form of the exception, that is, as something that is included solely through an exclusion" (Agamben 1998: 11).

We take up the arguments presented here to suggest that legal nationality neither provides proper "protection and recognition" nor does it prevent the Kurds from subjection to "bare life," as citizenship is substantially undermined by the systematic execution of the policies and practices of nonrecognition, domination, and external and internal displacement. Although we acknowledge structural vulnerabilities arising from statelessness, and recognize its severe consequences for the communities involved, we do not understand stateless communities as lacking agency and capacity of collective action.

Taken together, by statelessness we mean *a double dispossession*, that is, fundamental restrictions of freedom as a people and as citizens: being denied equal citizenship and minority rights within the "host" states of which they are nationals while simultaneously deprived of a statehood of their own (for a historical background see Bezwan 2018: 64). Statelessness as a double dispossession is understood as an underlying mechanism that links policies of domination, denial, and displacement conducted by the states ruling over Kurdistan with the Kurdish quest for self-rule in the homeland, and the search of the Kurdish migrant communities for belonging, identity, and collective action in receiving societies. As such, it is capable of explaining states' repressive policies in the countries of origin and the parameters within which Kurdish migrants navigate and negotiate their ways in the diaspora, and organize their transnational social relations and diasporic collective action.

This is where the question of identity and belonging acquires its specific relevance and meaning in the broader Kurdish context. In agreement with Stuart Hall, Nira Yuval-Davis, James Tully, and many other scholars, we do not understand identity as an essentialist concept, rather it is "a strategic and positional one" (Hall 1996: 3–4). As such, identities are neither unified nor singular "but multiply constructed across different, often intersecting and antagonistic, discourses, practices and positions" and therefore "are constantly in the process of change and transformation" (Hall 1996: 3–4).

As suggested by Tully (2008: 168 ff), as a "mutable and on-going construct of practical and intersubjective dialogue," identity is not a matter of "theoretical reason or unmediated ascription," but rather, it is a practical notion. As such, for one thing, it expresses self-awareness and self-formation. And for another, it remains relational and intersubjective as it is "acquired and sustained in relation with those who share it and those who do not" (Tully 2008: 169). Understanding identity in practical, intersectional, and relational terms implies that "the same positioning and identifications for all members of the group" cannot be assumed (Yuval-Davis 2010: 271). Framed in this way, the question then becomes how to shift the antagonistic boundaries set by homogenizing, essentializing, and hegemonic identities while avoiding delegitimizing the quest for different conceptions of belonging and identity, whether communal or individual.

Drawing on theoretical and empirical work presented in this article, we offer *inclusion through recognition* as a way of linking identity, belonging, and citizenship within multicultural and multinational societies. Inclusion based on recognition would allow us to see the "diverse cultural and national identities of citizens as overlapping, interacting and negotiated over time" (Tully 2008: 160). It would then more likely facilitate a sense of belonging to, and identification with, a polity, as it would provide citizens with having a say "in the formation and governing of the association" and seeing "their own cultural ways publicly acknowledged and affirmed in the basic institutions of their society" (Tully 1995: 197–98). Finally, inclusion through recognition would, it is hoped, create conditions of democratic diversity where identity and belonging can be reconsidered and renegotiated in a dialogical and participatory manner and where "difference without domination" can be achieved (Allen 2020: 45).

Conclusion

The causes of migration have significant bearings not only on the position of migrants vis-à-vis both the country of origin and that of settlement but also on the ways in which migrants are perceived by the receiving societies. In this chapter, we have therefore outlined the factors affecting internal and external displacements of the Kurds and how these shape the identity, self-understanding, and collective action of the emerging Kurdish diaspora. Based on our fieldwork and a growing body of literature, we have argued that the Kurdish diaspora can be seen as a politicized, plural, and hybrid formation that has been shaped not only by the fundamental experience of statelessness, violence, and displacement in the country of origin but also by opportunities and difficulties in the country of exile.

Given the importance placed on statelessness in the wider Kurdish context, we have presented a framework by conceptualizing statelessness as a political phenomenon that entails a simultaneous deprivation of the full enjoyment of citizenship rights and of self-rule. This dual deprivation, we argued, provides the key to understanding the Kurdish quest for recognition, belonging, and identity both in the homeland and the "abroad home."

The nature of the proverbial activism of Kurdish diaspora, its pronouncedly decolonial discourse, including its militant, and sometimes problematic, manifestations, cannot be properly understood without considering the background presented above. Approaching this activism in terms of "ethnic terrorism" or "terrorism" is both unjust and unproductive. Instead of labeling the collective action of migrants in such a way that their legitimate concerns and nonviolent activism are delegitimized, we have offered *inclusion through recognition* as a more democratic and inclusive way of achieving cohesion under conditions of heterogeneity. Based on both our academic work and long-standing personal experiences arising from our own history of emigration, we would like to emphasize the immense importance of a politics of recognition toward the Kurdish diaspora by host societies, which will facilitate both a deeper understanding of their conditions and a better integration into the country of settlement.

Notes

1 "Al-Anfal" literally means "the spoils of war" and refers to the eighth chapter of the Quran, which came to the prophet in the wake of his first jihad against non-believers (Hardi 2011: 13).

2 As indicated by the Office of the United Nations High Commissioner for Human Rights (OHCHR) in February 2017, "the Turkish Government forces have been conducting security operations in a number of provinces of South-East Turkey involving thousands of troops serving with combat-ready infantry, artillery and armoured army divisions, as well as the Turkish Air Force" (UN Report 2017: 6).

3 A UN Independent International Commission of Inquiry has qualified ISIS's violence against the Yazidis a case of genocide, indicating that "ISIS has sought to destroy the Yazidis through killings; sexual slavery, enslavement, torture and inhuman and degrading treatment and forcible transfer, causing serious bodily and mental harm; the infliction of conditions of life that bring about a slow death; the imposition of measures to prevent Yazidi children from being born, including forced conversion of adults, the separation of Yazidi men and women, and mental trauma."

4 Although there are a large number of the Kurds, most notably the Kurds in Syria, who have been deprived of citizenship or simply not recognised as citizens in their homeland (Chatty 2017: 133 ff.).

References

Adamson, Fiona B. (2013). *Mechanisms of Diaspora Mobilization and the Transnationalization of Civil War*. Cambridge: Cambridge University Press.

Agamben, Giorgio (1998). *Homo Sacer: Sovereign Power and Bare Life*. Stanford, CA: Stanford University Press.

Alexander, Claire (2017). "Beyond the Diaspora Diaspora: A Response to Rogers Brubaker." *Ethnic and Racial Studies* 40(9): 1544–55.

Allen, Danielle (2020). "A New Theory of Justice, in Difference Without Domination: Pursuing Justice," 27–57. In Danielle Allen and Rohini Somanathan, eds., *Diverse Democracies*. Chicago: University of Chicago Press.

Arendt, Hannah (2007). "We Refugees," 264–74. In Jerome Kohn and Ron H. Feldman, eds., *Hannah Arendt: The Jewish Writings*. New York: Schocken Books.

Ayata, Bilgin (2011). "Kurdish Transnational Politics and Turkey's Changing Kurdish Policy: The Journey of Kurdish Broadcasting from Europe to Turkey." *Journal of Contemporary European Studies* 19(4): 523–33.

Baser, Bahar, et al. (2015). "(In)visible Spaces and Tactics of Transnational Engagement: A Multi-dimensional Approach to the Kurdish Diaspora." *Kurdish Studies* 3(2): 128–50.

Bauböck, Rainer, and Thomas Faist (2010). "Diaspora and Transnationalism: What Kind of Dance Partners?," 9–34. In Rainer Bauböck and Thomas Faist, eds., *Concepts, Theories and Methods*. Amsterdam: Amsterdam University Press.

Bezwan, Naif (2018). "Addressing the Kurdish Self-Determination Conflict. Democratic Autonomy and Authoritarianism," 59–82. In Ephraim Nimni and Elçin Aktoprak, eds., *Democratic Representation in Plurinational States: The Kurds in Turkey*. London: Palgrave Macmillan.

Blitz, Brad, and Caroline Sawyer (2011). "Introduction," 1–19. In Caroline Sawyer and Brad K. Blitz, eds., *Statelessness in the European Union, in Displaced, Undocumented, Unwanted*. Cambridge: Cambridge University Press.

Brubaker, Rogers (2005). "The 'Diaspora' Diaspora." *Ethnic and Racial Studies* 28(1): 1–19.

Chatty, Dawn (2017). *Syria. The Making and Unmaking of a Refuge State.* Oxford: Oxford University Press.

Demir, Ipek (2017). "The Global South as Foreignization: The Case of the Kurdish Diaspora in Europe." *The Global South* 11(2): 54–70.

McAdam, Doug (1982). *Political Process and the Development of Black Insurgency 1930-1970.* Chicago: University of Chicago Press.

Eccarius-Kelly, Vera (2019). "Kurdish Studies in Europe," 22–34. In Michael M. Gunter, ed., *Routledge Handbook on The Kurds.* London: Routledge.

Eccarius-Kelly, Vera (2002). "Political Movements and Leverage Points: Kurdish Activism in the European Diaspora" *Journal of Muslim Minority Affairs* 22(1): 91–118.

Eliassi, Barzoo (2016). "Statelessness in a World of Nation-states: The Cases of Kurdish Diasporas in Sweden and the UK." *Journal of Ethnic and Migration Studies* 42(9): 1403–19.

Encyclopaedia Iranica (n.d.) "Qasemlu (Ghassemlou) Abd-al-Rahman," Accessed October 13, 2020, https://www.iranicaonline.org/articles/qasemlu%20accessed%20 October%2013

Grossman, Jonathhan (2019). "Toward a Definition of Diaspora." *Ethnic and Racial Studies,* 42(8): 1263–82.

Hall, Stuart (1996). "Who Needs Identity?," 1–17. In Stuart Hall and Paul Du Gay, eds., *Questions of Cultural Identity.* London: Sage.

Hardi, Choman (2011). *Gendered Experiences of Genocide: Anfal Survivors in Kurdistan-Iraq.* Farnham, UK: Ashgate Publishing Company.

Hassanpour, A. 1992. *Nationalism and Language in Kurdistan, 1918–1985.* San Francisco, CA, Mellen Research UP.

Hassanpour, Amir (1998). "Satellite Footprints as National Borders: MED-TV and the Extraterritoriality of State Sovereignty." *Journal of Muslim Minority Affairs* 18(1): 53–72.

Hiltermann, Joost (2008). "The 1988 Anfal Campaign in Iraqi Kurdistan." SciencesPo: Mass Violence and Resistance Network. Accessed September 18, 2020, https:// www.sciencespo.fr/mass-violence-war-massacre-resistance/en/document/1988-anfal-campaign-iraqi-kurdistan.html

Keles, Janroj Yilmaz (2011). "Transnational Media and Migrants in Europe: The Case of the Mediated Turkish-Kurdish Ethno-National Conflict." PhD thesis, Brunel University, London.

Keles, Janroj Yilmaz (2015). *Media, Diaspora and Conflict: Nationalism and Identity Amongst Turkish and Kurdish Migrants in Europe.* Houndmills, UK: I.B.Tauris.

Keles, J. Y. (2016). "Digital Diaspora and Social Networks." *Middle East Journal of Culture & Communication* 9(3): 315–333.

Klein, Janet (2010). "Minorities, Statelessness, and Kurdish Studies Today: Prospects and Dilemmas for Scholars." *The Journal of Ottoman Studies* 36: 225–37.

King, Russell, Thomson, Mark, Mai, Nicola, and Keles, Janroj Yilmaz (2008). "'Turks' in the UK: Problems of Definition and the Partial Relevance of Policy". *Journal of Immigrant and Refugee Studies* 6 (3): 423–34.

Kohn, Jerome (2007). "Preface: A Jewish Life: 1906-1975," xi–xxxii. In Jerome Kohn and Ron H. Feldman, eds., *Hannah Arendt: The Jewish Writings.* New York: Schocken Books.

Lyon, Alynna J., and Emek M. Uçarer (2001). "Mobilizing Ethnic Conflict: Kurdish Separatism in Germany and the PKK." *Ethnic and Racial Studies* 24(6): 925–48.

Mohammadpour, Ahmad, and Kamal Soleimani (2020). "'Minoritisation' of the Other: The Iranian Ethno-theocratic State's Assimilatory Strategies." *Postcolonial Studies.* doi: 10.1080/13688790.2020.1746157.

Romano, David (2002). Modern Communications Technology in Ethnic Nationalist Hands: The Case of the Kurds. *Canadian Journal of Political Science* 35 (1): 127–49.

Sawyer, Caroline (2011). "Stateless in Europe: Legal Aspects of De Jure And De Facto Statelessness in the European Union," 69–107. In Caroline Sawyer and Brad K. Blitz, eds., *Statelessness in the European Union: Displaced, Undocumented, Unwanted.* Cambridge: Cambridge University Press.

Sheffer, Gabriel (2003). *Diaspora Politics. At Home Abroad.* New York: Cambridge University Press.

Staples, Kelly (2012). *Retheorising Statelessness. A Background Theory of Membership in World Politics.* Edinburgh: Edinburgh University Press.

Syrett, Stephen, and Janroj Yilmaz Keles (2019). "Diasporas, Agency and Enterprise in Settlement and Homeland Contexts: Entrepreneurship in the Kurdish Diaspora." *Political Geography* 73: 60–69.

Tas, Latif (2016). "How International Law Impacts on Statelessness and Citizenship: The Case of Kurdish Nationalism, Conflict and Peace." *International Journal of Law in Context* 12(1): 42–62.

Tully, James (2008). *Public Philosophy in a New Key, Volume I: Democracy and Civic Freedom.* Cambridge: Cambridge University Press.

Tully, James (1995). *Strange Multiplicity. Constitutionalism in an Age of Diversity.* Cambridge: Cambridge University Press.

UN Report (2017). Report of the Office of the High Commissioner of Human Rights. Accessed September 13, 2020, https://www.ohchr.org/Documents/Countries/TR/OHCHR_South-East_TurkeyReport_10March2017.pdf

Vali, Abbas (1998). "The Kurds and Their 'Others': Fragmented Identity and Fragmented Politics." *Studies of South Asia, Africa and the Middle East* 18(2): 83–94.

Vali, Abbas (2020). *The Forgotten Years of Kurdish Nationalism in Iran.* London: Palgrave Macmillan.

van Bruinessen, Martin (1998). "Shifting National and Ethnic Identities: The Kurds in Turkey and the European Diaspora." *Journal of Muslim Minority Affairs* 18(1): 39–52.

Wahlbeck, Osten (2002). "The Concept of Diaspora as an Analytical Tool in the Study of Refugee Communities." *Journal of Ethnic and Migration Studies* 28(2): 221–38.

Yuval-Davis, Nira (2010). "Theorizing Identity: Beyond The 'Us' and 'Them' Dichotomy." *Patterns of Prejudice* 44(3): 261–80.

Zeydanlıoğlu, Welat (2012). "Turkey's Kurdish Language Policy." *International Journal of the Sociology of Language* 2012(217): 99–125.

14 What makes a place a home?

Syrian refugees' narratives on belonging in Turkey

Doğuş Şimşek

Introduction

I do not feel that I belong to Turkey. How can I feel that I belong here if I am not welcomed? How can I think this country as my home if I do not know how long I am allowed to stay in this country? How can I establish my life here and feel at home if I am not welcomed and find difficult to establish a life?

These are the words of Farid, who migrated to Turkey in 2014 from Aleppo. He escaped from the war to a safer place as did many others in Syria. As for many refugees who must leave their countries of origin, the process of establishing a new life in Turkey has not been easy for Farid. Through the study of Syrian refugee narratives of belonging and daily life in Istanbul, I aim in this chapter to explore the home-making processes of Syrian refugees within the political and social context of the settlement country, their individual trajectories, and their experiences of settlement.

The first movement of Syrian refugees to Turkey began in 2011. At the beginning of Syrian migration, Turkey adopted an "open door" policy for Syrians fleeing their country based on a policy of religion-oriented hospitality and began constructing tents in the southern provinces of Hatay, Kilis, Gaziantep, and Şanlıurfa, where Syrian refugees were called "guests," a status that granted them no legal rights and assumed a temporary and short stay. In 2013, the influx of Syrian refugees in Turkey peaked. By late 2014, 55,000 people were seeking asylum in Turkey every month (Içduygu and Şimşek 2016). In summer 2015, the movement of refugees from Turkey to Europe reached the highest level. By 2016, policies and practices were moving in the direction of integration after applying the Law of Foreigners and International Protection (LFIP) in April 2014. For example, Turkey issued a new regulation allowing registered Syrian refugees to apply for work permits, and it was announced that Syrians living in Turkey would be granted citizenship in 2016. Granting full citizenship is an important development but it is not clear whether it would include all Syrians under temporary protection. As of November 2020, according to the Turkish Directorate of General of Migration Management, 59,457 Syrian refugees reside in the camps out of 3.6 million Syrians.[1] The vast majority of Syrian refugees in Turkey are urban

DOI: 10.4324/9781003170686-17

refugees, as only 2.4% of Syrians reside in camps, with the majority of refugees remaining in Turkey's cities. Istanbul has the highest population of Syrians, with 496,561 in residence.[2] I focus on the experiences of urban refugees in Istanbul not only because of the high rate of population but also because of the heterogeneity among them.

The main points I address in this chapter concern how Syrian refugees negotiate questions of "home" and "belonging" within their everyday lives in Turkey, how a sense of home emerges among refugees, and what role migration policies play in shaping refugees' sense of belonging and feelings of home. I employed a narrative methodology to explore Syrian refugees' conception of "home" and sense of belonging in order to understand what makes a place a "home" in the context of forced migration. Through semistructured in-depth interviews, I collected stories of four Syrian refugees in Istanbul; I argue that the meaning of home and a sense of belonging for Syrian refugees in Turkey is articulated by the hierarchies of belonging, which are mainly related to class. Focusing on the stories of four Syrian refugees who reside in urban areas and come from different class, ethnic, religious, and gender backgrounds, I show that the meaning of home and a sense of belonging is intimately connected to class, a situation that has not received much attention in the literature on forced migrants and belonging.

The questions of home and belonging

The questions of home and belonging in the context of displacement have started to gain scholarly attention in the literature on forced migration since Helen Taylor (2013: 130) stated, "Though home is central [to] any understanding of displacement, the concept has not been explored as fully as possible in forced migration literature." Rather than placing peoples and cultures in defined geographical spaces contained within national borders, "home" should be defined as a set of social relations, socially and culturally constructed places focusing on a variety of experiences of refugees in the countries of settlement (hooks 1990; Massey 1994; Mallett 2004; Rosello 2020). Although the relationship between place and space is considered as an important factor in defining "home" and "belonging," the relationship within and through time also played a crucial role in their definition (Blunt and Dowling 2006; King and Christou 2008). Avtar Brah (1996: 190) poses important questions regarding how to define home: "When does a location *become* home? What is the difference between 'feeling at home' and staking claim to a place as one's own?"

In trying to understand the meaning of home, scholars aim to answer similar questions. For instance, Shelley Mallett (2004: 84) stated that "how home is and has been defined at any given time depends upon 'specification of locus and extent' and the broader historical and social context." In this sense, the definition of home also comprehends the situations and conditions of refugees in the receiving society—their feelings, practices, experiences, and relationships with the members of the receiving society. Doreen Massey (1994: 171) defines home as a "meeting place" where a set of social relations take place. In this case, refugees may feel the receiving country, the country of origin, or elsewhere as "home," or as a place where they "belong." But the receiving societies tend to disagree with the complexity of "home" when they argue that "refugees go home" (Rosello 2020). Rather than focusing on a geographical unit, the definition of home could include social networks, experiences, and even memories of refugees. According to Hall (1990), "home" and "belonging" are indicated through symbolic categorization of place and space. The questions of home and belonging in the context of displacement mainly focus on the relationship within and through time, and the sense of space; however, the ways in which refugees define the concepts of home and belonging might be different.

Taking into account the agents' perspectives, I aim in this chapter to present an understanding of how refugees define the concepts of home and belonging and whether their experiences in the receiving society have an impact on the ways they define these concepts. As I explore whether possessing economic resources are crucial in refugees' definition of "home" and "belonging," I focus on the class of Syrian refugees. My analysis of class in the case of Syrian refugees highlights the fact that the allocation of rights is based on refugees' economic resources, which influence refugees' sense of belonging. I also look at class as an analytical category that shapes one's positioning in society—in terms of current and future possibilities, life chances, and relationships—and thus examine it as a set of inequalities that affect how power works, how people get exploited, and how people are differently valued (Sayer 2005; Skeggs 2004).

Research methods

The primary methods I employed in my "narrative research" (Connelly and Clandinin 1990; Gudmundsdottir 2001) are participant observation and semistructured in-depth individual interviews with four refugees. By choosing to conduct narrative research, I aimed to understand the experience of Syrian refugees through my discussions with them. My approach involved providing a space for Syrian refugees' to tell their stories. In total I have collected stories from four Syrian refugees between the ages of 27 and 45—two women and two men—in Istanbul.

Syrians in Turkey are heterogeneous in terms of ethnicity, religion, gender, generation, and social class. There are Kurdish, Turkmen, Arab, Shi, Dom, Abdal, Armenian, Yazidis, Assyrian Syrian nationals; Palestinian and Iranian refugees coming from Syria; working-class, middle-class, and upper-class Syrians; and Syrians from diverse religious backgrounds, including Christians and Muslim Alawites and Sunnis, settled in various cities of Turkey. In selecting research participants, I have also considered ethnic, religious, and social class heterogeneity of Syrian refugees besides gender. Among four research participants, three were one Sunni Arab, one Kurdish and one Turkmen, and one Armenian. Some worked in the informal economy for very low wages; others were wealthier, running their own businesses in Istanbul. I selected research participants from different class, ethnic, religious, and gender backgrounds in order to understand whether the heterogeneity of Syrian refugees—their diverse identities—had an impact on their settlement processes and the ways they define "home," on how they developed a sense of belonging, and how they negotiated the questions of home and belonging within their everyday lives in Istanbul. Their class, ethnic, religious, and gender differences matter because these differences might affect their sense of belonging in the new place. The research participants' length of stay in Turkey varied; while some had migrated three years before the time of my observation, others had been living in Istanbul for six years. Pseudonyms are used when referring to all research participants.

I conducted the ethnographic research in Istanbul over a two-year period between 2016 and 2018. During the course of my fieldwork, I observed participants at a local NGO that offered services and organized events for refugees in Istanbul's Fatih neighborhood. Apart from participant observation, I also conducted semistructured in-depth interviews with my research participants to learn how they created a sense of belonging, their meanings of home and experiences in Turkey. I used qualitative content analysis to identify a set of common themes from the narratives, and then employed a thematic coding system with NVivo, which helped to create analytical categories. The next sections present the accounts of the four Syrian refugees and their definition of home.

"Home is in our memories"

When I met with Farid at the community organization, he was interested in sharing his settlement process in Turkey with me. Farid, a 27-year-old Syrian, male, Sunni Arab, who has been living in Turkey for five years and working in a textile workshop, answered the question of what he thinks about his life in Turkey by saying, "Which life? Is this a life?" While sitting at a café, he began talking about the early phases of migration and then his experiences during the settlement process. As Farid described:

> We (the Syrians) had to leave our home. I migrated to Turkey with my mother and father five years ago. We thought that we will stay in Turkey for some time and then we will be back to Syria when things get better. However, the situation was different than we thought.

His expectations of his life in Turkey differ from what he has been experiencing. His disappointment is related to not feeling secure about their migratory status and future in Turkey, to experiencing racism in everyday life, and to the lack of access to rights. He told me:

> Some of our relatives and my friends went to Lebanon, others migrated to Jordan. We decided to come to Turkey because my father's friend who live in Istanbul said that there are many Syrians in Istanbul and it would be easier to find a job and accommodation here. After staying at the hotel

in Gaziantep for two days, we went to Istanbul and stayed at my father's friends' house until we find an accommodation. We spend our saving on rent. Rent and bills are quite expensive in Istanbul, so I had to find a job as soon as possible. I started to work in a textile workshop informally because getting a work permit is not easy. I have had no social security, worked longer hours and, have not got my wages on time. As you know we are under temporary protection and not knowing whether it will be turned into a permanent residency affects our plans for the future.

Due to a lack of policy implementation, Syrian refugees have been struggling to survive (Şimşek 2019). Although Syrians have had the right to apply for a work permit according to the Regulation on Provision of Work Permits for Foreigners under Temporary Protection since January 2016. Farid has been working informally, as many Syrians do, because getting a work permit is difficult; it depends upon employers' willingness to offer contracts of employment and is accessible for refugees who have held Turkish identification documents for at least six months. In fact, since 2016, more than 132,497 work permits have been issued to Syrian nationals (including those under temporary protection and Syrians who have a residence permit).[3] This clearly shows that the majority of Syrians, working in an informal economy without social security, face exploitation, experience a lack of safe working conditions, and are overworked and underpaid, all of which causes the exclusion of many refugees from wider society (Şimşek 2020). When explaining his life in Turkey, Farid refers to struggles mainly related to the socioeconomic inequalities that highlight the class issue. Despite all these uncertainties about the future, experiencing racism in everyday life also makes it difficult for him to feel like Turkey is home. He said:

We do not know how long we will be allowed to stay in Turkey, and I am not sure whether I want to live here permanently. There are many reasons that make me feel insecure. I felt outsider every single day I have been in Istanbul. I have experienced racism in the street, at work, at the hospital and often been reminded that I can never be a part of the society. When you know that you are excluded from the society, you can never feel [like you] belong to that society. Then, you always think about the past, the society you'd grown up in, you feel that it is a part of you; you also start thinking about the meaning of home more than before.

Farid's account shows that the question of belonging is correlated with the feeling of being both insecure and excluded. The feeling of not being a part of the society makes him think about the meaning of home, which is something he did not question when he was living in Syria. When I asked him, "What does home mean for you?," Farid added:

I have not thought about the meaning of home before leaving Syria. I only lived in Syria and it was a place that is safe for me. Syria I am referring here is the country before the war. Syria is, of course, not a safe country for us now. I had had my childhood, social relationships; spent happy times in my life there. When I started to experience being an outsider, it made me think about the sense of belonging and safety as well. I still do not have a particular meaning of home. I think what makes a place home is the feeling of safety, and I believe that it would be easier to feel [like I] belong to somewhere if I would be welcomed and feel safe. I neither feel safe in Turkey nor in Syria. So, I do not know where my home is, it is probably in my memories.

"Home" for Farid is not associated with a particular place; rather it is about security. But he cannot find security either in Turkey (because of being under temporary protection, lacking access to rights, and experiencing racism) or in Syria (because of the country's insecure environment). Farid's life in Turkey as a Syrian refugee and the situation of his country of origin reveals how home is defined through ongoing experiences of forced migration. He did not think about what home means to him before experiencing forced migration. Home suddenly becomes a place in his memories rather than a geographical place. The meaning of home, therefore, is closely linked with the experience of forced migration as argued by Liisa Malkki (1995).

Home as a community

Eliza is a 39-year-old female, an Armenian Syrian who has been living in Turkey for four years and working in a shop that sells jewelry in Istanbul. She told me, "I have never felt this rootless before" when I asked her about life in Turkey. I had a conversation with her while we walked around Istanbul's Grand Bazaar, one of the oldest covered markets in the world, where she works. During the conversation, she motioned with her hand to show me a young man in the bazaar and said,

> Look at this young man, he came from Afghanistan a few years ago, he is carrying heavy bags, earning very low income, has no social security, and he does everything his boss tells him to do. He never says 'no' because he is in need of this job. I know that he hates what he is doing, this place and people he has to work for. I can see this from his eyes, because I went through the similar experience.

In response to my question of what she thinks about her life in Turkey, she related her experiences of living in Turkey with the experiences of the Afghan refugee, and said the following:

> In order to survive, refugees struggle a lot. My friend, Syrian, was carrying heavy bags like this Afghan guy when he first came to Istanbul five years ago. Then, he found another job. I was luckier and found this job straight away. My employer got a work permit for me because I was

speaking English and Arabic and communicating with tourists in both languages. We, refugees from Syria, are in a better situation compare to Afghan or Iraqi refugees as they do not have a right to work in Turkey. Although I can formally work and earn money in Turkey, I was happier within the Armenian community in Aleppo. I was running a jewelry shop in Aleppo where I was making craft. It was a family business. In Aleppo, we (Armenians) were supporting each other, shopping from each other's shops, and spending time together. When I was living in Aleppo, I was always around the Armenian community who largely consisted of the descendants of people who managed to escape the genocide in Turkey a century ago. We were always in solidarity especially when our churches and houses were destroyed during the war. Many Armenians in Aleppo have migrated to Armenia and Lebanon, none of us wanted to migrate to Turkey because of the history of the genocide. I had to come to Turkey because I have a friend here and I did not know anyone in Lebanon and Armenia.

When Eliza explained the process of finding employment in Turkey as a Syrian–Armenian refugee, she talked about the privilege of being Syrian compared to other forced migrants in terms of getting a work permit. But she also mentioned the difference between herself and her friend in relation to class. In fact, Eliza's and her friend's experience show that neoliberal migration policy creates a distinction between Syrian refugees that is based on their class and profession. Eliza's account indicates that the meaning of home is associated with a community. Compared to Farid, Eliza did not mention insecurity related to her migratory status, socioeconomic inequalities, or exclusion when defining home. When I asked her whether she feels belonging to Istanbul, she explained:

I do not have friends who support and help me in Istanbul. This makes me feel [like an] outsider, not part of the society. It is also very difficult for me to establish social relations with the receiving society, as it is not only because I am a Syrian refugee but being an Armenian also makes constructing social relations with the receiving society difficult. I am also not feeling safe to be here as an Armenian because of the memory and the history of the genocide. There are ongoing anti-Syrian sentiments which even turn into racist attacks against Syrians in Turkey. This makes it difficult to establish social relations with Turkish people. It is difficult to feel [like I] belong to the Turkish society knowing that you are not welcomed, supported. I do not feel at home here because I am not a part of a community which helps me to have a sense of belonging and, feel safe.

The meaning of home for Eliza is not related to a geographical place; rather, it is about community associated with a sense of solidarity and affinity. She does not refer to insecurity and socioeconomic inequalities but rather to not being a part of the society, a feeling linked to the lack of social bridges between Syrians and the receiving society. As stated by Alistair Ager and Alison Strang (2008), social bridges constructed with the members of the receiving society are crucial for refugees to be integrated into the receiving society and to provide aspirations to establish a new life in the receiving country. Eliza does not feel at home in Turkey because "the feeling of being 'at home' has not been built" as stated by Ghassan Hage (1997: 102).

Eliza's definition of home as beyond a geographical place is closely linked to Massey's definition of home as a set of social relations in a "meeting place" (Massey 1994: 171). When Eliza talks about the social relations she established in Aleppo, she refers to the practice of solidarity. What makes a place a home is not only related to social relations for Eliza, it also means sharing and solidarity that is left in memories. She highlighted the role of community when referring to belonging and home. In her work on the deconstruction of home in migration literature, Rosemarie Buikema (2005: 178) emphasizes the role of community as a major site of belonging and identity. Home is neither a place nor a space for Eliza: it is related more to a sense of solidarity, which makes her feel like she belongs to a society, than it is to the impact of neoliberal migration policy.

Home as a space of freedom

Solin, 41 years old, is a female Kurdish Syrian who has been living in Turkey for seven years and working in a textile workshop. She explained her life in Syria by highlighting her socioeconomic background and ethnic identity. She told me:

> I was living in Aleppo with my husband and two children. I was working in a textile manufacturer and my husband was working in building construction. We were living in a small flat, our children were carrying out their education, and we had our friends around. We did not have money, but we were happy. Then, we had to migrate to Turkey although we were not keen on going to Turkey as Kurds have been discriminated there. Although we did not want, we migrated to Turkey because we knew people who could help us to settle down. Since the day I migrated to Turkey, I have been feeling discrimination in every aspect of my life. It is not only in the workplace, but also in the street, in the institutions and in interacting with people. I have always been a minority in my whole life. I was a minority in Syria. When I migrated to Turkey, I became a refugee. Turkey is not a safe place for refugees, especially for Kurdish people.

Even before migrating to Turkey, she was worried about experiencing discrimination in Turkey because of her ethnic identity. Solin made assumptions about what she might be experiencing in Turkey based on her previous experiences as a minority in Syria. But she stated the fact that Turkey is not a safe place for Syrian-Kurdish refugees because of how the country has been treating Kurdish people. When I asked her about her experiences in Turkey, she referred to difficulties not only of accessing rights but also of the socioeconomic inequalities. She explained:

> When things get so violent for us, we decided to leave Syria in one day. Since living in Turkey, we have experienced difficulties in finding jobs and accommodation, as have many refugees. After some time, I started working in a textile workshop and, my husband in a restaurant. We were employed informally because getting a work permit is not easy for unskilled Syrians. I do not know whether we will have a future in Turkey. I feel insecure and an outsider in Turkey, and do not feel I belong to the Turkish society at all. I am not sure whether I was feeling at home in Syria. I have not thought about it before I started living in Turkey.
>
> I think I was freer in Syria because I was not a refugee. I did not need to get a work permit to work. Surviving is difficult in Turkey. When you work informally, you do not know whether you will get your wages or keep the job. There is no guarantee. You do not know whether they will deport you. We are insecure and cannot plan for our future under these circumstances. I also feel excluded from the society as Turkish people do not want Syrian refugees and I know that they do not like Kurdish people as well. So, I am very scared to speak Arabic and Kurdish in the street. Apart from being insecure and [experiencing] economic hardships, I lost the freedom of being able to speak in my mother tongue without any fear in Turkey, and this causes a lack of a sense of belonging.

Solin's account indicates that Turkey's neoliberal migration policy does not support the integration of all Syrians residing in Turkey, as only "selected" Syrians are to be propped up. Her settlement process is influenced by her class background. The lack of implementation of an integration policy causes unequal access to rights and participation of refugees in the receiving society and insecurity about her future. Her sense of belonging is influenced by the way she was treated by the receiving society, which is related to her class and to being a refugee and a Kurd. Her relationship with home is not only linked to being able to survive and having security but also identified with having spaces of inclusion, which can be understood as a response to racism.

Home is where I earn money

Berdi, 43 years old, male, a Turkmen Syrian, has been living in Turkey for seven years and running his own business in Istanbul. He smiled as he explained his life in Turkey to me. He started the conversation by referring to his Turkish identity. He said, "I am also Turkish, like yourself," and continued:

> I am a Syrian Turkmen, Syrian citizen of Turkish origin. My grandads were living in Turkey; they arrived in Syria during Ottoman rule. Our

roots are in Anatolia. I am happy to be back in Turkey. I came to Turkey in 2013 with my wife and two children. I was living in Aleppo and owning my own restaurant. I had a good life in Aleppo. I have a few friends in Turkey who could help me with establishing a business. Knowing Turkish has been an advantage for me in communicating with people and understanding the regulations. We have not experienced an adaptation problem in Turkey. In the beginning, we, of course, found it hard to get used to our new environment but the help of people around us, and knowing the language and culture, helped us to overcome the issues of adaptation.

I know that our Turkmen Syrian situation is much better compared to Sunni, Kurdish, and Christian Syrians due to cultural similarity and language efficiency. With the help of my friends in Istanbul, we found a decent flat to rent and a shop to run my business. Before coming to Turkey, I sold out my business and brought some investment with me that helped me to establish my own restaurant in Aksaray, Istanbul. When you asked me the meaning of home, I first think about the livelihood. I have established my business, been earning a good income and bought a flat. Our life quality has not changed at all. The meaning of home is a place where I earn money, have a quality of life, understand the language and culture of the society. I, therefore, feel that Istanbul is my home. Of course, having historical past and our roots in Turkey also make me feel that Turkey is my home.

Berdi highlights different aspects of his identity which influence his sense of belonging and feeling of home. In Berdi's account his class background, social networks, and language efficiency have an impact on his sense of belonging and his feeling of home. Compared to Farid and Solin, he does not experience financial hardship as he brought investments with him to establish his business. He also does not feel insecure about his future as other participants do. Citizenship offered to Syrians to be a "selective citizenship," which targets investors and highly skilled individuals and is not extended to less skilled individuals, laborers, and small-wage earners (Şimşek 2020; see also Akçapar and Şimşek 2018). As for being invested in Turkey, it is easier for Berdi to be granted Turkish citizenship and establish his future in Turkey.

Berdi's definition of home refers to a place where he can earn money and establish a business. The feeling of loss in relation to "home," language, and occupation, which Hannah Arendt (2007) articulated in her well-known 1943 essay "We Refugees" is not part of Berdi's experience. When he talks about the meaning of home, he does not refer to a loss related to the culture, community, and unfamiliarity with the language. Although Salman Akhtar (1998) stated that the grief associated with losses as a result of migration can cause disturbance to the individual's identity and sense of belonging, in Berdi's case migration did not cause a particular loss or a serious disturbance to his individual identity or a sense of belonging; he is proud to define himself as a Turkmen who has a Turkish root. For him, migrating to Turkey refers to being back to the homeland of the previous generations. Being able to speak Turkish and share a cultural similarity are also advantageous for Turkmen-Syrians when building a sense of belonging with the receiving society. The meaning of home is not only associated with the essentialist understanding, it also highlights the aspect of wealth, being able to survive and carry out the similar lifestyle and habit.

Conclusion

By focusing on the stories of four Syrian refugees who are from different class, ethnic, gender, and religious backgrounds, this chapter shows that the meaning of home and a sense of belonging are intimately connected to class, a finding that has not received much attention in the literature on forced migration and belonging. The accounts of Farid, Eliza, Solin, and Berdi show various meanings given to the concept of "home" and the ways these concepts shaped the refugees' sense of belonging. The meaning of home for Syrian refugees is complex and refers to their past and present experiences. Home means a different thing depending on who defines it. In the case of Farid, Eliza, Solin, and Berdi, although being a refugee has an impact on their definition of home, their class background matters more in their conception of home and belonging. The accounts of Farid and Solin indicate that their sense of belonging and the feeling of home are influenced by the neoliberal migration policy, which highlights the fact that the allocation of rights is based on refugees' economic resources. Their class background is reflected in their experiences in the receiving society especially in relation to accessing rights that influence refugees' sense of belonging. In the case of Farid and Solin, home is not associated with a particular place; rather it is about security, which they cannot find either in Turkey (because of being under temporary protection, lacking access to rights, and experiencing racism) or in Syria (because of the country's insecure environment).

Although Farid and Solin referred to struggles mainly related to socioeconomic inequalities that indicate the class issue, Eliza and Berdi did not show that insecurity and socioeconomic inequalities influence their sense of belonging and feeling of home. While Berdi referred to Turkey as his home because of his smooth settlement process—which is mainly related to his class status and language efficiency, as those make him a "good refugee" in the eyes of the government and the society—Eliza mentioned the importance of having a sense of solidarity that is often linked with community rather than her experiences during the settlement process.

The difference in the refugees' conception of home is linked to their experiences during the settlement process in Turkey and how members of the receiving society perceive them. As Mireille Rosello (2020) states, refugees trying to rebuild a home depend on citizens' willingness to shape the function of home and make it more welcoming for refugees. Therefore, the refugees' concept of home is also identified with which refugees are welcomed by the receiving society—in the case of Syrian refugees in Turkey, this is mainly related to class. The stories of urban refugees reflect class heterogeneity and how it affects their sense of belonging. This chapter shows that Farid, Eliza, Solin, and Berdi articulate the meaning of home through their experiences in the receiving society. Those experiences are mainly related to their settlement processes and the hierarchies of belonging, especially regarding the emphasis on class that the receiving society applies to Syrian refugees as a result of the neoliberal migration policy. Thus, their experiences in the receiving society play a crucial role in refugees' sense of belonging and feeling of home.

Notes

1 Directorate General of Migration Management's recent statistics on demographics of Syrians under temporary protection in cities and at camps. Accessed April 2020, https://en.goc.gov.tr/temporary-protection27.
2 Ibid. (accessed April 2020).
3 Regional Refugee and Resilience Plan (3RP) in Response to the Syria Crisis: Country Chapter—Turkey (January 2020), Reliefweb Report. Accessed April 2020, https://reliefweb.int/report/turkey/regional-refugee-and-resilience-plan-3rp-response-syria-crisis-country-chapter-turkey.

References

Ager, Alastair, and Alison Strang (2008). "Understanding Integration: A Conceptual Framework." *Journal of Refugee Studies* 21(2): 166–91.

Akçapar, Şebnem, and Doğuş Şimşek (2018). "The Politics of Syrian Refugees in Turkey: A Question of Inclusion and Exclusion through Citizenship." *Social Inclusion* 6(1): 176–87.

Akhtar, Salman (1998). *Immigration and Identity: Turmoil, Treatment and Transformation*. Northvale, NJ: Jason Aronson.

Arendt, Hannah (2007). "We Refugess." In Jerome Kohn and Ron H. Feldman, eds., *The Jewish Writings*. New York: Schocken Books.

Blunt, Alison, and Robyn Dowling (2006). *Home*. London: Routledge.

Brah, Avtar (1996). *Cartographies of Diaspora*. London: Routledge.

Buikema, Rosemarie (2005). "A Poetics of Home," 177–87. In Sandra Ponzanesi and Daniela Merola, eds., *Migrant Cartographies*. Lanham, MD: Lexington Books.

Connelly, Michael F., and Jean D. Clandinin (1990). "Stories of Experience and Narrative Inquiry." *Educational Researcher* 19(5): 2–14.

Gudmundsdottir, Sigrun (2001). "Narrative Research on School Practice," 226–40. In Virginia Richardson, ed., *Fourth Handbook for Research on Teaching*. New York: Macmillan.

Hage, Ghassan (1997). "At Home in the Entrails of the West: Multiculturalism, 'Ethnic Food' and Migrant Home Building," 99–153. In Helen Grace, Ghassan Hage, Lesley Johnson, Julie Langsworth, and Michael Symonds, eds., *Home/World: Space, Community and Marginality in Sydney's West*. Sydney: Pluto.

Hall, Stuart (1990). *Cultural Identity and Diaspora: Identity: Community, Culture, Difference*. London: Lawrence and Wishart.

hooks, bell (1990). *Yearning: Race, Gender, and Cultural Politics*. Boston, MA: South End Press.

Içduygu, Ahmet, and Doğuş Şimşek (2016). "Syrian Refugees in Turkey: Towards Integration Policies." *Turkish Policy Quarterly* 15(3): 59–69.

King, Russell, and Anastasia Christou (2008). "Cultural Geographies of Counter-diasporic Migration: The Second-generation Returns 'Home.'" Brighton: Sussex Centre for Migration Research, University of Sussex, Migration Working Paper 45.

Malkki, Liisa H. (1995). "Refugees and Exile: From 'Refugee Studies' to the National Order of Things." *Annual Review of Anthropology* 24: 495–523.

Mallett, Shelley (2004). "Understanding Home: A Critical Review of the Literature." *The Sociological Review* 52(1): 62–89.

Massey, Doreen (1994). *Space, Place and Gender*. Cambridge: Polity Press.

Rosello, Mireille (2020). "Autobiography of a Ghost: Home and Haunting in Viet Thanh Nguyen's *The Refugees*," 519–32. In Emma Cox, Sam Durrant, David Farrier, Lyndsey Stonebridge, and Agnes Woolley, eds., *Refugee Imaginaries: Research Across the Humanities*. Edinburgh: Edinburgh University Press.

Sayer, Andrew (2005). *The Moral Significance of Class*. Cambridge: Cambridge University Press.

Şimşek, Doğuş (2019). "Transnational Activities of Syrian Refugees: Hindering or Supporting Integration." *International Migration* 57(2): 268–82.

Şimşek, Doğuş (2020). "Integration Processes of Syrian Refugees in Turkey: Class-based Integration." *Journal of Refugee Studies* 33(3): 537–54.

Skeggs, Beverly (2004). *Class, Self, Culture*. London: Routledge.

Taylor, Helen (2013). "Refugees, the State and the Concept of Home." *Refugee Survey Quarterly* 32(2): 130–52.

15 "This is about making family"

Creating communities of belonging in schools serving refugee-background students

Shawna Shapiro

Introduction

In recent decades, the numbers of students from immigrant and refugee backgrounds in US public schools have risen steadily (Bialik, Scheller, and Walker 2018). While this rise means an increase in cultural, racial/ethnic, and linguistic diversity, many school districts tend to view this population of students, most of whom are English Learners (EL),[1] as a problem, rather than an asset (Noguara 2004; Gutiérrez and Orellana 2006; Shapiro 2014). Deficit narratives about EL students and families—that they are "difficult" to teach and work with—result in school cultures in which these students are treated as second-class citizens (Valenzuela 1999; Valdés 2001). A number of scholars have documented how this phenomenon plays out in particular ways with refugee-background (RB) students (Kia-Keating and Ellis 2007; Roy and Roxas 2011; Edgeworth 2015). At schools in communities of resettlement, students whose families were forcibly displaced from their countries of origin often experience new forms of othering and exclusion (Kanu 2008; Keddie 2012; Taylor and Sidhu 2012).

Yet few studies have focused on how to counteract these conditions—that is, how to create school cultures of belonging—particularly at the secondary level. As Kathryn Edgeworth (2015: 363) explains:

> The task for future research is to understand how to work effectively with teachers and students to challenge these inequalities in schools. This entails attention to both explicit and latent understandings of difference and (un)belonging to interrupt prevailing discourses of exclusion.

My research responds to Edgeworth's (2015) call, by articulating how deficit discourse shapes students' educational experiences and how alternative discourses can contribute to increased educational equity. This project is informed by critical race theory (CRT), an approach to educational research that "challenges traditional research paradigms, texts, and theories used to explain the experiences of students of color," in pursuit of "liberatory or transformative solution[s] to racial, gender, and class subordination" (Solórzano and Yosso 2002: 24). One methodology used prevalently by CRT scholars is critical

DOI: 10.4324/9781003170686-18

storytelling, which seeks to identify and challenge the dominant "master narratives" about marginalized communities—in particular communities of color—by foregrounding "counter-stories" that are often overlooked within those dominant discourses (Ladson-Billings 1998; Solórzano and Yosso 2002).

Similar to research by Laura Roy and Kevin Roxas (2011), my work focuses on the "master narrative" of deficit discourse about refugee-background students and families (see also Montero et al. 2012; Edgeworth 2015). Deficit discourse is often used to justify exclusionary and inequitable educational policies and practices, shifting the focus away from ineffective school structures and casting blame on students and families (Roy and Roxas 2011; Shapiro 2014). Documenting these dynamics is only one part of the work, I am committed to finding ways to resist deficit discourse. Through interviews with students, I have begun to outline alternative discourses ("counter-stories") that are rooted in an understanding and appreciation of the assets that refugee students and their families bring to schools and communities (Shapiro and MacDonald 2017; Shapiro 2019). This asset orientation has salient implications on how researchers and practitioners think of school culture—specifically for our understanding of what hinders and contributes to students' sense of belonging. In this chapter, I draw on case study research in one particular school district in order to answer the question: *What might a school culture of belonging look and feel like for refugee-background (RB) students?*

Questions of exclusion and belonging in this community, which I call "Laketown," first came to my attention one day in April 2012, when a student-led protest at Laketown High School (pseudonym) brought to light the many ways in which RB students felt alienated and underserved. Dozens of students walked out of class that morning to gather in front of their school building, holding signs with messages such as "End Racism!" and "Justice for All!" That afternoon, students, a few teachers, and local reporters gathered for a video-recorded conversation. Much of that conversation was spent discussing an article that appeared in the newspaper the month prior, which included the following passage: "Annual standardized test scores ... show a yawning achievement gap between high-income and low-income students in the district. English language learners, many of whom are African refugees, have even lower scores."

As the conversation circled back again and again to this passage, the reporters reacted with puzzlement: in their view, they were simply stating the facts and trying to hold schools accountable for educational disparities. But students saw the article differently—as reinforcing negative stereotypes about RB students. As one student put it:

> Most of us interpret [the article about low test scores] as you guys calling us stupid. Like, 4-year-olds. ... Because they already call us stupid, and we don't do anything about it, because we know we're smart. ... But her publishing that article just, like, makes people ... say, "Oh wow, their stupidity is even published in the newspapers!"

Students articulated a number of ways in which these negative stereotypes shaped their experiences in school and the community. They described feeling "stuck" in EL classes, without access to a more rigorous academic curriculum, and noted the absence of RB students in Honors and advanced placement (AP) courses. They described hurtful and offensive comments from classmates and community members, including "Go back to your country!"—a refrain that is particularly jarring to students who have spent most or all of their lives in the United States. At one point, a student turned away from the reporters and addressed her teachers directly:

> I want to know if you guys really accept us. Do you guys accept us—that we're different, that we're here, and we're actually trying to succeed? You guys brought us to America, because our country wasn't giving us the chance to have success in life. We came here to have success in life!"

At the time of the protest, I had been conducting interviews with RB students in the Laketown community about their college transition processes. Many of the protesters' concerns were echoed in my interview data. But I realized that in my own analysis I had focused much more on curriculum than on climate. I had underestimated the nuances and impact of students' lived experiences at school and in the broader community—experiences that gave them the message that they did not (do not) belong. I began to focus more of my work on understanding the nature and impact of deficit discourse, as well as on highlighting alternatives to that discourse. This chapter provides an opportunity to share that work with an audience beyond educational or applied linguistics researchers. Specifically, I illustrate how asset discourse, as a form of counter-storytelling (Solórzano and Yosso 2002), can undergird the creation of a culture of belonging for RB students.

Discourse, exclusion, and (un)belonging in literature on RB students

I argue throughout this chapter that a culture of belonging requires countering the deficit narratives prevalent in much of the educational discourse about RB students and families. These narratives are prominent in public media as well, whether presenting migrant families as a political, cultural, or economic threat—what Heaven Crawley and colleagues call the "villain" frame—or recognizing them as deserving of pity but reliant on public assistance (i.e., a "victim" frame) (Crawley et al. 2016; see also Loring 2015; Ludwig 2016).

In educational settings, RB students are often constructed as the neediest among EL populations and expected to be less successful in school. Discussions of RB students often assume the existence of psychological trauma, for example, as an inevitable result of displacement and forced migration (McBrien 2005; Roy and Roxas 2011; Shapiro et al. 2018). Teachers, therefore, tend to attribute RB students' academic and/or social struggles to conditions such as post-traumatic stress disorder (PTSD), whether or not students have received such a diagnosis (Matthews 2008; Joyce et al. 2010). Well-meaning educators may, in fact, lower their expectations for RB students, in an effort to "protect" students from the potential psychological stress of academic rigor (Hirano 2011; Shapiro 2015).

RB students are also seen as having educational deficits, as is represented by the increasingly prevalent label "students with limited or interrupted formal education, or SLIFE" (DeCapua and Marshall 2011). Although past educational experience—including literacy levels in their home language(s)—does play an important role in the education of RB students after resettlement, the SLIFE label obscures the knowledge and skills that students might have gleaned from informal learning networks (Browder 2018). Moreover, this label often reinforces the assumption that students' families (especially family members who may themselves have had limited access to formal schooling) are an obstacle, rather than a resource (Roy and Roxas 2011; Shapiro 2019).

Deficit discourses about race and class also shape how RB students are seen and treated—particularly in settings where white, middle-class students are viewed as the "norm" (Edgeworth 2015). Constructing RB students as a racialized "other" makes it easier for schools to shift blame for disparate outcomes onto students and families, preventing a critical examination of how school policies and practices perpetuate structural racism (Roy and Roxas 2011). As explained by Gloria Ladson-Billings (2013), by focusing on the student "achievement gap," educators and policymakers often overlook the "opportunity gap" within school systems.

Ultimately, the "master narrative" told by educational discourse presents RB students and families as a burden and assumes that educational disparities are inevitable (Roy and Roxas 2011; Keddie 2012). Yet there is little discussion in existing literature about alternative discourses and their impact on educational policy and practice. By foregrounding voices and counter-stories from RB students, in keeping with the CRT approach discussed earlier, I hope to pick up where many studies leave off, so as promote a sense of agency among scholars and practitioners looking to resist deficit discourse and to contribute to rectifying educational inequalities.

Context and methods

For this study, I draw primarily on interview data from three particular students—Fardowsa, Madina, and Najib (pseudonyms)–from the larger set of 22 in my study of college transitions. These three participants, all of Somali Bantu background, spoke most at length about issues of exclusion and belonging in their schooling experiences. Recruitment for this Institutional Review Board–approved study took place through contacts at schools and community organizations. Later participants were identified via snowball sampling (Noy 2008). Analysis of interview data began inductively, with open coding (Thomas 2006), followed by secondary coding according to dominant themes that emerged from the data and were echoed in existing literature. I eventually clustered these themes under "deficit" and "asset." (For more on methods, see Shapiro 2014; Shapiro and MacDonald 2017; Shapiro 2019.) For this particular chapter, I conducted an additional round of coding of data from these three participants, looking for indicators of belonging/unbelonging.

Before delving into the data, I wish to provide a bit of context about Laketown School District, and about my own positionality as a researcher. Laketown is a midsized public school district in the northeastern United States, within which approximately 20% of students are classified as English Learners (ELs) or former ELs—that is, students whose primary or home language is not English. Most of the ELs in the district came to Laketown through refugee resettlement, and nearly all RB students receive free or reduced-price lunch—a key marker of socioeconomic class in the United States.

I have been involved in a number of school and community initiatives in and around Laketown. But I am white, US-born, and middle class. I do not share a cultural, linguistic, or racial background with most of the students I interview. A recognition of my own privilege as a researcher is, in part, why I focus on counter-storytelling and educational reform, as this reflects the desire of many students and families I have worked with. It is impossible for me to understand fully the experiences of RB students, and I have tried where possible to share findings and conclusions with participants to ensure that I represent them accurately. I have also worked to forge partnerships between my institution and the school district, which have resulted in some resources feeding back into Laketown schools and community organizations.

Findings

In this section, I organize my findings according to the negative effects of deficit discourse on belonging and the characteristics that define communities of belonging.

How deficit discourse creates a culture of unbelonging

From the first day that RB students enroll in Laketown schools, they are subject to deficit discourse. Each student undergoes an "intake" process, involving information gathering and assessment, in order to make placement decisions. Educators often place students by default into the lowest level of classes—not just for English but for other subjects, such as science and mathematics, because of low English-language proficiency and/or print literacy. Students are, therefore, set on an academic track in which much of their day is spent in "sheltered" classes taught by EL specialists. While this approach gives new students a "soft landing," as teachers often put it, students usually remain in this lower track for several years—sometimes more—and many come to resent being in separate (EL-only) classes year after year. Drawing on a metaphor used by Valdés (2001), the "shelter" begins to feel like a "ghetto." In other words, some of the school structures intended to support RB students may in fact contribute to those students' marginalization in the long run.

Indeed, Fardowsa, Madina, and Najib—like many other participants in my study—expressed frustration that they were not "pushed" more academically. For example, Najib said his original math placement was "so easy ... I was like acing it all the time." He was eventually moved into a more advanced class, but he never made it into college preparatory math by his senior year and, therefore, had to take (and pay for) remedial math courses in college. Najib also said he was not given enough critical feedback: after rereading some work he had completed in his first two years, he reported, "I can't even read it. ... How did they even understand? How did I get an A?" Najib recognized that grade inflation was a well-intentioned practice, since "teachers don't want to discourage their students," but he also saw how this practice hindered his growth: "I don't mind the grade, but I want to know what I can change. ... What did I learn from that?" (see Shapiro 2015, for more on grade inflation as an outgrowth of deficit discourse).

Fardowsa said similarly that much of her high school experience was "a waste of time" because "they kinda baby you so much." When she made it to college, intending to pursue a pre-medicine track, Fardowsa was academically "lost"—especially in lab sciences courses. Thus, the withholding of academic challenge, which teachers likely justified using deficit discourse (i.e., "This is the best we can do with this particular population"), was experienced by Najib and Fardowsa as a type of academic "othering" that overlooked their many aspirations and capabilities.

Many RB students at Laketown High School advocate repeatedly for placement into more challenging classes—contesting the decisions of their EL teachers, who become the gatekeepers. (See Callahan 2005, for more on academic tracking as a form of structural othering for ELs at the secondary level.) But students who make these requests are usually told that their standardized test scores are not high enough for advancement. This is one reason that the student protestors reacted so strongly to the newspaper's reporting on test scores: RB students know that those scores have a tangible impact on educational opportunity.[2] Madina, who had been assigned to EL classes since elementary school, said "I had to prove myself" before being allowed to take more rigorous high school classes. Participants also noted that spending most of their day with other ELs affects their motivation and social well-being: Fardowsa said "I got distracted" in EL classes, because "they were not pushing me."

One reason for the lack of academic "push" at Laketown was that until recently, high school graduation—not college readiness—was seen as the goal for RB students at Laketown (see Shapiro and Ehtesham-Cating 2019, for a discussion of ongoing changes). As long as students attained the credits needed to graduate—and most "sheltered" EL classes were credit-bearing— teachers assumed they had done their jobs. Yet this approach did not take into account students' academic and professional goals, which usually involve some form of postsecondary education. Notably, all but two of the students I have interviewed wanted to pursue a career in a "helping" profession, such as medicine, psychology, social work, or education—professions that all require some form of postsecondary education. (This finding has been echoed in other research on RB students—Hannah 1999; Naidoo et al. 2017).[3]

This commitment to "giving back" is another counter-story rarely discussed within the Laketown schools and community. Students' aspirations toward altruism are not usually discussed during the intake process, and they are not a major focus of academic advising. In fact, many RB students do not realize until their senior year that they have not taken the coursework needed to be ready for college. As Najib put it:

> I asked them what are the classes I need to graduate, and that's what they were giving to me. … But my dream was to go to college. … Someone should have talked to me about college stuff and what I need to do and how hard it is and the things I need to do to prepare for it.

Many RB students also miss out on opportunities for career-oriented coursework (e.g., technical education programs), as well as for service learning, internships, and other cocurricular experiences. Deficit discourse is part of what prevents educators and administrators from making these connections. RB students are seen as the ones "needing help" rather than as those who can and will "help others" (Shapiro and MacDonald 2017). Sadly, some students' aspirations were not just ignored but mocked: Madina said that when she told one of her teachers about her plans to attend college, the teacher's response was, "You're just going to get married." Madina said this message was given to "a lot of students from my culture." This is just one example of the marked cultural othering that took place regularly in students' experiences, in which their families and coethnic community members were seen as an educational obstacle, rather than an asset (see also Roy and Roxas 2011).

Given these dynamics, it is not difficult to see why many RB students conclude that their teachers do not see them as intelligent and capable. Similar trends were evidenced in students' descriptions of social dynamics as well, in which racist and xenophobic harassment were prevalent. Fardowsa said that (non-RB) students often asked her hurtful and ignorant questions, such as "Do you wear the headscarf because you're bald?" At her middle school, "There was a lot of racism stuff going on. … Students would tell us 'go back to your country'… [or] would hold their noses … because people spread rumors about African students." When she asked teachers to intervene, "They won't do anything. They usually just take the white kids' side."

Najib echoed these comments, recounting that at one point, the EL teachers at Laketown High School convened a meeting with ELs to talk about personal hygiene. For weeks afterward, Najib and other RB students were ridiculed in the hallways, usually as they were coming to or going from EL classes: "People see you and they start holding their nose. They're saying, 'Oh my god! They're smelly!'" Najib was angry not just at peers, but also at his teachers for the way they handled the situation: "They're having a conference with you and they're not talking about your education—they're talking about your cleanness."

Madina had similar experiences. She described an incident in which two female students of Somali background had misbehaved on a field trip but instead of addressing the issue just with those two students, "all of the [Somali] girls were called into the office." During that meeting, moreover, the administrator demanded that she look him in the eye. "I got in trouble," she explained, "because in my culture, you can't look guys in the eye." Although she had participated in some conversations about changing the school culture, she felt that they had resulted in very little: "In order to make change, you have to do something—actions. I've seen a few changes, but it's more talk than action."

Hence, the structural othering described by students here extended across their academic and social experiences, reinforcing the message that teachers and peers saw RB students as culturally, racially, and intellectually inferior. This message was amplified, in turn, when students tried to address problems with teachers and administrators but felt unheard and ignored. These sorts of experiences caused all three students to question whether they belonged in school. This dynamic took a particularly heavy toll on Madina, who came to the conclusion that "teachers and students didn't really understand me."

One way teachers might have challenged some of these deficit narratives would have been through an antiracist curriculum (Affolter 2019). Yet while the school curricula in Laketown did include some units on multiculturalism and diversity, it generally avoided addressing these topics from a critical perspective, and rarely if ever invited students to connect to their lived experiences at school and in the community. One exception was the ninth-grade humanities curriculum, which, around the time of my data collection, was being revised to focus on global citizenship and social justice, including a unit on race and racism. Sadly, because most EL students were in "sheltered" humanities classes that separated them from students from other backgrounds—and because the EL teachers decided not to integrate these themes into their curricula—RB students had no chance to engage these topics among themselves or with their non-EL peers. Hence, even if teachers had been open to facilitating conversations about institutionalized racism and other forms of oppression at school as part of their social justice curriculum, school structures would have prevented many RB students from taking part in those conversations.

These research findings show how discourse and school culture are interdependent. The student protesters at Laketown High School knew this all too well. Their protest was a manifestation of their awareness that that *how they are seen and talked about* is closely tied to *how they are treated*. The "master narrative" in which RB students are seen unlikely to succeed became a self-fulfilling prophecy, as those students were given little if any access to a challenging and meaningful education. Given these experiences of othering and exclusion, it is unsurprising that many students would ask: "Do you accept us—that we're different, that we're here, and we're actually trying to succeed?"

What communities of belonging look like

Just as students have helped me to understand the impact of deficit discourse, they have also helped me envision what a school system that "accepts" them—a school community in which they experience a sense of belonging— would look like. Such a system would involve an orientation in student assets throughout all aspects of the education system. For example, advising and placement processes should focus not just on what students *lack* in terms of educational background or language proficiency but also on what they (and their families) *bring* to schools and communities. One such asset is "aspirational capital"—that is, "the ability to maintain hopes and dreams for the future, even in the face of real and perceived barriers" (Yosso 2005: 77; see also Shapiro and MacDonald 2017). Families who have successfully navigated the slow, bureaucratic process of refugee resettlement tend to believe strongly in the potential for social mobility and societal change through education. Madina's mother, for example, frequently reminded her how "lucky and fortunate" she was to have the opportunity to attend school in the United States. Similarly, when Najib suggested that he might seek full-time employment instead of going to college, he said his mother "doesn't care if there's nothing to eat at home—she wanted me to go to school."

This counter-story of aspirational capital within RB families and communities, however, is largely overlooked by schools (Roy and Roxas 2011; Shapiro 2019). Creating a culture of belonging, therefore, requires that teachers take students' aspirations seriously and work with students and their families to achieve those aspirations. This means having conversations early on—even during the intake process—about RB students' educational and professional goals, outlining pathways for achieving those goals.

Taking student and family aspirations seriously also requires more focus on academic rigor and course placement. When RB students are under-challenged, as evident in this study, they begin to question whether their teachers and classmates see them as intelligent and capable. In other words, lack of access to academic rigor reinforces deficit narratives about and among RB students. Academic "push," in contrast—when provided within a supportive environment—serves as an enactment of "critical caring" (Valenzuela 1999) that challenges deficit discourse, foregrounding the alternative narrative that students can and will succeed. Indeed, while lack of rigor was a general concern, as discussed here, all three students mentioned at least one notable exception—a class with a teacher who "pushed" in a loving way. As Najib, for example, said of his senior English teacher:

> I still remember her, of course. I can still see her face helping me, talking with me right now—it's fresh. She helped me a lot. … She pushed me

hard. ... At the end of the year, I earned an A in her class, but I also
learned something from that class. It was hard, but it pushed my brain.

Yet increasing academic rigor cannot be done on an ad hoc basis; it requires
RB students to be integrated and supported throughout the curriculum—
not just in separate EL courses. This commitment to inclusion requires fore-
thought in course placement and scheduling, as well as opportunities for
collaboration between EL teachers and their colleagues in other departments (see Honigsfeld and Dove 2019 for more on these sorts of collabora-
tions). Within this more inclusive model, EL teachers could be positioned
less as gatekeepers and more as advocates for RB students. Indeed, there are
some indications that Laketown High School is shifting in this direction
(Shapiro and Ehtesham-Cating 2019; see also Keddie 2012; Taylor and
Sidhu 2012).

A school culture of belonging also requires a culturally relevant curriculum. Although Laketown School District has articulated a commitment to
global citizenship as part of its mission, global learning is not well integrated into the current curriculum. The three students I discuss here would
have benefitted tremendously from opportunities to engage curricular
themes in which they could draw on their lived experience—and of course,
their (non-RB) peers would have learned much from these opportunities as
well! Near the end of our interview, for example, Fardowsa spoke enthusi-
astically about an English course she was taking in college, in which
students were "reading plays from different parts of Africa and watching
films ... based on how different the culture is from here." When I asked if
she felt that she could draw on her personal experiences in that course, she
replied, "Yeah, a lot—I love that!" In Laketown, however, Fardowsa had
few if any opportunities to make these connections—"not even in my history classes."

Critical storytelling could, in fact, be a central feature of a culturally relevant curriculum, inviting accounts of both struggle and agency from RB
students and their families. Najib was particularly adept at foregrounding
these stories. In a memoir that he began in college and eventually expanded
and shared publicly, he wrote: "I have always had a good mind for figuring
things out. I always have been a problem solver. I knew if I worked hard and
put my mind to something, that I could make a difference." In our interview,
similarly, he described himself as a "street hustler kid, but in a good way,
helping people."

These comments invite the question: *How might Laketown schools have
better leveraged these assets?* Najib himself articulated a desire to serve as a
role model for other students—especially those with immigrant or refugee
backgrounds:

You gotta be a role model for somebody, but in order for that to be, you
gotta give somebody the credit. When you see a person—someone—a
guy that came from Asia or Brazil, or anywhere in the whole place, when

he do something good, make sure that you award him with that. ... He's gonna be a role model, and that will encourage a lot of kids and they'll want to be in the same position. ... [Schools] don't understand how important this stuff is.

Fardowsa and Madina, who were eldest daughters in large families, also saw themselves as role models. Fardowsa said that part of the reason she was "determined to get ahead" was that she wanted her younger siblings to "look up to me as a big sister." Schools can build and draw on this social capital through cocurricular opportunities such as peer mentoring and leadership training programs. While she was at Laketown High School, in fact, Madina became involved with a grant-funded, community-school partnership initiative. "It really helped me to speak in front of a group of people," she told me. "Now I have a lot of connections with teachers and in the community." This transformative experience was one of the main reasons Madina chose to study teacher education in college (see Montero et al. 2012 for more on asset-oriented cocurricular opportunities for RB students).

Conclusion

The research presented in this chapter has offered interconnected insights into discourse, perception, and treatment of RB students. I have illustrated how deficit discourses (i.e., "master narratives") about RB students and families can contribute to othering and exclusion. I have also drawn on student counter-stories to suggest ways that Laketown might shift toward a culture of belonging in which student and familial assets are better recognized. I hope that this case study has resonance not just for educational researchers, but for scholars and practitioners working in other institutional contexts. To aid readers in applying these insights to their own settings, I provide Table 15.1, which lists the most salient features of deficit versus asset-oriented discourse and practice.

The below framework could be useful to migration researchers in a variety of disciplines. Scholars in critical media studies, for example, could employ

Table 15.1 Features of deficit versus asset approaches

Deficit	Asset
Emphasizes what individuals/families lack	Emphasizes what individuals/families bring
Oriented in the past	Oriented in the present and future
Narratives of victimization/trauma	Narratives of agency/empowerment
Surviving is the goal	Thriving is the goal
Invokes pity/sympathy	Cultivates pride/aspiration
Constructs individuals/families as recipients of help	Constructs individuals/families as also able/wanting to help others
Contributes to sense of alienation	Contributes to sense of belonging

this approach not only to expand the body of research on deficit discourse in mainstream media (Loring 2015; Crawley et al. 2016) but also to open up new lines of inquiry focused on the types of asset discourse prevalent in alternative media. An asset approach in fields such as anthropology, similarly, might encourage researchers to focus not only on the impact of trauma on migrant communities but also on how those communities cultivate resilience, aspiration, and belonging.

I wish to conclude this chapter with a quote from the final few minutes of my interview with Najib, in which he described his vision for Laketown School District, where he now works, in fact, as an interpreter and cultural liaison. His words capture best, I think, what is at the core of an educational culture of belonging:

> I'm gonna just say, this is about making family—making *family*. ... You bring all kids together and you're telling them that 'This is what we need to do. This is what we want you to be ... all positive stuff. ... We want you to have a family, a future, and we'll be working hard on you.

Notes

1 "English Learner" (EL) is the most common label used by K–12 schools in the United States for students who use a language other than English at home and who are in the process of acquiring English as an additional language. This descriptor is also used for courses designed specifically for English Learners.
2 It is important to note that there are federal and state laws that often prevent EL teachers from "exiting" students from EL courses before their test scores reach a particular level. However, parents/guardians can "opt-out" of EL instruction for their children if they wish—an option that many families are unaware of.
3 To understand why so many RB students are looking to "give back," it is important to keep in mind that less than 1% of all families living in refugee camps will be given the opportunity to resettle to a third country such as Australia, Canada, or the United States (UNHCR n.d.). Thus, students who have had this extremely rare opportunity often wish to "pay it forward" to others, and are encouraged to do so by their families and coethnic communities.

References

Affolter, Tara (2019). *Through the Fog: Towards Inclusive Anti-racist Teaching.* Charlotte, NC: Information Age Publishing.

Bialik, Kristen, Alissa Scheller, and Kristi Walker (2018). "6 Facts about English Learners in U.S. Public Schools." Pew Research Center. Accessed January 20, 2021, https://www.pewresearch.org/fact-tank/2018/10/25/6-facts-about-english-language-learners-in-u-s-public-schools/

Browder, Christopher (2018). "Recently Resettled Refugee Students Learning English in US High Schools," 17–32. In Shawna Shapiro, Raichle Farrelly, and Mary Jane Curry, eds., *Educating Refugee-Background Students: Critical Issues and Dynamic Contexts.* Bristol, UK: Multilingual Matters.

Callahan, Rebecca M. (2005). "Tracking and High School English Learners: Limiting Opportunity to Learn." *American Educational Research Journal* 42(2): 305–28.

Crawley, Heaven, Simon McMahon, and Katherine Jones (2016). "Victims and Villains: Migrant Voices in the British Media." Accessed January 20, 2021 from Centre for Trust, Peace and Social Relations, Coventry Universiety, https://pure. coventry.ac.uk/ws/portalfiles/portal/8263076/Victims_and_Villains_Digital.pdf

DeCapua, Andrea, and Helene W. Marshall (2011). "Reaching ELLs at risk: Instruction for Students with Limited or Interrupted Formal Education." *Preventing School Failure: Alternative Education for Children and Youth* 55: 35–41.

Edgeworth, Kathryn (2015). "Black Bodies, White Rural Spaces: Disturbing Practices of Unbelonging for 'Refugee' Students." *Critical Studies in Education* 56(3): 351–65.

Gutiérrez, Kris D., and Marjorie Faulstich Orellana (2006). "The "Problem" of English Learners: Constructing Genres of Difference." *Research in the Teaching of English* 40: 502–07.

Hannah, Janet (1999). "Refugee Students at College and University: Improving Access and Support." *International Review of Education* 45: 151–64.

Hirano, Eliana (2011). "Refugees Negotiating Academic Literacies in First-year College: Challenges, Strategies, and Resources." Unpublished doctoral dissertation, Georgia State University, Atlanta.

Honigsfeld, Andrea M., and Maria G. Dove (2019). *Collaborating for English Learners: A Foundational Guide to Integrated Practices* (2nd ed.). Thousand Oaks, CA: Corwin Press.

Joyce, Andrew, Jaya Earnest, Gabrielle De Mori, and Genevieve Silvagni (2010). "The Experiences of Students from Refugee Backgrounds at Universities in Australia: Reflections on the Social, Emotional and Practical Challenges." *Journal of Refugee Studies* 23(1): 82–97.

Kanu, Yatta (2008). "Educational Needs and Barriers for African Refugee Students in Manitoba." *Canadian Journal of Education* 31(4): 915–40.

Keddie, Amanda (2012). "Pursuing Justice for Refugee Students: Addressing Issues of Cultural (Mis)Recognition." *International Journal of Inclusive Education* 16: 1295–310.

Kia-Keating, Maryam, and B. Heidi Ellis (2007). "Belonging and Connection to School in Resettlement: Young Refugees, School Belonging, and Psychosocial Adjustment." *Clinical Child Psychology and Psychiatry* 12(1): 29–43.

Ladson-Billings, Gloria (1998). "Just What Is Critical Race Theory and What's It Doing in a Nice Field Like Education?" *International Journal of Qualitative Studies in Education* 11(1): 7–24.

Ladson-Billings, Gloria (2013). "Lack of Achievement or Loss of Opportunity," 11–23. In Prudence L. Carter and Kevin G. Welner, eds, *Closing the Opportunity Gap: What America Must Do to Give Every Child an Even Chance*. Oxford: Oxford University Press.

Loring, Ariel (2015). "Positionings of Refugees, Aliens, and Immigrants in the Media," 21–34. In Emily Feuerherm and Vaidehi Ramanathan, eds., *Refugee Resettlement in the United States: Language, Policy, Pedagogy*. Bristol, UK: Multilingual Matters.

Ludwig, Bernadette (2016). "'Wiping the Refugee Dust from My Feet': Advantages and Burdens of Refugee Status and the Refugee Label." *International Migration* 54(1): 5–18.

Matthews, Julie (2008). "Schooling and Settlement: Refugee Education in Australia." *International Studies in Sociology of Education* 18(1): 31–45.

McBrien, J. Lynn (2005). "Educational Needs and Barriers for Refugee Students in the United States: A Review of the Literature." *Review of Educational Research* 75(3): 329–64.

Montero, M. Kristina, Hany Ibrahim, Colleen Loomis, and Sharon Newmaster (2012). "'Teachers, Flip Your Practices on Their Heads!' Refugee Students' Insights into How School Practices and Culture Must Change to Increase Their Sense of School Belonging." *The Journal of Multiculturalism in Education* 8(3): 1–28.

Naidoo, Loshini, Jane Wilkinson, Misty Adoniou, and Andrew Langat (2017). *Refugee Background Students Transitioning into Higher Education: Navigating Complex Spaces.* New York: Springer.

Noguara, Pedro (2004). "Social Capital and the Education of Immigrant Students: Categories and Generalizations." *Sociology of Education* 11(2): 180–83.

Noy, Chaim (2008). "Sampling Knowledge: 'The Hermeneutics of Snowball Sampling in Qualitative Research.'" *International Journal of Social Research Methodology* 11(4): 327–44.

Roy, Laura A., and Kevin C. Roxas (2011). "Whose Deficit Is This Anyhow? Exploring Counter-Stories of Somali Bantu Refugees Experiences in 'Doing School.'" *Harvard Educational Review* 8: 521–42.

Shapiro, Shawna (2014). "'Words That You Said Got Bigger': English Language Learners' Lived Experiences of Deficit Discourse." *Research in the Teaching of English* 48(4): 386–406.

Shapiro, Shawna (2015). "A 'Slippery Slope' of Too Much Support? Ethical Quandaries Among College Faculty/Staff Working with Refugee-Background Students," 118–34. In Emily Feuerherm and Vaidehi Ramanathan, eds., *Refugee Resettlement: Language, Policies, Pedagogies.* Bristol, UK: Multilingual Matters.

Shapiro, Shawna (2019). Familial Capital and the College Transition for Refugee-Background Students. *Excellence & Equity in Education* 51(3–4): 332–46.

Shapiro, Shawna, and Miriam Ehtesham-Cating (2019). From Comfort Zone to Challenge: Toward a Dynamic Model of English Language Teacher Advocacy In Secondary Education. *TESOL Journal* 10(4): e488.

Shapiro, Shawna, Raichle Farrelly, and Mary Jane Curry, eds. (2018). *Educating Refugee-Background Students: Critical Issues and Dynamic Contexts.* Bristol, UK: Multilingual Matters.

Shapiro, Shawna, and Michael T. MacDonald (2017). "From Deficit to Asset: Locating Discursive Resistance in a Refugee-background Student's Written and Oral Narrative." *Journal of Language, Identity & Education* 16(2): 80–93.

Solórzano, Daniel G., and Tara J. Yosso (2002). "Critical Race Methodology: Counter-storytelling as an Analytical Framework for Education Research." *Qualitative Inquiry* 8(1): 23–44.

Taylor, Sandra, and Ravinder Kaur Sidhu (2012). "Supporting Refugee Students in Schools: What Constitutes Inclusive Education?" *International Journal of Inclusive Education* 16(1): 39–56.

Thomas, David R. (2006). "A General Inductive Approach for Analyzing Qualitative Evaluation Data." *American Journal of Evaluation* 27(2): 237–46. doi: 10.1177/1098214005283748.

UNHCR (n.d.) "Resettlement." Accessed January 20, 2021, from https://www.unhcr.org/en-us/resettlement.html

Valdés, Guadalupe (2001). *Learning and Not Learning English: Latino Students in American Schools*. New York: Teachers College Press.

Valenzuela, Angela (1999). *Subtractive Schooling: Issues of Caring in Education of US-Mexican Youth*. Albany: State University of New York Press.

Yosso, Tara J. (2005). "Whose Culture Has Capital? A Critical Race Theory Discussion of Community Cultural Wealth." *Race, Ethnicity and Education* 8(1): 69–91.

Part IV

Gender, sexuality, age, and belonging

16 "I am not alone"

Rohingya women negotiating home and belonging in Bangladesh's refugee camps

Farhana Rahman and Nafay Choudhury

Introduction

There's more! There's more! They're coming! They're coming!

These words echoed throughout Bangladesh's Rohingya refugee camp one early morning in September 2017. It was just after dawn—the sand had muddied from the torrential monsoon rains the night before, the yellow-orange hue of the sun shining over the soggy terrain. In the distant horizon beyond the sandy expanse and just over the rice paddies, crowds of people emerged out of nowhere, first in tens, then in thousands, until their presence carpeted the entire landscape. Their sunken faces revealed the exhaustion of a prolonged journey, escaping unspeakable calamity only to be faced with an uncertain future. The refugees were hungry and drained, pacing through the scorching heat, some carrying the weight of their children, others their elderly family members. Some were able to gather a few belongings at the last minute, which they wrapped inside larger knotted shawls that they carried on their shoulders. Many others brought nothing except the clothes on their backs. Many Rohingya women later recounted that they did not know their exact destination, but they ran nonetheless, out of necessity, following one another across dense forest lands until they had reached the camps.

In the days that followed, the urgency of flight from Myanmar translated into a frenzy of disorder in the camps that swelled exponentially as the numbers of refugees increased. The early days of their arrival were the most difficult; one by one, families were grouped and regrouped by NGOs and aid organizations that operated around the clock as thousands more refugees appeared in the quiet of the night, confused and disoriented. Rohingyas were "settled into the camps" (the words used by a Rohingya woman named Zannat), with little time to think about how daily life would unfold in the coming days.

The Rohingya refugee camps are situated in the Cox's Bazar district of Bangladesh, which borders Myanmar to the east (see Figure 16.1). Myanmar consists of over 130 different ethnic groups. Rakhine State, located in southwest Myanmar and bordering Bangladesh, is home to the Rohingya people.

DOI: 10.4324/9781003170686-20

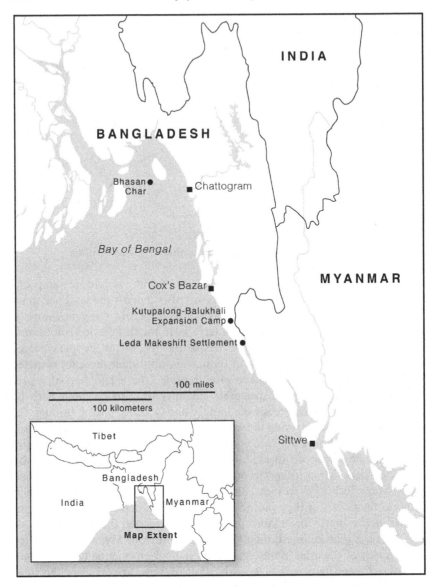

Figure 16.1 Rohingya refugee camps in Bangladesh.

Data & Sources: Natural Earth Data; "Cox's Bazar, Bangladesh" (Human Rights Watch 2018).
Cartographer: Gregory T. Woolston.

A small number of Rohingya refugees have been living in camps on the
Bangladesh–Myanmar border since the 1990s as a result of violence during
that period. In 2017, a fresh wave of attacks by the Burmese military in
Rakhine State caused the Rohingya people to flee for their lives, with roughly
a million taking shelter in Bangladesh. These refugee camps—the largest in

the world—now host over a million Rohingya who have little prospects of returning home in the foreseeable future.

Within this context, Rohingya women have had to negotiate and reforge community ties as well as renew their understandings of "self." Though estranged from their native homeland of Myanmar, a sense of "home" has begun to emerge in the camps from the shared experiences of displacement, the social and cultural interactions that constitute the camps, and the routine activities that provide meaning in their everyday lives. This chapter explores the arduous journey of Rohingya women after being uprooted from Myanmar and forced into refugee camps in Bangladesh. While having lost their homes in Myanmar, Rohingya women have been able to reestablish "community" and traditions in the refugee camps, thereby creating a feeling of home and belonging despite their predicaments. By exploring the lived experiences of Rohingya refugee women, this chapter reveals that even in the direst of circumstances, refugees maintain a productive capacity to transform their setting, thereby producing a new sense of "home."

Gender and belonging in forced migration

Few events cause as much upheaval in an individual's life as forced migration, as it entails intense rupture and dislocation. A refugee camp is a profoundly new environment for refugee women, who must navigate the new spatial and social dynamics of a cramped, overpopulated environment. As Linda McDowell (1999: 5) writes, "places are defined, maintained and altered through the impact of unequal power relations." Refugee women are forced to rebuild their lives in a space where their personal autonomy is greatly limited. In such settings, it is particularly important to pay attention to the tremendous human capacity to not only survive—but also to "thrive," as Nancy Scheper-Hughes' (2008) contends, despite experiences of violence, deprivation, and profound trauma. As Lawrence Grossberg (2000: 154) notes, belonging involves the active sentiment "of identification, of involvement and investment, of the line connecting and binding different events together." Refugee women infuse new meanings into their lives in the camps, despite the limitations of their immediate setting, by continuing practices that appropriate meanings from their past. It is through the appropriation and reshaping of deeply held personal, communal, and social meanings that Rohingya women are able to recreate a sense of order and belonging in their lives in displacement, thereby creatively recasting their environment as a familiar setting with intimate meanings.

To understand and analyze the way in which "home" and "belonging" are experienced by Rohingya refugee women, this chapter employs the notion of "placemaking" that Julie Peteet (2005) uses to explain the relationship between space and society. Studying Palestinian refugees in Lebanon, Peteet (2005: 93) contends that refugees engage in the "social production of place" whereby new meanings and understandings are created "by the regular, patterned activities and social relationships that unfold in it and the cultural rules governing them." The issues relating to space, place, and home are a primary locus of discussion for refugees dislocated from their countries of origin and forced into camps. As David Turton (2005) argues, making a place for oneself involves the production of individual localized meanings, and thus the emergence of a sense of community within refugee camps follows patterns that reflect both existing familiar and wider social structures and the disruption to these structures resulting from displacement. Peteet (2016) further contends that the profound experiences of displacement play a defining role in the spatial and temporal meanings that animate life and social relations in the camp. While refugees face "social paralysis" resulting from the general lack of opportunities and livelihoods, as well as extensive dependence on outside parties, they nonetheless are capable of adapting to their environment (Peteet 2016). This may entail the transformation of their environment and thus the circumstances that they inhabit. Alice Corbet (2016) demonstrates that refugees, in adapting to their new settings, may bring into existence "new social forms and opportunities" through their everyday practices. Thus, despite the setbacks and limitations brought about by circumstances in the camp, refugees possess the capacity to reorient their environment by adopting new meanings and, ultimately, a renewed sense of self. The remainder of this chapter engages with Rohingya women's profound, productive actions as they craft a feeling of "home" and belonging—and in the process adapt their gendered identities and subjectivities.

Feminist ethnographic fieldwork in the refugee camp

The ethnographic research for this chapter took place in the Kutupalong-Balukhali Rohingya refugee mega-camp an hour outside of Cox's Bazar, Bangladesh, and involved spending time in refugees' homes, sharing and cooking meals together, and observing and participating in their daily routines and interactions. It was important to observe not only their actions but also their inactions—the moments of solitude, silence, and reflection that provided texture to the routines of *everyday* life.

Throughout the course of fieldwork, research, and writing, we employed a feminist methodology, paying particular attention to power relations that existed in the field but were unspoken or hidden behind tacit understandings of the social environment. In studying forced migration, feminist ethnography "is an essential tool for narrating the experiences and perspectives of people living with the unspeakable consequences of mass atrocities" (Abusharaf 2009: 5). In the case of Rohingya women, feminist ethnography emphasizes narrating personal experiences and practices, which involves focusing on Rohingya women's voices as the starting point of analysis. The urgency of this task is highlighted by the gaps in academic and policy discussions on forced migration concerning the lived experiences of refugee women as gendered bodies. Most importantly, through a feminist ethnographic approach, women "emerge as agents of change," thus avoiding traditional categories of merely being "victims" of forced migration, and instead emphasizing their transformational capacity even within dire circumstances (Greenhouse, Mertz, and Warren 2002; Abusharaf 2009). As the rest of this chapter shows, often it is the seemingly mundane actions in daily life that provide meanings that help reconstitute "home" and "place" for Rohingya women.

Fleeing from Myanmar, settling in the camps

It was shortly after the first large wave of refugees in late August 2017 when Zomila arrived with her brother and elderly mother, their eyes glazed over from the horrific acts they had witnessed. Her husband and son were killed in Myanmar during the brutal onslaught, and many of her neighbors were also killed when the Burmese army descended upon their village without forewarning. The group she arrived with had traveled for two weeks through thickets and rivers—many in the group now unable to stand upright out of sheer exhaustion. Zomila and her group were received by the UN Human Rights Commission (UNHCR) and other NGOs, local communities, and the Bangladesh army. She recounted her experience of those initial days of fleeing from Myanmar, traversing the countryside by foot, and arriving into the camps:

> After the *mog* (Rakhine Buddhists) came to our village they took my husband, brother, and 10-year-old son. My brother managed to escape somehow and came to find me and my mother and told us the *mog* was coming to rape the women. It was torture. There was so much pain everywhere and some of my neighbors, my two aunts and their children, and a few others from the village, we started running together. We didn't know where we were going—we just saw others from other villages also running so we followed them. When nighttime came, it was pitch dark and we heard there were *mog* waiting to catch Rohingyas, so we hid for many hours behind some trees.
>
> It was 15 days that we were hiding in the trees, trying to make our way to Bangladesh. We tried to cross the river at the border to escape, but the boat's engine failed and it started taking on water. One of my neighbours from my village, her mother, son, and daughter all were in the boat and they all drowned. And we tried again the next night—we had to move only at night to avoid being shot by the *mog* and Burmese military. We had no water to drink, no food, nothing. My feet became very swollen but there was no chance to stop. I had to help carry my mother during many parts since she was unable to keep up. It was horrible and scary. We were so exhausted and hungry. In the dark while we were crossing, we could hear noises; and every time we would think the *mog* is coming after us again.
>
> After 15 days we arrived on Shahpuri Island, which is at the mouth of the river. We stayed there for a few days. That was the first point for us

when we arrived to Bangladesh. We stayed in a madrassa (Islamic religious school) for many days. At that time, we were all confused about where to go next. I was with my family and some of the families from my village. Many others were there too from other villages and we were all waiting to see what is the next step. Because we reached Shahpuri Island at night, it was very dark and quiet and we were not sure if we were even in the right direction. But at the madrassa they gave us food and some shelter for a few days, and then we started moving again. Another friend from my village was with us—she lost her two babies who drowned when our boat first capsized in the river. One elderly lady could not continue and collapsed and died from the running when we were trying to reach the hills in Myanmar. There was so much death all around us. I cannot bear it.

When the refugees first arrived in the camps, the landscape was cloaked by jungle. A small number of Rohingya refugees had been in Bangladesh since the 1990s and had been officially registered and lived in well-coordinated and well-managed camps. But the 2017 wave of refugees was exponentially greater—in those initial days after arrival, the atmosphere was chaotic. Convoys of overfilled UNHCR trucks were rolling in every few hours with support from the Bangladeshi army, both working around the clock to build dirt roads to allow greater aid access. Construction agencies had been called in to stabilize the landslide-prone hills where incoming refugees would eventually settle. UNHCR with the cooperation of local communities in surrounding villages, as well as with the help of newly arrived Rohingyas themselves, carved away large swaths of forest—including parts of a protected nature preserve—to be able to provide building materials and wood as fuel for cooking. They broke away large mounds of sand to set up the makeshift tarpaulin shelters.

As Zomila's narrative illustrates, initially families and neighbors from the same village fled together and tried remaining together during the arduous journey across the Myanmar countryside. But many families were separated from their group when hiding in the pitch-black of the night. Others broke off as they progressed more slowly to accommodate the elderly amongst them. Other communities began to splinter when crossing the Naf River into Bangladesh on rickety, over-packed boats. Arrival into the camps was varied and the "settling in" process depended on the time of arrival as well as the location of available shelters. UNHCR and local organizations worked to settle refugees into various "zones" as they arrived, with an appointed (unelected) Rohingya member (referred to as a *majhee*) as the head of each zone and as a camp leader, responsible for acting as a liaison with various officials. These *majhees*—all young men, mostly in their late 20s or early 30s—were installed by the Bangladeshi army as a stopgap measure, though their role as "block leaders" continued thereafter. They served as intermediaries with aid organizations and dealt with local issues or community governance matters within their specific zone.

On the evening that Zomila and her family arrived at the camps, she was gathered with 100 other refugees in the receiving area to await news of where they would be placed next. A UNHCR worker looked after the group of newly arrived refugees and told them that the *majhee* would take them to their designated area and set them up there. Another lady, Aleya, a 23-year-old woman who was Zomila's neighbor in their village in Myanmar, had been part of the same group that journeyed to the camps and later mentioned how the "distribution" of refugees took place:

> When we came to the camp, it was chaos. We were standing for hours in line waiting to be put into locations, and then we were given bamboo sticks and tarp to set up our own shelters. There were so many of us. I was still with my family but some of my neighbors who we came with were separated when we were walking for days in the rice paddies. Many of them died. Altogether from my village there were four families—though most had missing family members—and we were put into this area in this group of shelters with some of the families that we met along the way.
>
> Everything was just crazy then and we didn't know what was going to happen to us. We followed the *majhee* who brought the group of us to this area in the camp and now we are settled here. It was better for us to be with some families that we know—at least one or two—it is better than knowing nobody. I heard later on that more families from my village came to the camps a few weeks after us but they are settled elsewhere. So now there are many people from different villages here. But we are all from neighboring villages as many of us came from the same area in our township.

After being taken by the *majhee* to their designated zone, the refugees' first task, as Aleya notes, was to begin the difficult work of constructing their shelters. The very act of establishing the camps and constructing the dwelling—the *jupri* (bamboo) shelter—for individual families was, as Peteet (2005: 111) asserts, the "most basic features of placemaking," because it serves as the endpoint of an arduous journey and the beginning of a new episode of life. The shelter serves as the foundation from which new possibilities could emerge. In the case of Palestinian refugees, Peteet's (2005) research revealed that camps were organized according to the village of origin as well as neighborhoods within a specific village, a settlement pattern that attempted to socially recreate "lost villages." The process allowed for "placemaking" that integrally depended on existing villages and clan units. In the case of Rohingya refugees however, settling in was more haphazard. As some families managed to travel together, they were already clustered, thus allowing many families from the same village to end up in the same zone. Of the families under a single *majhee*, however, only a handful would end up from the same village—the others being strangers to one another, though possibly coming from the same township (and all sharing the harrowing experience of forced displacement). With the sudden arrival of such a significant volume of refugees in such a short period of time, the

Bangladeshi government and official organizations prioritized the act of providing any feasible place of shelter to these refugees—a formidable task in itself—rather than focusing on keeping communities intact. Majed, the *majhee* in charge of Zomila and Aleya's zone, explained:

> When the crisis began in August 2017, much of the migration happened in waves. So, you will find that let's say an attack took place in a village in Myanmar at the end of August, at the time maybe 4 or 5 families left their village, which is about on average 5 members per family. So, you can say from one village about 25 members from the same village came during that wave and once they reached the camp they were waiting in lines or in groups with 4 or 5 families from another village, or others they met along the way. So, like that different families came in waves, and because the migration was very chaotic and so many people came at one time, we tried our best to divide them as best as we can. These are not registered refugees so there is no concept of registering and then nicely putting everyone in their preferred location with all their neighbors from their home village or even their relatives. So, you will see that many families have relatives in another makeshift camp 10 or 30 minutes away. When the crisis happened that time with the thousands and thousands of refugees coming every day, our main—and only—goal, was to give them shelter right away.

With the haphazard nature of arrival, government officials and NGOs worked alongside *majhees* with the principal objective of providing shelter. While members from the same village would have ideally preferred to be housed in the same area, the settlement process did not lend to this level of reorganization. Within any given area, a *majhee* would be responsible for 90–120 refugees (or 20–25 families), and among them, four or five families may have come from the same village in Myanmar. In some cases, families may have met during flight across the countryside. While families attempted to create a semblance of "home" in exile by remaining connected to and settling alongside families from the same village, the chaos of flight, and death and dispersal of relatives, caused each section of the camp to be dotted with families from various villages.

Over time, as Rohingya refugees settled in, they began looking for missing family members, relatives, and fellow villagers. These efforts provided the initial steps of reestablishing social ties and creating a semblance of community in the camps. Through these patchwork efforts of community-building, fragmented lives were woven together through a renegotiation of gender identities and social norms in an attempt to find belonging and ultimately "home."

Re-creating home, re-creating community

It was late December 2017, and the weather was cool and windy in the camps as the evening settled over the dusty valley of orange tarpaulin dwellings. In the days after being "allocated" to their respective zone of the camp and setting up their shelters, refugees began looking for their lost families. Locating family members occurred by word of mouth—most refugees did not bring cellphones with them when they fled Myanmar, and thus they were considered a real luxury as very few in the camps—and even fewer women—possessed one. Furthermore, as the camps were still in the process of being developed, formidable challenges existed in navigating the labyrinth of unmarked, undifferentiated shelters without getting lost.

After weeks of inquiring with everyone she knew, Shofika received news that her older brother Kobir was at the other end of the Kutupalong-Balukhali mega-camp. At the time, Shofika was living with her aunt and her family, as Shofika's family members were all killed in Myanmar. With the news that Kobir was alive, Shofika suddenly regained a glimmer of hope that had all but disappeared since escaping Myanmar:

> I need to find him. I have nobody left. When we were all together under one roof in one country, what a beautiful time it was. I didn't see him for so many months now. But if I can find him then it will be good. In our culture, we like to be with all our family and all our relatives. I am now at least with my aunt but my mother and father and siblings—everyone is gone. There is nothing left—only memories of them. You are at peace when you have everyone in your family with you. I am happy to be with my aunt—at least I know I am not alone. She and my little cousins are all I have now. If I can find my brother that will make me happy.

A few weeks after, Shofika and her brother were finally reunited, and they now shared the one room in her aunt's shelter. Another woman, Maleka, who was in search of her son after he was separated when they fled by boats, shared her worries, which were echoed by many:

> We need to be with our family members. Our culture is like that. Home is where everyone remains together—all our brothers, sisters, aunts, cousins, grandparents. It is nice to have everybody together. My daughter was killed in the massacre in our village, then my 15-year-old son was separated when we were fleeing. He was put into another boat from us and now I'm still looking for him. My main priority is to find him so we can be together again.

Rohingya culture is based strongly on communal and kinship ties. Thus, one of the main tasks of many families after fleeing to the camps was attempting to reunite with lost family members and reestablish their social networks: for Shofika and Maleka, that was an important and necessary process of re-creating home. While "home" denoted a sense of familial belonging, oftentimes, it did not refer only to one's immediate family; rather, it also alluded to the meaning of "community" consisting of a wider set of social ties and connections. Those unable to find immediate family and kin members, or whose family members died or disappeared, had to rely on social relations with distant relatives or with members of their village in Myanmar. These ties radiated across the camps since individuals arrived at different times and were settled into a variety of zones. These social ties with individuals that refugees knew in Myanmar were overlaid by new social relations forged with those within one's immediate community in the camps. While previously held social ties could be useful for locating family and maintaining a sense of connection with the past, new social ties within one's immediate physical locality allowed for women to carry out their daily tasks in the camps—whether looking after children, preparing meals, or otherwise. Close physical proximity thus helped bridge social proximity by creating new communities that struggled with common day-to-day issues as well as by sharing the collective pain of being displaced from Myanmar.

In this way, "home" had been recast in the form of what Naohiko Omata (2016: 28) refers to as a "relational home" that "emerges from the lived experiences of refugees through interactions and connections which they forge with other people in a new place." The notion of home was a vital part of the way Rohingya women understood their connection to others, and the construction of a community of new fictive kin relations based on common experiences of displacement and exile, which were a pivotal step in establishing oneself in the settlement process. Although one of the outcomes of forced displacement is the "dismantling of communities" (Cernea 2000), refugees sharing no previous personal relations retain the capacity to forge new social bonds with one another that lead to the formation of new communities (Hammond 2004; Omata 2016). For many refugee women like Shofika and Maleka, community assumes a new meaning as it entails a combination of preexisting social ties radiating across the camp, alongside new ties with immediate community members who can be relied upon to support them with their everyday lives. The combination creates a sense of social belonging.

Negotiating belonging, feeling at home

As the weeks and months wore on for Rohingya women, life assumed intimate meanings through mundane everyday practices that included performing group prayers, sharing food with neighbors, forging friendships, attending the mosque, collecting wood, and frequenting the market. Domesticity saturates the lives of Rohingya women, and in many ways, they attempted to recreate that daily rhythm of life in the camps. In their everyday actions, they reconstructed their social worlds replete with intimate meanings. Maleka reminisced about life in Myanmar while vividly reflecting on the nuances of everyday life in exile:

> We were happy back in Burma and our village was very beautiful and green. My house had lots of furniture and other nice things—many special items and beautiful clothing for my children. I had a lot of gold and jewellery. I could only bring a few of those things with me, which I had to sew into my *bazu* [blouse] all of which I had to sell once we reached here because we had no money. We had lots of friends in Burma and a

good life. But we did not have peace—we were worried for our lives every day and so we ran before the *mog* came to our village. Here in the camps it is more or less a better existence than I expected. Things are very difficult and there are so many women, men, children—everyone together in one place so close to each other. But I take care of my children and play with them and we try to be happy. What's the use of holding on to pain? If it comes to mind or I feel bad, I try to forget. I try to keep myself busy and occupied with housework and taking care of my children to take my mind off the pain I feel. Especially when I talk to the women in the shelters around me—my neighbors—we sit together in the afternoon and sew together, sometimes we laugh, help each other to cook. This is how we can go on.

Maleka's narrative provides a glimpse of the routines of everyday life in the camps where, after facing the trauma of violence and displacement, a steady rhythm of normalcy has begun to settle. Women sit for hours in front of their shelters exchanging stories and banter, decorating themselves and their houses with the adornment available to them—activities that continue from life before exile. Another woman, Zolekha, remarked:

It is not Burma but we are making a life here. We don't know what the future will be like so we try our best to survive. What else can we do? We don't have our cows and poultry here like we had in our villages or my vegetable garden, so things are more difficult here but we are still living our life based on the values and traditions we had back home. Every day we are cooking and cleaning and being with our children—in many ways it is so different than life in Burma but sometimes I also feel that it's not very different. We don't know how long we will be here, but for now we are still making a life here. I like to decorate my shelter with the small things we brought with us—with those same decorations and traditions.

The spatiality of the camp has dramatically transformed the way interactions take place. Privacy became an issue as everyone now lives in close proximity with one another. In Myanmar, the Rohingya had homes with plots of land as well as animals they took care of and also depended on for their livelihood. The camps are thus a radical rupture from the past as women now face a significant reduction in privacy as well as an inability to sustain themselves with their own resources. Despite these changes, Rohingya women have been able to establish a sense of "home" through the routines of everyday life. Home becomes a setting where daily activities provide a sense of self-worth and meaning, where certain traditions of the past continue, while mixing with new practices in the present. The very ability to maintain such routines helps transform a foreign camp into a familiar environment with intimate meanings. For Rohingya women, there is food to be made, cleaning to be completed, chores to be finished. These simple, everyday household tasks are a means of coping and "carrying on" with life (Gren 2015). Routines, order, and predictability, as Gren (2015) suggests, help individuals deal with crises and difficult circumstances. This mundanity captures the micro-level process that provides Rohingya women a meaningful sense of attachment to their new setting. Despite the hardship of living in the camps, these daily routines—the "repeated actions, those most travelled journeys, those most inhabited spaces that make up, literally, the day to day" (Highmore 2011: 1)—make the camps feel like "home."

Part of the "home-making" process also entails cementing bonds with those who share similar experiences of exile despite the lack of prior personal connection. These new community bonds provide a source of friendship, company, and support in the camps. The significant reduction in private space in the camps also reduced opportunities to engage privately in domestic activities such as cooking and food preparation. But food preparation has now become a community event: often, women sit outside together on the front steps of their shelters cleaning and cutting vegetables, providing an opportunity to bond with other women. These newly forged social ties became especially important for parenting, as women come to rely on other women in their vicinity to help look after their children. Because of the higher proportion of men killed by the Burmese army, women outnumber men in the camp and thus carry a particularly high burden in ensuring that children are taken care of, as they may be the only parent of a child. Zolekha noted the importance of community support in helping to raise her young children in the absence of family members.

> In our village in Myanmar we had all our family members and the women used to take care of the children. The men would do the work but now that my husband died I have to go to pick up food in the bazaar or pick up wood for cooking. I worry about my babies because I cannot carry them in the heat and walk so long to get the items. But at least I am relieved to have the support of my neighbors who look after my children when I have to go out.

Rohingya women are thus able to create a sense of belonging by developing daily routines that help infuse the camp with intimate meanings, by forging new social ties with immediate community members based on common daily activities and shared experiences in exile. Another specific way in which Rohingya women have been able to maintain past traditions is through participation in *taleems*, which has become an important space of solace and belonging—a replica of a custom held in Myanmar that continues to operate in the camps.

Taleem: a sanctuary of belonging

The *jumu'ah* prayer (Friday afternoon congregational prayer) had just ended, and the relentless sunrays provided little afternoon solace. Past rows of bamboo shelters carved into the steep, sandy hillside, carefully climbing up doughy clay steps still damp from flashes of heavy rain earlier that morning was Zomila's shelter where she would be holding a *taleem*—a women's prayer circle.

The women slowly entered the large, open room covered by a single piece of tarpaulin until every space on the floor was occupied. The two dozen of them sat in concentric circles around the *alima* (a learned woman/teacher of Islam and Arabic). The *alima*'s eyes had a certain depth to them and all of the women carried themselves with a heightened sense of humility in her presence. She began by making a prayer supplication in solitude, and then recited the supplication aloud. Within a few minutes, a new supplication begun, and this time, the other women joined in, praying in a subdued voice that floated across the room like a murmur. All the women partook in this communal prayer, though each maintaining their solitude, many of the women weeping as they supplicated. After about 45 minutes, the *alima* made her closing prayers, by which time there was not a dry eye remaining in the room. The heat of the afternoon summer sun made the room increasingly stuffy and so, upon completing her supplication, Zomila quickly drew open the curtain in the middle of the room to allow a gust of fresh air to rush into the space. A few of the women got up and quickly dispersed into the late-afternoon sun to carry on with their daily activities, trotting along with children in their arms. Some lingered, taking a few moments to pray on their own; others sat in quiet solitude. Twenty-two-year-old Ismat Ara walked up to shake hands with the *alima* and convey her *salaam* (greetings, this one goodbye) to the group before quickly rushing off with her one-year-old son.

Soon after her marriage in 2017, Ismat Ara and her husband found themselves fleeing for their lives; the two eventually finding their way to the camps. She shared that one of the reasons she left quickly from the *taleem* was to keep peace with her husband. She had to balance, on the one hand, her desire to attend the *taleem*, a space that provided her solace, and, on the other hand, the expectations from her husband. She said:

> I go there [to the *taleem*] as early as possible, but I have to leave quickly because I know my husband does not like for me to stay on longer than necessary. So, I go home as fast as possible. My husband does not let me go anywhere. But I told him I have to go to the *taleem* because it is only for women and we are doing religious things. You know, we used to have *taleem* in Burma? It reminds me of better times in Burma. It is hard for us Rohingya women—we don't have any opportunities and nothing to do here in the camp.

Ismat Ara's remarks reveal the sense of alienation pervasive in the camp and the importance of continuing past practices, like attending the *taleem*, and thus provide a caution against too quickly assuming that a new setting may be transformed into a place of belonging. While daily routines and a new sense of community may help create a feeling of belonging, the new environment may still raise challenges that make it difficult for women to feel at home. Ismat's experience in the camps has been greatly confining and her movements restricted, thereby causing her feelings of depression (Rahman 2019). As Robyn Sampson and Sandra Gifford (2010: 117) note, "gendering of place relates to feeling at home," and while the dynamics of the camp with many people living within a small area has caused more interactions among people, it has also led to greater gender segregation and thus greater isolation than in the past.

Despite this difficulty, Ismat insisted on attending the *taleem*. For her, the *taleem* provided a sense of relief, a means of connecting with a past that has been destroyed, but that is being recreated—even if imperfectly—in the present. The "personal dimension" of the *taleem*—the convening of women who have suffered and who supplicate communally as a means of support—gives it meaning and provides Ismat and others a sense of belonging, even in an otherwise estranged land (Sampson and Gifford 2010).

The women who attended the *taleem* expressed an overwhelming sense of peace and contentment. This "temporary space" that the women created provide a momentary microcosm of a home that once was, as Zomila articulated the evening after the *taleem*:

> I really wanted to hold it [the *taleem*] here because in Burma we used to have these gatherings. I know other areas here in the camps are also doing it. I informed the *majhee* about it and set it up. It felt it was important to have something like this here in the camps. When I was growing up, in our house in our village in Burma my mother used to hold *taleems*. It was just a few people—the women in my family that lived with us—me, my mother, my sisters, aunt, and a few female cousins. Though we have experienced so much suffering and pain, when I think of those memories now I still look back fondly. My heart aches for my village. This gathering I hold here is just a small reminder of those happy times when my whole family used to be together—us girls we used to laugh, pray, and have lots of fun together. It's not the same here, but I am trying my best.

Zomila's words provide an incisive opening into the emotional, cognitive, and psychological attachment to spaces like the *taleem*. For Rohingya women like Zomila, the *taleem* is a means of creating continuity in their lives through actions that encode memories of "happy times" during their life in Myanmar. Clare Twigger-Ross and David Uzell (1996: 207) suggest that the act of trying to replicate such spaces in displacement is a deliberate attempt at placemaking, to maintain "a link with [a place that] provides continuity to a person's identity." While Rohingya women have been forced into a new environment, they preserve the practice of convening weekly *taleem* gatherings, thereby maintaining continuity despite profound rupture (Hartsock 1983). Their feelings of longing and attachment evolve out of the complex experience of a home lost (in Myanmar) but also a sense of belonging preserved (in the *taleem*) through their collective organizing (Gren 2015).

The *taleem* had thus become a way to reestablish the social structures of lost communities and a way of bringing together dispersed lives through the shared customs of prayer and remembrance. For Rohingya women, it plays an important role in forging a sense of belonging in displacement.

Conclusion

While forced migration entails the uprooting of individuals from their native countries into a foreign land, refugees carry with them their practices, memories, and meanings when they relocate. Old practices reemerge over time and can help provide normalcy and continuity despite the experience of profound dislocation. These practices can transform an unfamiliar environment into a familiar setting attached to a time immemorial (Rahman 2019).

Within the new setting of the refugee camp, Rohingya women are forging a sense of home by re-creating community, establishing new social ties, sustaining familiar routines in daily life, and continuing meanings and practices from the past. Preexisting personal ties are overlaid by new social relationships with women in one's immediate community, the latter of whom may not be known acquaintances from Myanmar. This creates a feeling of belonging and also allows women to adapt to new challenges in the camp, such as social isolation and parenting. Furthermore, everyday routines help build a sense of self by providing women tasks for which they maintain their individual autonomy and regain control of daily life. Certain traditions such as the *taleem* help to bring past practices into the present, thereby facilitating continuity despite tremendous rupture. In this way, Rohingya women negotiate home and belonging in Bangladesh's refugee camps: once strangers in a foreign land, they now creatively adapt traditions and craft new meanings within the limitations and contingencies of the camps. Even while faced with these various constraints, Rohingya women remain capable of transforming their circumstances.

References

Abusharaf, Rogaia M. (2009). *Transforming Displaced Women in Sudan: Politics and the Body in a Squatter Settlement.* Chicago: University of Chicago Press.

Cernea, Michael M. (2000). "Risks, Safeguards, and Reconstruction: A Model for Population Displacement and Resettlement," 11–55. In Michael M. Cernea and Chris McDowell, eds., *Risks and Reconstruction: Experiences of Resettlers and Refugees.* Washington, DC: The World Bank.

Corbet, Alice (2016). "Community after All? An Inside Perspective on Encampment in Haiti." *Journal of Refugee Studies* 29(2): 166–86.

Greenhouse, Carol J., Elizabeth Mertz, and Kay B. Warren, eds. (2002). *Ethnography in Unstable Places: Everyday Lives in Contexts of Dramatic Political Change.* Durham, NC: Duke University Press.

Gren, Nina (2015). *Occupied Lives: Maintaining Integrity in a Palestinian Refugee Camp in the West Bank.* Cairo: American University in Cairo Press.

Grossberg, Lawrence (2000). "History, Imagination and the Politics of Belonging: Between the Death and the Fear of History," 148–64. In Paul Gilroy, Lawrence

Grossberg, and Angela McRobbie, eds., *Without Guarantees: In Honour of Stuart Hall*. London: Verso.

Hammond, Laura (2004). *This Place Will Become Home*. Ithaca, NY: Cornell University Press.

Hartsock, Nancy C. M. (1983). "The Feminist Standpoint: Developing the Ground from a Specifically Feminist Historical Materialism," 283–310. In Sandra Harding and Merrill B. Hintikka, eds., *Discovering Reality*. London: D. Riedel Publishing.

Highmore, Ben (2011). *Ordinary Lives: Studies in the Everyday*. London: Routledge.

McDowell, Linda (1999). *Gender, Identity and Place: Understanding Feminist Geographies*. Cambridge: Polity Press.

Omata, Naohiko (2016). "Home-Making During Protracted Exile: Diverse Responses of Refugee Families in the Face of Remigration." *Transnational Social Review* 6(1–2): 26–40.

Peteet, Julie (2005). *Landscape of Hope and Despair: Palestinian Refugee Camps*. Philadelphia: University of Pennsylvania Press.

Peteet, Julie (2016). "Camps and Enclaves: Palestine in the Time of Closure." *Journal of Refugee Studies* 29(2): 208–28.

Rahman, Farhana (2019). "'I Find Comfort Here': Rohingya Women and *Taleems* in Bangladesh's Refugee Camps." *Journal of Refugee Studies*. Accessed January 20, doi:10.1093/jrs/fez054.

Sampson, Robyn, and Sandra M. Gifford (2010). "Place-making, Settlement and Well-being: The Therapeutic Landscapes of Recently Arrived Youth with Refugee Backgrounds." *Health and Place* 16(1): 116–31.

Scheper-Hughes, Nancy (2008). "A Talent for Life: Reflections on Human Vulnerability and Resilience." *Ethnos* 73(1): 25–56.

Turton, David (2005). "The Meaning of Place in a World of Movement: Lessons from Long-term Field Research in Southern Ethiopia." *Journal of Refugee Studies* 18(3): 258–80.

Twigger-Ross, Clare L., and David L. Uzell (1996). "Place and Identity Processes." *Journal of Environmental Psychology* 16(3): 205–20.

17 Journeys of belonging

Latina migrant lesbians in Long Beach, California

Sandibel Borges

Introduction

Creating a sense of belonging is often a necessity for queer migrants in the diaspora, both within and beyond the United States (Manalansan 2003; Acosta 2013; Carrillo 2017). In the US context, nativist, capitalist, and homophobic narratives propagate a logic that portrays immigrants as public charges and "invaders" who are disposable, and queers either as "diseased" or as Western, "modern," and white (Puar 2007; Potts 2011; Ngai 2014; Chang 2016). Given these narratives, which either erase the existence of queer migrants of color or else make them hyper-visible and criminalized, a sense of belonging—not to the nation-state but to queer communities of color—can be deeply affirming.

In this chapter, I discuss the ways a group of Latina migrant lesbians living in Long Beach, California, practiced community building in the context of structural violence. I argue that the narrators used processes that were unique to their multidimensional locations as lesbians, migrants, and Latinas to create community through, despite, and against the displacement, homophobia, and racism they faced. While community can serve many purposes, in this chapter, I focus on the role it plays in a context of intersecting oppressions. To do this, I look at two spaces of belonging: lesbian bars and physical homes. To be clear, I am not arguing that Latina/o/x queer migrants always and unquestionably belong to Latina/o/x or LGBTQ spaces. Rather, I propose that creating and maintaining a sense of belonging was an important tool that the narrators used to survive in a context where queer migrants are targets of violence.[1]

I interviewed the narrators in fall 2013 in Long Beach, conducting the interviews in English, Spanish, and Spanglish at two of the narrators' homes. To honor the narrators' words and their preferred language, I offer the original quotes followed by English translations when necessary.[2] At the time of the interviews, all of the narrators in this chapter were in their late 30s or early 40s, identified as lesbian and women, and were friends who lived in Long Beach. They held gatherings in their homes on a regular basis with other Latina migrant lesbian friends. But they are not a monolithic group: their similarities demonstrate that oppression is systemic, while each

DOI: 10.4324/9781003170686-21

individual story provides insight into the ways social locations influence their sense of belonging.

The narrators' names and ages (at the time of the interviews) were Rocío, 40 years old; Claudia, 39; Deyamira, 37; Berenice, 39; and Victoria, 44. All names (but one) are pseudonyms chosen by the narrators. Claudia is from Honduras and the other four are from Mexico. Except for Rocío, all of the participants identified their socioeconomic status as working class or working poor. Rocío defined her status as middle class because she owned a house with her partner Deyamira, and they were paying a mortgage. Finally, I use the term *Latina* to refer to the self-identified female narrators, and Latinx to refer generally to people of Latin American heritage in the United States, with the "x" indicating gender neutrality.

This chapter is part of a larger oral history project on Latinx queer migration. As a method, oral history creates public knowledge (Roque-Ramírez and Boyd 2012) and demonstrates that individual stories matter and expose how systems of power work. Following the lead of other oral historians, I call the participants "narrators" to credit them as authors of their own narratives.

Intersecting violence in the lives of the narrators

Migration studies have tended to overlook the experiences of queer migrants, focusing on men and/or women within nuclear heterosexual families. These important works have contributed significantly to our understanding of borders, neoliberalism, and exploitation. But the intersections of gender, sexuality, and migration are palpable in the narrators' stories. Following Kimberlé Crenshaw (1989), I take an intersectional—not an additive—approach to analyzing the experiences of Latina migrant lesbians. This chapter is neither a study of sexuality that incorporates migration nor a study of migration that incorporates sexuality and gender. As will become evident, migration and sexuality are intimately linked in the lives of the narrators.

The field of queer migration studies understands sexuality, gender, race, and class as inseparable from migration and citizenship. Sexuality has "shaped *all* migration in its practice, regulation, and study" (Cantú 2009: 26).[3] For instance, queer immigrants were barred from entering the United States until 1990, having been constructed as "psychopathic inferiors" in 1917, "psychopathic personalities" in 1952, and "sexual deviates" in 1965 (Luibhéid 2002: xii–xii). Having a queer perspective then reveals the heteronormativity within US immigration policy as well as the study of migration itself (Luibhéid 2002; Cantú 2009). Furthermore, researchers have found that queer immigrants often negotiate their sexual identities in ways that allow them to remain connected to their migrant communities (Decena 2011; Acosta 2013; Carrillo 2017).

I build on queer migration scholarship to position migrant lesbian belonging as a refusal to the isolation created by intersecting violence. In a context where anti-immigration policies, racism, exploitation, homophobia, and patriarchy oppress queer migrants, building a sense of belonging is a powerful act. It pushes back against the message that as migrants and/or lesbians and racialized, they do not belong anywhere; and it involves negotiating where they can safely be themselves.

The topic of isolation came up in several of the narratives. Rocío, for example, explicitly described feeling a deep sense of loneliness in Mexico as she was coming to terms with her sexuality. Having witnessed homophobia, she feared for her safety.

> A mi me daba pavor, "Ah, es tortillera!" "Dicen que le gustan las viejas!" Y entonces cuando oías eso—cuando tú sabías que te gustaban las mujeres y escuchabas cómo trataban a las mujeres que les gustaban las mujeres […] eso me causaba como [gasp], lo mismo van a decir de mi.

> It was terrifying. "She's a dyke!" "People say she likes women!" So, when you heard that—when you knew that you liked women and heard others treat women who liked women that way […] it made me feel like [gasp]; people will say the same about me.

Rocío migrated to the United States at age 30, partly to live her sexuality with less fear. Like Héctor Carrillo's (2017) participants, Rocío migrated not because the United States is more "liberated," but rather because she would be away from family, friends, and other people who knew her. She also migrated for economic reasons: she could make more money in the United States than in Mexico. Moreover, migrating allowed her to escape drug and alcohol abuse. From ages 22 to 28, she used drugs and drank alcohol excessively, recognizing her own loneliness coupled with anxiety and fear of rejection as reasons for her substance abuse. Because she knew no other lesbians, she thought she was the only one. In 2000 or so, she migrated to Los Angeles with a tourist visa, intending to live as a lesbian and stop using drugs. Years later, she made an arrangement with a close friend, a gay Latino man born in the United States, to marry and apply for permanent US residency. Rocío expressed that it was necessary to use the system to navigate its hurdles.

Claudia was familiar with feeling like "the only one" and, like Rocío, struggled to explore her desire amid homophobia and widespread gender conformity. Social pressure to conform resulted in fear, anxiety, and anger. From an early age, Claudia felt attraction toward other females, including one friend and a teacher. Because she liked dressing "like a boy," other children, she recalled, "would tease me and would call me 'marimacha' [dyke]. I remember feeling this anger when they used to call me that, and it's because I was! But I wanted to hide it so much." Back then, Claudia was still living in Honduras with her father and siblings, as she did from age 2—when her mother first migrated to the United States in search of economic opportunities—until age 14. By the time she was 14, her mother had married a Mexican man, become a US citizen, and petitioned for her two daughters to reunite with her. But before their immigration appointment, her mother went back to Honduras to bring Claudia and her 17-year-old sister to the United States. They entered Mexico with passports and Mexican visas; from Mexico City, they flew to Tijuana, and from Tijuana, their mother entered the United States alone, and the two sisters crossed the border with the help of *coyotes* (smugglers). Their mother picked them up in Los Angeles. Claudia detailed the journey:

> Ellos [coyotes] nos pasaron a nosotros por el cerro—corriendo [...]. Pude ver también los riesgos porque, a los catorce años yo me miraba como una niña todavía pero mi hermana a los diecisiete, casi dieciocho, ya era una adolescente y estaba desarrollada y todo. A nosotros nos dijeron, "no se separen, siempre juntitas," pero como te digo, nunca habíamos salido de nuestro pueblo en sí. Pues estás a la merced de estas personas, ¿no? Recuerdo que estuvimos ahí esperando por mucho tiempo. Este señor [coyote] estaba tomando, tomando, tomando, y de repente fue por mi hermana, la tomó de la mano y se la llevó. Estábamos en el cerro y estaba oscuro. Se la llevó a lo oscuro. En poquito tiempo—su esposa de él también era parte del grupo, estaba embarazada—ella corrió a buscarlos y pues regresaron ya con mi hermana. Te digo, no pasó nada, gracias a Dios. Ya cuando regresaron ellos discutieron un poco, después parece que les avisaron que era hora de que podíamos pasar, so, nosotras y muchas otras personas a correr no sé cuánto. No sé cuánto corrimos; había helicóptero y todo. Nos dijeron, "si la luz del helicóptero te alumbra, no te muevas, no te muevas." Eso hicimos, cruzamos, llegamos a una casa, estaba completamente deshabitada, sin ningún mueble. En un cuarto grande nos pusieron a todos las mujeres y los niños, y en otro a los hombres. [...] Al siguiente día ya nos llevaron aquí a Los Ángeles, mi mamá fue por nosotras. Por cierto, era un viernes. El lunes me llevaron a la escuela [laughs]. Así de rapidito.

They [smugglers] helped us cross through the mountains—running [...]. I saw the dangers because, at 14, I still looked like a girl but my sister who was 17, almost 18, was already developed and all. They said to us, "Don't split up, always together," but like I was telling you, we had never left our town before. You're at the mercy of these people. I remember we were there waiting for a long time. This man [smuggler] was drinking and drinking and drinking, and suddenly he went to get my sister; he grabbed her by the hand and took her away. We were up on a mountain and it was dark. He took her to the darker side. Shortly afterward—his wife was also in the group, she was pregnant—she ran to look for them and they came back with my sister. Like I said, nothing happened, thank God. When they came back, they argued. Then they were informed that it was time to cross, so we, and a lot of other people, started running who knows how long. I don't know how long we ran; there was a helicopter and everything. They [the smugglers] told us, "If the helicopter's light is on you, don't move, don't move." That's what we did; we crossed, got to a house that was completely empty, without any furniture. They put all the women and children in one big room and the men in a different room. [...] They brought us here to Los Angeles the next day; my mother went to pick us up. By the way, that was a Friday. The following Monday they took me to school [laughs]. It was that fast.

I quote Claudia at length because she captures the dangers to which she and her sister were exposed as they made the journey across the US–Mexico border without trusted adults. The risks for teenage girls were many: they could fall victim to sexual violence, be apprehended by the border patrol, or get lost or kidnapped, all experiences that have been widely documented (Falcón 2016; Fernández 2019; Leyva-Flores et al. 2019).

Upon arriving in the United States, Claudia was faced with adjusting to a new environment and culture while feeling alone and disconnected from her mother. She lived in a two-bedroom apartment with 11 other people; her mother had a partner and a small child; and at school, she struggled to learn English. Her sister, whom Claudia considered her best friend, married and moved out within a few years. Mirroring Lilia Soto's findings on Mexican migrant teenage girls in Napa, California, Claudia "discover[ed] that [she] [had] come to a place that ha[d] already marginalized [her]" (Soto 2018: 178). A new layer of isolation and marginalization as a migrant was placed on top of the isolation that came from feeling desire for other women.

Like Claudia, Victoria migrated with her family at age 14, escaping poverty from Jalisco, Mexico, to Los Angeles. At the time of our interview, she had been undocumented for 28 years. Living in constant fear of deportation, she trained herself to be her own "detective"; she learned to observe her own driving skills as if she were a police officer. She also became a detective to hide her sexuality from her family. Victoria explained that as an undocumented immigrant and a lesbian, she often lived in two closets, which she negotiated depending on the safety of the space. Similarly, Berenice negotiated "out-ness" as an immigrant, lesbian, racialized, working-class mother who had been married to a man for 20 years. Berenice first migrated to the United States at age 16. A few years later, she married, became a legal resident, and eventually became a US citizen. Throughout her 20 years of marriage, she maintained romantic and sexual relationships with women. Like Rocío, Berenice learned to use the system to her advantage, and like Victoria, she became her own detective so that her ex-husband and family would not find out about her affairs. In 2013, Berenice was in the middle of a divorce and was in a relationship with a woman; no one in her family knew about her sexuality.

The narrators, as Latinas, lesbians, and migrants, most of them working-class, faced homophobia, racism, displacement, family separation, and complex family dynamics, all fueled by heteronormative and white supremacist logics. This experience of living within multiple systems of oppression was isolating, leading the narrators to think they were "the only ones." The multiplicity and multidimensionality of their social locations, which operated within interlocking oppressive systems, made community building all the more necessary.

"And that's where my world opened up": gay clubs as sites of belonging

Scholars have identified bars and nightclubs as "the quintessential gay space" (Manalansan 2003: 70), where the dance floor is "a stage for queer performativity that is integral to everyday life" (Muñoz 2009: 66). The clientele of these bars and clubs include queer immigrant people of color (see Álvarez 2019). Here, I do not differentiate between bars and nightclubs or between gay clubs and lesbian bars, as the narrators used these terms interchangeably and the lines are often blurred when such spaces serve queer and trans communities of color. In research that documents the ambiguity of queer of color spaces in the immigrant neighborhoods of Woodside and Jackson Heights in Queens, New York, Martin Manalansan (2003) found that some Filipino family restaurants became gay clubs after 7 p.m., showing that queer of color spaces are not separate from their migrant communities. These spaces become a kind of home to those who attended, allowing them to see themselves reflected in others, eat familiar foods, and listen and dance to familiar music.

Similarly, the narrators in this chapter named bars and clubs as spaces where they felt whole and seen, and where they could meet other Latina lesbian migrants with whom to build community. Rocío's introduction to lesbian spaces, for example, changed her life. Shortly after moving to Long Beach, she became friends with a woman from her home city who was also a lesbian:

> Y pues ella fue la que me empezó a llevar a lugares, a presentar amistades. Y ahí fue donde empecé a ver que la vida era normal; porque yo pensé que ser lesbiana era como una vida … por eso tenía miedo de salir del closet, porque sentía que salir del closet o aceptar que era lesbiana era como marginarme yo sola con mi familia. […] Cuando conocí a mi amiga esta que me paseó por aquí por Long Beach, pues conocí dos-tres personas, entonces me empecé a sentir cómoda de conocer muchachas y que se tocara el tema de que uno era lesbiana porque le gustaban las mujeres.

> She was the one who started taking me places, introducing me to other people. And that's where I began to see that life was normal, because I used to think that being a lesbian was like … I was afraid of coming out of the closet because I felt like coming out and accepting that I was a lesbian was like marginalizing myself within my family. […] When I met this friend who took me places here in Long Beach, well, I met two or three people [lesbians] and started feeling comfortable meeting women, talking about being lesbians and liking women.

Meeting other Latina lesbians introduced Rocío to a world where being a lesbian was not associated with shame, and where she could still hold on to her Latina roots. The first time she went to a gay bar was momentous: "entonces ahí fue cuando se me abrió el mundo" [and that's where my world opened up]. When Deyamira went to a gay bar for the first time, her eyes too were opened to something new:

Y le dije, es que yo nunca he ido a un gay bar. Tenía muchas inquietudes. No conocía el scene. [Me dijo], "Ah vamos." Me llevó a Coco Bongo que está en la Victoria, creo, pero tiene muchos nombres. Pero está así como en el ghetto. Entonces entramos y yo iba con los nervios, como que no sabía qué esperar. Haz de cuenta que parecían puros muchachitos de la high school. Eran mujeres pero parecían vatitos y yo estaba en la barra como de, "Oh my god." Me la pasé platicando con la bar tender y bailamos un rato. Yo estaba a gusto... empecé a ver la variedad de nuestro género.

And I told her that I had never been to a gay bar. I was very curious. I didn't know the scene. [She said], "Come on, let's go." She took me to Coco Bongo on Victoria, I think; it has different names. But it's like in the ghetto. So we went in and I was nervous, like, I didn't know what to expect. They looked like high school boys. They were women but they looked like young men and I was at the bar thinking, "Oh my god." I spent the whole time talking with the bartender and we danced for a while. I was happy ... I started to see the variety of our gender.

Being at a lesbian bar introduced Deyamira not only to a space where she was not alone, but also to a space where she could witness different gender expressions. Seeing herself as a lesbian and finding other lesbians were liberating experiences, especially because migrating to the United States had made her "como cohibida, vergonzosa y muy introvertida" [very timid, shy, very introverted]. In her mid-20s, however, she gained a sense of identity when she recognized her attraction to other women: "Nunca me había sentido tanto en mi cuerpo, nunca me había sentido tan identificada; me dio identidad" [I had never felt so much in my own body. I had never felt that I had an identity; it gave me an identity]. Her lesbianism allowed Deyamira to feel connected to her sexuality despite feeling out of place as an immigrant. It gave her confidence, which grew as she connected with others with similar experiences. Deyamira admitted that gay bars were not entirely her "scene" because it was hard to establish relationships with people there, especially because she did not drink much. These bars, however, served as a space of representation that affirmed her existence.

Victoria, in contrast, was clear that gay clubs were places where meaningful connections were created.

The only places that I knew I could find other people like me was at the gay clubs—the Executive Suite, el Que Sera—at the time, because I live in Long Beach and that was our point of reunion. Actually, I built my social circle from there. I haven't been there in a long time ... and throughout the years, my social circle expanded, and the points of reunion weren't the clubs anymore; it was somebody's house, a barbeque ... I found that I really enjoy barbeques among friends. That's how I found [a community].

Victoria knew she would not be alienated in gay clubs. Like the scholar Juana María Rodríguez's description of her own experience in gay bars—where she discovered her "queer latinized body" and her "dancing body" found a "home" (Rodríguez 2003: 153)—Victoria found a home and formed "community and *familia* through music and dance" (155). Gay clubs hold deep significance for queer racialized Latinxs like Victoria, for whom "it matters to get lost in [queer] dance or to use dance to get lost: lost from the evidentiary logic of heterosexuality" (Muñoz 2009: 81). At these clubs, the narrators found the potential to become queer *familia*. Nevertheless, these sites of belonging were not free of complexity:

> If you look Latina and you have an accent, you don't belong. Even if they look Latina but they don't speak Spanish, you don't mingle with them. Yeah, it was like that. But again, there's a lot of different [groups] … like white girls, we used to call them "white girls"—Caucasians—who mingle with everybody. […] I also noticed that most of [my] friends at the time, they wanted to keep themselves separate from men, from gay men. I never liked [these divisions]. (Victoria).

Gay clubs were certainly not perfect—and belonging and safety are not clear-cut. Victoria voiced the often dichotomous tensions within Latina/o/x spaces (migrants vis-à-vis US-born) and lesbian and gay spaces (women vis-à-vis men). She emphasized that she never liked these distinctions, but her closest friends were Latina lesbian migrants, including the other four narrators in this chapter. This apparent contradiction is not surprising, because internal disagreements and conflicts often occur within communities where marginalized individuals find refuge (Ahmed and Fortier 2003; Kunstman 2009). This happens because "communities are affected by relations of power in the very way in which they involve some bodies and not others" (Ahmed and Fortier 2003: 255). Linguistic markers—having an accent or not speaking English or Spanish—determined who belonged where at these clubs. This phenomenon cautions us against romanticizing the concept of community as though it were free of conflict, power dynamics, or divisions. But while it is undeniably necessary to recognize these conflicts, they did not diminish the sense of belonging that the narrators felt when they were with other Latina migrant lesbians.

When the home becomes home

Berenice, like the other narrators, enjoyed going to gay bars, but unlike them, she had to be very careful about where she went: "No podía tener el lujo de andar en los bares pero sí salíamos de vez en cuando. Sí íbamos a gay bars y eso, allá en Santa Mónica" [I didn't have the luxury of going to the bars but we did go out every once in a while. We did go to gay bars, in Santa Monica]. Throughout her 20 years of marriage, Berenice maintained relationships with women but never disclosed her sexuality to her family; fearing that someone would recognize her at bars, she preferred social gatherings at friends' homes.

Home plays an important role in the lives of migrants, and of queer migrants in particular. Migrants often recreate a sense of home by holding on to their traditions, cultures, and values (Zavella 2011; Levin 2016; Sahney 2016). In doing so, however, they sometimes reproduce heteropatriarchy and nationalism, rendering women and queers invisible (Espiritu 2003; Gopinath 2005). For this reason, queer migrants often create their own homes, which is part of their strategy for surviving violence as queer, racialized migrants (Manalansan 2003). In this section, I discuss how the physical home became a space where the narrators created an emotional home that fostered belonging. The physical home was a place to gather, build community, and strengthen their connections to one another.

Scholars have found that immigrants often construct and decorate their homes in ways that recreate a feeling of being "at home" (Levin 2016; Sahney 2016). The connection between home as a physical and an emotional space was clear at a Día de los Muertos celebration hosted by Rocío and Deyamira. Día de los Muertos is a Mexican tradition that takes place on November 1 and 2 and honors those who have passed. People set up an altar, or *ofrenda*, in their homes to offer food, drinks, sweets, and flowers to the spirits of loved ones who are no longer physically present. The celebration is "vibrant with spiritual and emotional meaning for the people who participate in it" and "full of reverence, sorrow, and prayer" (González et al. 2015: 28). The five narrators and other friends gathered to enjoy food, drinks, music, laughter, and conversation. When I arrived, their *ofrenda* was already set up, taking up most of the wall adjacent to the kitchen; it had food, skulls, candles, and photos of loved ones. Spanish music was playing in the background. Deyamira and Rocío served homemade Mexican food and Mexican beer from Baja California, and their home had colorful Mexican decorations. Some of the conversations that evening revolved around coming out (or not), migration stories, and lesbian jokes. With this celebration, Rocío and Deyamira embraced their Mexican heritage while making their physical home a lesbian space for themselves and their friends. They used decorations, food, music, and spiritual objects, beliefs, and practices to make their home *home*, a space where they could feel safe and welcome.

Such gatherings nurtured friendly and sometimes romantic relationships. Perhaps because Latina lesbian migrants were a relatively small community in Long Beach, the lines that separated the two kinds of relationships were often blurred. Rocío and Deyamira were partners, as were Victoria and Claudia. Berenice met Rocío, Victoria, and Claudia through Deyamira, whom she dated for a brief period, when Deyamira invited Berenice to a *carne asada* at a friend's house on the Fourth of July. Deyamira also met Rocío on a Fourth of July weekend. There, Rocío invited Deyamira to a barbeque with her friends the next day. It is unclear whether Berenice and Deyamira were referring to the same Fourth of July weekend when they told these stories. Regardless, Deyamira recalled meeting Rocío and her "crew":

Yo así como, "It's a trap or … ?" como que nadie hace eso de conocer a sus amigas. Yo eso pensaba, ¿no? Nadie te invita a conocer a sus amigas en el segundo día de conocerla, no me conoces bien … y yo así como, "Uhh, it could go very well or very wrong." Dije, voy a ir. Y esa noche estábamos con unas amigas de todas nosotras and I was like, "Oh my god, I'm home"—puras mujeres lesbianas con sus parejas, algunas solteras pero … por primera vez I didn't have to do anything. I didn't have to be anything. I was just there. I was home. […] And it branches out porque ella conoce a ella, y ella a otra. Entonces hay gente que todavía no conozco pero la comunidad se va haciendo más grande. Somos todas muy diversas pero sí, somos todas Latinas.

I was like, "It's a trap or … ?" Like, nobody does that; nobody takes you to meet their friends that quickly. That's what I thought. Nobody introduces you to their friends the very next day after having met them … you don't really know me. I was like, "Uhh, it could go very well or very wrong." I said, I'll go. And that night we were with our friends and I was like, "Oh my god, I'm home"—all lesbian women with their partners, some single but … for the first time I didn't have to do anything. I didn't have to be anything. I was just there. I was home. […] And it branches out because she knows her, and she knows someone else. So there are people that I still don't know but the community just gets bigger and bigger. We're all very diverse but yeah, we're all Latinas.

Deyamira used the word "home" to describe being in a community with other Latina lesbians. Her feeling of being at home while at a friend's home disrupted her isolation. As I argue elsewhere, homebuilding, or the active verb "homing," allowed the narrators to survive systemic violence, as home became a space of strength where they were embraced and welcomed without judgment. As a tool for surviving systemic violence, homing is a "resistance that pushes against patriarchal, cis-hetero-patriarchal, capitalist, and ablest systems that build and maintain borders, deeming racialized, gendered, and sexualized bodies disposable" (Borges 2018: 70). Deyamira's words capture the way that being part of a community affirmed her.

Moreover, some of the narrators invoked the term "family" to describe their connections. At a different home gathering, one of the narrators' friends said to me: "It's more than just a community; it's like family. It feels right that we have something in common." In Berenice's case, having this *familia* reassured her that she was not alone and she had their support: "Me siento apoyada y en comunidad. Me siento a gusto. Es otra nueva experiencia para mi esto. Sí, estar viviendo y tener más contacto con ellas, me siento más a gusto, más apoyada" [I feel supported and in community. I feel happy. This is a new experience for me. Yes, living and having more contact with them, I feel happy, supported.]. I asked, "Where there are other lesbians?" to which Berenice responded, "Yes, exactly. You don't have to hide anything. I feel happy, I feel free; I feel at peace." The narrators built what Giancarlo Cornejo calls "queer friendship," which has the power to "create affective spaces that heal wounds inflicted by social norms" (Cornejo 2014: 360). The narrators created, maintained, and nourished these relationships, which brought material benefits as the friends shared information and resources. Their home-building processes, moreover, did not reproduce heteropatriarchal values but instead allowed the narrators to live uncompromisingly as Latina lesbians and migrants.

Beyond comfort: challenging violence

Belonging, Adi Kunstman argues, "is constituted *through* and not against violence" (Kunstman 2009: 24). Although I agree, I propose that a sense of belonging also has the potential to challenge the products of violence. With the exception of Victoria, who was starting to become involved in politicized spaces, the narrators did not engage in activism or move in activist spaces. Nevertheless, these five oral histories demonstrate that a sense of belonging confers far more than comfort. Finding, building, maintaining, and nourishing a welcoming community allowed the narrators both to live through and to challenge isolation, rejection, displacement, invisibility, and hyper-visibility, all created by systemic violence.

In the face of oppression, the narrators built community while re-signifying the concepts of home and family through activities such as going to gay clubs, gathering for barbeques at each other's homes, blurring the lines between friends and lovers, having lesbian relationships outside heterosexual marriage, and embracing cultural traditional practices. Their processes of community building were unique to their multidimensional locations as lesbians, Latinas, and migrants. In these locations, among themselves, they were able to live without pretending or accommodating others. The spaces of belonging that they created humanized and affirmed their existence.

Notes

1 While it is beyond the scope of this paper, it is worth highlighting that a sense of belonging can, and does in many cases, spark activism against structures of power.
2 The bracketed ellipses in the interview indicate material that I left out, whether for brevity or because of relevance or repetition; the unbracketed ellipses indicate that the interviewee paused for a moment before continuing to speak.
3 Due to Cantú's passing in 2002, he was unable to complete his manuscript, which was then published by Nancy A. Naples and Salvador Vidal-Ortiz in 2009.

References

Acosta, Katie L. (2013). *Amigas y Amantes: Sexually Nonconforming Latinas Negotiate Family*. New Brunswick, NJ: Rutgers University Press.

Ahmed, Sara, and Anne-Marie Fortier (2003). "Re-imagining Communities." *International Journal of Cultural Studies* 6(3): 251–59.

Álvarez, Eddy Francisco Jr. (2019). "Finding Sequins in the Rubble: The Journeys of Two Latina Migrant Lesbians in Los Angeles." *Journal of Lesbian Studies*. Accessed January 19, 2021, https://doi.org/10.1080/10894160.2019.1623600

Borges, Sandibel (2018). "Home and Homing as Resistance: Survival of LGBTQ Latinx migrants." *Women's Studies Quarterly* 46(3–4): 69–84.

Cantú, Lionel Jr. (2009). Nancy A. Naples and Salvador Vidal-Ortiz, eds., *The Sexuality of Migration: Border Crossings and Mexican Immigrant Men*. New York: New York University Press.

Carrillo, Héctor (2017). *Pathways of Desire: The Sexual Migration of Mexican Gay Men*. Chicago: University of Chicago Press.

Chang, Grace (2016). *Disposable Domestics: Immigrant Women Workers in the Global Economy* (2nd ed.). Chicago: Haymarket Books.

Cornejo, Giancarlo (2014). "For a Queer Pedagogy of Friendship." *TSQ: Transgender Studies Quarterly* 1(3): 352–67.

Crenshaw, Kimberlé (1989). "Demarginalizing the Intersection of Race and Sex: A Black Feminist Critique of Antidiscrimination Doctrine, Feminist Theory and Antiracist Politics." *University of Chicago Legal Forum* 1 (article 8): 139–67.

Decena, Carlos (2011). *Tacit Subjects: Belonging and Same-sex Desire Among Dominican Immigrant Men*. Durham, NC: Duke University Press.

Espiritu, Yen Le (2003). *Home Bound: Filipino American Lives across Cultures, Communities, and Countries*. Berkeley: University of California Press.

Falcón, Sylvanna (2016). "'National Security' and the Violation of Women: Militarized Border Rape at the US-Mexico Border," 119–29. In INCITE, ed. *Color of Violence: The INCITE! Anthology*. Durham, NC: Duke University Press.

Fernández, Manny (2019, March 3). "'You Have to Pay with Your Body/: The Hidden Nightmare of Sexual Violence on the Border." *New York Times*. Accessed January 19, 2021, https://www.nytimes.com/2019/03/03/us/border-rapes-migrant-women.html

González, Rafael Jesus, Chiori Santiago, Meoy Gee, and Oakland Museum of California (2015). *El Corazón de la Muerte: Altars and Offerings for Days of the Death*. Berkeley, CA: Heyday Books.

Gopinath, Gayatri (2005). *Impossible Desires: Queer Diasporas and South Asian Public Cultures*. Durham, NC: Duke University Press.

Kunstman, Adi (2009). *Figurations of Violence and Belonging: Queerness, Migranthood and Nationalism in Cyberspace and Beyond*. Bern: Peter Lang.

Levin, Iris (2016). *Migration, Settlement, and the Concepts of House and Home*. London: Routledge.

Leyva-Flores, René, Cesar Infante, Juan Pablo Gutierrez, Frida Quintino-Perez, MariaJose Gómez-Saldívar, and Cristian Torres-Robles (2019). "Migrants in Transit Through Mexico to the US: Experiences with Violence and Related Factors, 2009–2015." *PLoS One* 14. Accessed January 19, 2021, https://doi.org/10.1371/journal.pone.0220775

Luibhéid, Eithne (2002). *Entry Denied: Controlling Sexuality at the Border*. Minneapolis: University of Minnesota Press.

Manalansan, Martin F. (2003). *Global Divas: Filipino Gay Men in the Diaspora*. Durham, NC: Duke University Press.

Muñoz, José Esteban (2009). *Cruising Utopia: The Then and There of Queer Futurity*. New York: New York University Press.

Ngai, Mae M. (2014). *Impossible Subjects: Illegal Aliens and the Making of Modern America* (updated edition). Princeton, NJ: Princeton University Press.

Potts, Michelle C. (2011). "Regulatory Sites: Management, Confinement, and HIV/AIDS," 99–112. In Nan Smith and Eric A. Stanley, eds., *Captive Genders: Trans Embodiment and the Prison Industrial Complex*. Chico, CA: AK Press.

Puar, Jasbir (2007). *Terrorist Assemblages: Homonationalism in Queer Times*. Durham, NC: Duke University Press.

Rodríguez, Juana María (2003). *Queer Latinidad: Identity Practices, Discursive Spaces*. New York: New York University Press.

Roque-Ramírez, Horacio N., and Boyd, Nan A. (2012). "Introduction: Close Encounters. The Body and Knowledge in Queer Oral History," 1–22. In Horatio N. Roque-Ramírez and Nancy A. Boyd, eds., *Bodies of Evidence: The Practice of Queer Oral History*. Oxford: Oxford University Press.

Sahney, Puja (2016). "*Darśan,* Decoration, and Transnational Hindu Homes in the United States." *Asian Ethnology* 75(2): 279–302.

Soto, Lilia (2018). *Girlhood in the Borderlands: Mexican Teens Caught in the Crossroads of Migration*. New York: New York University Press.

Zavella, Patricia (2011). *I'm Neither Here nor There: Mexicans' Quotidian Struggles with Migration and Poverty*. Durham, NC: Duke University Pres.

18 The welfare state and women's citizenship in Buchi Emecheta's *Second Class Citizen*

Ben Suzuki Graves

Buchi Emecheta's fictional and autobiographical accounts of black womanhood in postcolonial London frequently address questions of migrant belonging in terms of access to healthcare and other forms of aid emanating from the postwar British welfare state. Set in the 1960s, her novel *Second Class Citizen* (1974) offers a memorable picture of the National Health Service (NHS) seen from the perspective of Adah, a young working-class Nigerian mother in the throes of an abusive marriage. The army of medical professionals who intervene in Adah's life (ranging from primary care providers and OB/GYN specialists to emergency room doctors, pediatricians, nurses, counselors, and even ancillary workers) register the welfare state's ambition to guarantee protections and benefits for British citizens, including what were then called "commonwealth immigrants." Emecheta figures the intimate space of an NHS maternity ward as a potential site of national belonging and mutual recognition for mothers of different races and backgrounds. Yet she ultimately describes how the welfare state's reassertion of colonial hierarchies of race—and its stigmatization of women who are seen to fall short of prescribed standards of normative womanhood—preempt that sense of belonging. While Emecheta critiques the racism and punitive gender expectations underwriting state health services, she also identifies alternative possibilities for belonging and solidarity among black women. Contrary to critics who charge that, apart from Adah, Emecheta negatively represents the novel's black Nigerian characters, I argue that in *Second Class Citizen* she discovers meaningful structures of mutual aid and assistance in the recurring flashbacks to Adah's mother and her involvement in unofficial communities of care among women in late-colonial Nigeria.[1]

My analysis of the novel brings into focus how the regulatory functions of the welfare state establish criteria of national belonging for nonwhite women. For Emecheta, "belonging" has multiple valences and is often defined negatively in scenes depicting how Adah feels alienated and excluded as a black woman in the white metropolis. Much of the novel's emotional power derives from Adah's experience of the racist "colour bar," which denied housing to black tenants in many areas of London in the 1950s and 1960s.[2] But while the novel dwells on the humiliating *feelings* prompted by racism and xenophobia,

DOI: 10.4324/9781003170686-22

it also explicitly treats the migrant's vexed pursuit of belonging in terms of denying what T. H. Marshall (1950) has called "social citizenship"—the right to access state welfare provision and other protective services. In *Second Class Citizen*, belonging is not just a feeling or cosmetic cultural experience but a matter of access to state health services, council housing, contraception, the interventions of social workers, maternal leave, and so on. In the imagination of the novel, the individual's experience of belonging coexists with an administered sense of belonging whose terms are dictated and policed by the state. Indeed, as we will see, Emecheta's novel shows how the individual's affective experience of belonging can actually signal submission to colonial norms and thus stand as an obstacle to the achievement of meaningful social citizenship. By emphasizing these material coordinates of belonging, I aim to offer an alternative to the often idealizing and de-historicizing treatment of belonging that currently enjoys favor in Anglophone postcolonial literary theory.[3] By drawing attention to the caring institutions of the welfare state as they intersect with issues of racial formation, my aim is not to distract from Emecheta's portrayal of race and migrancy, but rather to place my own emphasis where the author's lies, in the impinging structures that alternately enable or thwart forms of migrant belonging.

Because my main concern is the complex *representation* of migrant belonging in Emecheta's work, I should begin with a brief description of the novel's plot and stylistic features. *Second Class Citizen* functions in some ways as a coming-of-age story that tracks Adah's movement from the Nigeria of her girlhood to a newly postcolonial London. Both sides of the colonial divide are depicted as sites undergoing profound social and demographic changes— Nigeria as it prepares for decolonization in the years leading up to its achievement of independence in 1960, and England as it begrudgingly adjusts to the arrival and settlement of new communities of color hailing from the West Indies, Africa, and the Indian subcontinent. The novel's treatment of the migrant condition is partially autobiographical; the life of the fictional character Adah resembles, with some exceptions, that of Emecheta herself. In its capacity as a realist novel attempting to capture the lived experience of a Nigerian migrant in London, *Second Class Citizen* stands alongside contemporary examples of life writing and first-person testimony such as that of Hazel Carby of the Center of Contemporary Cultural Studies in Birmingham, the often memoiristic pieces collected in anthologies such as *Black British Feminism: A Reader* (1997), and of popular autobiographical accounts of early black British life by Mike Phillip in volumes such as *Windrush* (1999).

Significantly, the novel connects the young Adah's struggle to define her role within her family and community in Nigeria to her later attempts to carve out a place for herself within the imagined community of the British nation. Although the opening chapters of the novel dwell on Adah's girlhood in Nigeria, its third chapter sees her migrating to London as a young woman to join her husband Francis, who is pursuing a law degree there. Central to the plot is her disastrous marriage to Francis, who serially abuses her in the

London phase of the narrative. The novel offers a sympathetic portrait of Adah's attempt to secure housing, employment, reproductive health care, child care, and ultimately a divorce from Francis. Many of its most moving episodes involve the emotional and physical violence she endures within the space of the home. Never a place of sanctuary, the council flat where her family ekes out its existence denies her the sense of belonging she seeks, sending her in search of alternative sites of meaningful affiliation.

Early parts of the novel set outside Lagos emphasize the colonial conditioning of Nigerian structures and institutions. From the very beginning of the novel, Emecheta foregrounds Adah's scramble for access to these institutions, namely a school to which she is initially denied entrance based on gender. Like many elements of Adah's characterization, her self-fashioning as a responsible, upwardly mobile student carries with it an intriguing ambiguity. On one hand, her determination to receive an education exhibits a hard-won agency that Emecheta links to a mystical "presence" connected with the memory of her father's mother. Yet the novel also presents Adah's aspirations in terms of a problematic imperial identification. The novel's narrator reflects critically on the occasional tone of derision and superiority that her education affords her, and which she mostly directs toward her in-laws. Anticipating the representation of agencies of the British welfare state later on in the novel, Emecheta also codes the village school as the province of charismatic male teachers who implicitly demand rituals of deference from their students. Although her education enables Adah's self-possession and material security, it also embroils her in a hierarchical structure whose rules she is continually forced to observe. Emecheta invites the reader to ask what is at stake in "belonging" to such an institution. In the London phase of the narrative, national belonging is also figured in terms of Adah's ceding of autonomy and agency to the custodial institutions of care associated with the NHS.

Like other black British novels of the postwar decades, such as Sam Selvon's *The Lonely Londoners* (1956), George Lamming's *The Emigrants* (1954), or Joan Riley's *The Unbelonging* (1985), *Second Class Citizen* portrays the fitful and uneven process by which black migrants in mid-century Britain came to view themselves as a community forged through a common experience of displacement, urban alienation, and white racism.[4] In these texts, the problem of belonging is linked intimately to questions of form and style. In Selvon's novel, for instance, the fracturing of the larger storyline into disjointed episodes focusing on individual characters registers the difficult process of fusing together a composite black British identity out of separate ones then defined in terms of country of origin. A more important similarity with *The Lonely Londoners*—and with the Australian expatriate novelist Colin MacInnes's *City of Spades* (1957) and *Absolute Beginners* (1959)—lies in Emecheta's extensive representations of the social democratic programs (council housing, the NHS, public casework and counseling) charged with protecting the safety and well-being of Britain's newest arrivals. Selvon's and MacInnes's treatment of dole queues and welfare bureaucracy, as well as vernacular West Indian welfare alternatives such as "pardner" or *sousou* systems of group savings, offer a useful point of comparison to Emecheta's treatment of the welfare state's caring functions. For scholars of black British culture, these literary works bridge two adjacent if seldom-connected histories: the rise and fall of the postwar welfare state and the making of Britain as a multicultural society.[5]

The bonds of motherhood

The British welfare state receives its fullest treatment in the novel's depiction of Adah's multiple pregnancies and experiences of childbirth. Emecheta describes her protagonist's experience of childbirth and both ante- and postnatal care in considerable detail (Adah is the mother to two children at the novel's opening and four at its close). *Second Class Citizen* gives readers a privileged view of the 1960s' obstetric practices, which tended to curtail women's reproductive choices and echoed the patriarchal structures that elsewhere thwart Adah's agency. The case I want to pursue here is that the NHS performs an important symbolic function as a signifier of British national culture. Adah's interactions with medical professionals register in great detail the way in which Adah, as a black migrant, is forced to demonstrate a palatable femininity in order to fulfill the full promise of British citizenship.

What takes shape in these episodes is a narrative of acculturation where the black migrant woman is recruited into British national culture but simultaneously racialized as an unassimilable cultural other, both on account of her race and practices of homemaking deemed unfit for meaningful civic inclusion. These scenes show how Adah negotiates overlapping forms of oppression as a migrant in a racist culture but also as a woman navigating a patriarchal health care system.

Emecheta's misgivings about NHS birthing services take shape in a lengthy sequence set in the maternity ward of a hospital, where Adah recovers after giving birth via cesarean section to her third child, a boy named Titi. The operation itself is depicted as a terrifying experience in which Adah has control over neither her body nor the conditions of the birth. Incapacitated by anesthesia, Adah stands at a certain hallucinatory remove from what is happening to her and her baby during the procedure. In her dream, she is delivered five years into an imaginary future, where her husband has emerged as "lord and master of several farms" and proclaims Adah to be his "virtuous wife" (Emecheta 1974: 108). Francis's smug patriarchal image is eclipsed by another that serves to amplify the male presence dominating the operating theater; Adah awakens to the sight of the male doctor who had "cut her open, to take out the funny baby that had lain across her, instead of lying straight like every other child" (108). The British specificity of these medical procedures helps explain Adah's terrifying Kafka-like confusion. Adah is repeatedly made to submit to the white doctor who embodies the dominant racial order that she has learned to venerate.

After a period of recuperation, she awakens again to find herself effectively bound to a bed with intravenous tubes.

> She was ashamed of herself, because somebody, she did not know who, had decided to make a fool of her. She was lying there, all tied up to the bed with rubber cords, just like the little Lilliputians tied Gulliver.
>
> (Emecheta 1974: 110)

A breathing tube also notably renders her unable to speak. As Emecheta structures the scene, Adah intuitively cedes control over herself to the charismatic figure of the male surgeon. Although he is presented in idealized rather than negative terms, the reader is invited to think of Adah's deference toward him as a sign that she has internalized the hospital's patriarchal and imperial value system. Thus, Adah identifies him in ingratiating terms as a "handsome dark man. ... A great man. A man who knew how to handle his knife. ... The man's confidence never left him" (113).

Emecheta's disturbing portrayal of obstetric medicine accords well with women's oral histories of the postwar decades. In *Modern Motherhood: Women and Family in England, 1945–2000*, historian Angela Davis synthesizes first-person accounts of a generation of mothers who testify to the distressing nature of childbirth in the hands of forbidding, distant male gynecologists (Davis 2014). Several factors conspired to rob women of autonomy in the birthing process in this period: the increasing medicalization of childbirth through intervention and anesthesia, the normalizing of induction and accelerated labor, and the shunning of home birth in favor of hospital deliveries. The animating figure here was the lab coat–wearing male gynecologist. It was his relationship with the baby that mattered, not the relationship between mother and child. "In those days of course," reports one of Davis's interviewees, "people paid such reverence to somebody with a medical coat" (Davis 2014: 93). Others recall factory-like conditions in which expectant mothers lying on gurneys were lined up in corridors awaiting their turn to give birth. In *Second Class Citizen*, Emecheta situates the antifeminist thrust of obstetric practices within a broader denial of reproductive rights to women.

There is, however, an intriguing tonal complexity to the maternity ward sequence. Emecheta repeatedly emphasizes its seductive allures as seen from the perspective of a nonwhite outlier figure. The disorientation and anxiety shaping the event of the birth give way to a soft-focus, dream-like atmosphere issuing from Adah's warm interactions with other convalescing mothers. Because of this focus on the women's mutually therapeutic exchanges, the reader finds it difficult to tell if the hospital is a site of emancipation or discipline, or both.

The lavish, seemingly unexceptionable representation of the maternity ward is worth noting because it gestures toward Adah's intuitions about what she construes as a benevolent and welcoming white Britishness. From Adah's own perspective, the women of the maternity ward represent a community of peers that initially appears welcoming and therapeutic. This camaraderie among women conveyed as being in opposition to, rather than continuous with, the therapeutic culture of the medical professionals. The ward is duly filled with an assemblage of kind-hearted women, new and expectant mothers whose fellow feeling helps ease Adah's physical recovery. Pointedly undesignated in terms of race, the women signify as white in ways that attach their benevolent, tutelary relation to Adah to a certain idealized Englishness.

> They were kind, those women in the ward. For the first few days, when Adah was deciding whether it was worth struggling to hold onto this life, those women kept showing her many things. They seemed to be telling her to look around her, that there were still many beautiful things to be seen, which she had not seen, that there were still several joys to be experienced which she had not yet experienced.
>
> (Emecheta 1974: 111)

The company of these women gives Adah a much-needed reprieve from Francis's verbal and physical violence, not to mention his philandering, selfishness, and chronic laziness. The narrator dwells on Adah's increasing emotional investment in the white mothers around her, and on her subsequent soul-searching about the state of her own marriage. For each of Adah's new acquaintances seems to enjoy a happier marriage and more attentive husband than she. And, significantly, each openly embraces her new role as a mother. While gently modeling and guiding Adah toward a happier domesticity, these women simultaneously shame Adah and stigmatize her on account of her uncouth appearance (her stained medical gown is repeatedly mentioned) and comparably unhappy marriage with Francis. Adah's provisional acceptance in this order of women signals her identification with and internalization of colonial values.

One woman, clutching her "cherished baby," has finally given birth after 17 years of marriage to her devoted husband (Emecheta 1974: 112). This image of forbearance justly rewarded is accompanied by that of another woman, two weeks overdue, seen cuddling with her "film-star-like husband" as if they were "lovers in the cheap movie pictures Adah had seen at home" (115). This woman, identified only as the "sleek girl" in bed number 11, is herself an orphan, and views her pregnancy as a path to the close-knit biological family she has always wanted. "She was determined to make a happy home for herself, where she would be loved, really loved, and where she would be free to love. ... It seemed as if her dream was coming true" (114). Yet another "gorgeous" Greek woman engrosses Adah with her fashionable nightdress from Marks and Spencer's, an uncomfortable reminder of Adah's shabby, stained hospital-issue gown. Adah listens eagerly to each woman's story, at once humbled and inspired by her tale of hardship overcome and happiness achieved. The healthy bonds of attachment enjoyed by the couples around accentuate the love deficit in her own marriage; Francis, unlike the other husbands, only begrudgingly visits her in the hospital and comes bearing neither flowers nor the clean nightdresses she requests. As she gets to know these women, Adah finds her initial bitterness and annoyance, as when "she got tired of admiring this baby with thick brown hair" (111), dissolving in a flood of her own tears. "Coming to have her baby in this hospital had opened her eyes a great deal" (120). Specifically, she begins to view herself as a deserving recipient of the care and compassion directed at the other mothers: "Why was it she could never be loved as an individual," the narrator asks, slipping into free-indirect discourse, "the way the sleek woman was being loved?" (115).

How are we to understand Adah's appetite for these love stories? On the one hand, Emecheta renders the maternity ward a crucial, if precarious, site of shared struggle and solidarity among fellow mothers. Adah's open, intimate verbal exchanges with the women in neighboring beds stand starkly opposed to the coldness and privacy elsewhere associated with England, described earlier as a "society where nobody was interested in the problems of others" (Emecheta 1974: 66). The maternity ward, as an emblem of Britishness, outwardly appears welcoming and hospitable to the figure of the black migrant. The temporary sisterhood emerging in the hospital offers an important emotional resource as Adah builds up the courage to assert greater rights within her marriage. However, the maternity ward is simultaneously a site of exclusion and stigmatization. The medical professionals and female patients alike establish an imposing set of behavioral standards that effectively stigmatized Adah on account of her racial difference. The title of the chapter containing the hospital episode is "Learning the Rules," bracketed by ones titled "Role Acceptance" and "Applying the Rules." This somewhat managerial nomenclature evokes willed acquiescence rather than emancipation, thwarting the narrative of self-awakening elsewhere suggested. Adah's self-consciousness about her medical gown and the shabby, blood-stained shawl with which she swaddles Titi register a complex sense of racial self-loathing encouraged by the idealized therapeutic culture of the maternity ward, which constantly reminds her of how far short she falls of its civilizational standards. A self-conscious politics of respectability shadows Adah's attempt to disguise her class and racial origins by mimicking the dress and behavior of the white women surrounding her. Here again, "belonging" encodes imperial identification.

But it also interpellates Adah as the kind of caring, responsible mother enshrined in what Nikolas Rose (1999) calls the then-emerging "psy" disciplines around childbearing and parenting. Although Adah is stigmatized on account of her race, she is also increasingly held accountable to the patriarchal assumptions of a male-dominated medical culture. Emecheta asks readers to think critically about Adah's emotional over-involvement in the idealized picture of motherhood enshrined by the maternity ward. The debilitating effects of Adah's newfound maternal orientation become especially pronounced after the birth of her fourth child, who gives Adah a new opportunity to demonstrate the values of the fit mother. While on maternity leave caring for this child, Adah attempts to defuse tensions between her and

Francis by briefly, and disastrously, reinventing herself in the mold of the "real housewife" of women's magazines and pulp romance.

> She had been reading a great number of women's magazines, and was surprised to read of mothers saying that they were bored just being housewives. She was not that type of woman. There were so many things she planned to do, and she did them. She knitted endless jumpers and cardigans for everybody, including thick big ones for Francis. It was a way of telling him that that was all she asked of life. Just to be a mother and a wife.
>
> (Emecheta 1974: 163)

In this passage, Emecheta gives memorable shape to the happiness imperative found in mass-mediated women's fictions and in the maternalist assumptions of the welfare state. A matrix of prescriptive discourses about and for women lead Adah to "tame her emotions" and reign in an intuitive frankness and aggressivity that Emecheta connects to her Nigerian past (Emecheta 1974: 50). The psychic space abdicated by these ugly feelings must be taken up with a range of positive affects (civility, personality, fellow feeling, maternal care) holding forth the promise of fuller inclusion in the national culture.[6] As the novel proceeds, Adah's pursuit of happiness in the domestic sphere ironically constitutes one of the biggest hazards to her well-being. It raises false hopes of a conjugal reconciliation with Francis—the responsibility for which rests on the shoulders of the supplicant wife and not those of her abuser—and prolongs what is clearly an unsustainable, violent marriage.

A related scene that brings Adah's maternal ambitions into painful focus involves the birth of her next (fourth) child. Adah takes steps to ensure Francis will not embarrass her, as he has before, by failing to visit or bring flowers and gifts like the other husbands.

> She addressed twenty greeting cards to herself, gave three pounds to Irene, the girl, and told her to post three cards a day after the baby was born. Two big bunches of flowers were to be sent her, one on her arrival, with Francis's name attached to it *with sentimental words*.
>
> (Emecheta 1974: 160, my emphasis)

In this act of authorial impersonation, Adah effectively writes into being an idealized birth scenario, with a starring role for Francis. She ventriloquizes his "sentimental words" and even signs off on his behalf. Viewed in this way, the greeting cards constitute a somewhat desperate act of over-compensation. By recentering the voice and stabilizing presence of the male breadwinner, these cards emerge as extensions of, rather than critical alternatives to, Francis's patriarchal excesses. Whereas an earlier act of forging Francis's signature on the consent form for birth control suggests Adah's emerging self-possession and agency, the forging of the birth cards suggests relinquished empowerment and continued entrapment in a dangerous domestic reality. Adah's sentimental conception of writing and literature forms one of the most enduring aspects of her colonial education.

Emecheta explicitly links Adah's maternal devotion to a middle-brown tradition of popular psychological texts about proper parenting. The idealized maternal identity Adah aspires to bears a strong resemblance to the one championed by the psychologist John Bowlby, in influential mid-century treatises such as *Child Care and the Growth of Love* (1953) and *Maternal Care and Mental Health* (1950). Bowlby joined a growing chorus of voices placing the onus of national reconstruction on the mothers who bear the nation's children. With his prescriptions for "normal motherhood," Bowlby strongly influenced the postwar British welfare state's profound investment in motherhood and the family. He did so by seizing upon and giving scientific sanction to popular anxieties about "absent mothers." He popularized the figure of the childcare-shirking mother as a kind of folk devil through influential studies of child evacuees during World War II. Bowlby argued that without a "warm, intimate and continuous relationship with mother," the developing child would spiral into antisocial behavior and delinquency (Bowlby 1950: 11). His solution to the problem was to advance a set of moralizing prescriptions for proper parenting. Two such prescriptions appearing in *Second Class Citizen* are first, a command for mothers to speak to children (even preverbal infants) as much as possible and second, an instruction that enjoins the mother to take pleasure in the services she offers her children.

Emecheta's novel extends the feminist critique of Bowlbyism by situating it in the colonial endeavor in which it participated. As a black migrant, Adah is described repeatedly as being browbeaten by parenting techniques whose supposed universality bespeaks a decidedly Eurocentric set of values and concerns.

> At home in Nigeria, all a mother had to do for a baby was wash and feed him and, if he was fidgety, strap him onto her back and carry on with her work while that baby slept. But in England she had to wash piles and piles of nappies, wheel the child round for sunshine during the day, attend to his feeds as regularly as if one were serving a master, talk to the child, even if he was only a day old!
>
> (Emecheta 1974: 46)

In *Second Class Citizen*, becoming the kind of parent recommended by Bowlby requires Adah to undergo a dehumanizing, and implicitly imperialist, affective transformation that obliges her to police and contain her unsavory or antisocial emotions. As Adah prepares to leave the maternity ward after giving birth to Titi, she reflects on the successes and failures of the ethical education she has gained there.

> She must learn to thank people, even for their smiles, and kindly nods. This consoling conclusion, this new code of conduct Adah learned from

the hospital and from staying together with other women for thirteen days, was to be with her for a long time.

(Emecheta 1974: 126)

This "code of conduct" appears innocuous enough at first, but it takes on a coercive dimension when we consider the kinds of self-abnegation it demands of Adah. *Second Class Citizen* is full of scenes in which Adah must hold her tongue or tame her emotions in the name of English decorum, collegiality, and restraint.

This demand for the racialized citizen's emotional compliance exemplifies the governing strategies that Anne-Marie Fortier describes as targeting the "affective citizen." Discussing in an interview the neoliberal British state's drive for "community cohesion" in recent years, Fortier locates this mode of governmentality in the "policing of the kinds of public feelings that are acceptable and unacceptable—protesting is bad for cohesion, talking about racism is bad for cohesion, meeting your neighbours in 'meaningful exchanges' is good for cohesion" (Fortier 2011: 11). In the imagination of *Second Class Citizen*, this policing of affect is increasingly enforced by Adah herself; she is both the subject and object of the policing strategies Fortier describes. One of the striking features of the novel is Emecheta's sustained attention to what we might call the "affective" criteria of national belonging. Emecheta frames Adah's somewhat cautious and erratic attempts at self-fashioning in terms of the emotional and behavioral criteria ethnic minorities were forced to meet in order to demonstrate fitness for belonging in the imagined community of Britain. Beyond the formal, legal criteria of citizenship, black migrants in Emecheta's writings are pictured negotiating a kind of cultural obstacle course designed to differentiate between those deserving and those undeserving of full civic membership in the nation.

Women's support networks and alternative forms of belonging

In this context, more promising sites of belonging and mutual assistance in the novel begin to come into focus. If the therapeutic culture of the maternity ward reveals itself as a coercive apparatus that polices the behaviors of black mothers, then Emecheta identifies alternative sites of female solidarity in the caring strategies modeled by Adah's mother and the women's support networks pictured in the Nigeria phase of the novel. This dimension of the novel stands apart from conventional understandings of "belonging" that construe it only in terms of the nonwhite migrant's access to or mastery of white metropolitan norms. Instead, Emecheta treats "belonging" in terms of mutual, lateral forms of care and community that need not attach to the self-legitimizing mythologies of any one national culture. In the many moments of confrontation shaping Adah's adult life in Britain, the memory of her mother frequently authorizes her defiance and allows her to articulate dissent. Significantly, the defiance inspired by Adah's mother runs counter to the very English restraint and reticence expected of her as a wife and mother. Emecheta pictures Adah's mother in terms of her temper, her appetite for confrontation, and her refusal of imperial standards of modesty. These qualities, it is worth noting, initially render her a somewhat forbidding character to the young Adah. As a girl, she, in fact, much prefers the company of her father; he enables his daughter's education (it is his slate board that Adah uses at school) whereas her mother actively discourages it. But the memory of her mother, retrieved in adulthood, takes on a very different valence in scenes where an older Adah grasps for emotional resources in confronting metropolitan racism.

In one of these scenes, Adah confronts a negligent white babysitter named Trudy, who has been minding her children. Adah arrives unannounced and discovers the children cavorting in a "slum" of a backyard "filled with rubbish, broken furniture, and ... an old type of toilet with faulty plumbing, smelly and damp" (Emecheta 1974: 51). Already enraged, Adah enters Trudy's flat to discover her in a compromising position, in a state of partial undress with a male lover.

> There was a pause, during which Adah could hear her heartbeat racing. She was finding it more and more difficult to control her temper. She remembered her mother. Ma would have torn the fatty tissues of this woman into shreds if she had been in this situation. Well, she was not Ma, but she was Ma's daughter, and, come what may, she was still an Ibo. She screamed.
>
> (51)

Inspired by her mother, Adah's violent oath implicitly rebukes the standards of female decorum upheld by the welfare state's maternalist ideologies. Adah immediately curbs her anger. "Where are my children? You pro—" (51). Adah's cut-off locution registers at the level of the sentence the kind of bottling up of emotions required of the aspiring British citizen.

In a later scene, however, Adah unleashes the anger that here she keeps in check. After finding out that Francis himself has been having an affair with Trudy, Adah throws a broom at her and, vowing vengeance, has to be restrained by a neighbor:

> Adah spat, foaming in the mouth, just like the people of her tribe would have done. Among her people, she could have killed Trudy, and other mothers would have stood solidly behind her. Now, she was not even given the joy of knocking senseless this fat, loose-fleshed woman with dyed hair and pussy-cat eyes. She belonging to the nation of people who had introduced "law and order."
>
> (Emecheta 1974: 66)

In this passage, Adah justifies her aggression by noting its compatibility with what Emecheta identifies as Nigerian (and particularly Ibo) modes of justice and conflict resolution. Adah claims kinship and belonging with the women's support communities that "would have stood solidly behind her." As the scene unfolds, Emecheta uses the counterexample of these support communities in order to launch a scathing critique of the emotional restraint that structures British imperial culture.

Englishness, despite being rhetorically grounded in an ideology of "welfare," is in fact portrayed in the novel in terms of a failure of mutuality, community, and support. The failure of communal assistance in England is repeatedly contrasted with the Nigerian communities of sympathy that Adah remembers from her childhood. The narrator derides welfare state caseworkers and counselors as "paid listeners" who "make you feel that you are an object to be studied, diagnosed, charted and tabulated," and describes England as a "society where nobody was interested in the problems of others" (Emecheta 1974: 66). Emecheta juxtaposes the cynical, condescending "listening" practices of British caseworkers with the genuine understanding and willingness to intervene that Adah attaches to remembered Nigerian women from her childhood.

> In England, she couldn't go to her neighbor and babble out troubles as she would have done in Lagos. ... You don't have the old woman next door who, hearing an argument going on between a wife and husband, would come in to slap the husband, telling him off and all that, knowing that her words would be respected because she was old and experienced.
>
> (66)

In an essay titled "Feminism with a Small 'f'," Emecheta offers a useful picture of the Nigerian models of community that Adah recalls. "In the villages the woman will seek the company of her age-mates, her friends, and the women in the market, and for advice she goes either to her mother or to her mother-in-law" (Emecheta 1988: 176). Recognizing the value Emecheta places in combative but supportive female communities of care helps us put pressure on John McLeod's argument, in his book *Postcolonial London*, that Emecheta tends to pathologize the Nigerian characters in the novel, portraying them as obstacles to Adah's happiness. "The bulk of the horrors of living," McLeod argues, "are created by Nigerians in the neighbourhood who demand that women live according to the gender restrictions which Adah has been keen to leave behind in Lagos" (McLeod 2004: 102–03). It is true that Adah's female neighbors occasionally delight in her misfortune, singing songs to ridicule her. But just as often Nigerian women come to her rescue and offer emotional resources for the struggles she undergoes. McLeod elaborates Emecheta's interest in spaces of "transcultural" connection that exceed conventional notions of national identity—for example, the conspicuously cosmopolitan space of the Chalk Farm Library where Adah works, replete with a Canadian boss and coworkers hailing from Ireland and the West Indies, respectively (107).

It would be a mistake, however, to dismiss Emecheta's enthusiasm for the specifically black diasporic affective community that binds together Adah and the memory of women's support communities in Nigeria. Adah's edgy, antagonistic relationships with women such as Trudy challenge many of the assumptions of universal sisterhood driving Western feminism, and in this sense, Emecheta's black postcolonial feminism aligns with theoretical work in black cultural studies such as Hazel Carby's influential essay "White Woman Listen!" (Carby 1997). The term "belonging" doesn't quite do justice to the aggression and antisocial energies *Second Class Citizen* seems to celebrate. Yet in the imagination of the novel, black diasporic women's support networks prove to be a far more sustainable resource of hope than the caring institutions of the British welfare state.

Notes

1 See McLeod (2004) for an example of this position.
2 See Procter (2003) on housing and racial formation in black British literature.
3 See Lazarus (2011) for a critique of the seeming indifference to the state in the dominant strains of postcolonial theory.
4 On this early period of black diasporic literary production in Britain, see Procter (2003), McLeod (2004), Premnath (2002), and Dawson (2007).
5 See Natarajan (2013), Webster (1998), and Bailkin (2012) for historical accounts of racial formation in the context of the British welfare state.
6 For queer and feminist perspectives on the coercive "promise" of happiness and contentment for women, see Ahmed (2010) and Berlant (2011).

References

Ahmed, Sara (2010). *The Promise of Happiness*. Durham, NC: Duke University Press.

Bailkin, Jordanna (2012). *The Afterlife of Empire*. Berkeley: University of California Press.

Berlant, Lauren (2011). *Cruel Optimism*. Durham, NC: Duke University Press.

Bowlby, John (1950). *Maternal Care and Mental Health*. London: Penguin.

Carby, Hazel (1997). ""White Woman Listen!" Black feminism and the boundaries of sisterhood." *Black British Feminism: A Reader*. New York: Routledge.

Davis, Angela (2014). *Modern Motherhood: Women and Family in England, 1945–2000*. Manchester: Manchester University Press.

Dawson, Ashley (2007). *Mongrel Nation: Diasporic Culture and the Making of Postcolonial Britain*. Ann Arbor: University of Michigan Press.

Emecheta, Buchi (1974). *Second Class Citizen*. New York: George Braziller.

Emecheta, Buchi (1988). "Feminism with a Small 'f.'" In Kirsten H. Petersen, ed., *Criticism and Ideology: Second African Writer's Conference, Stockholm 1988*. Uppsala: Scandinavian Institute of African Studies.

Fortier, Anne-Marie (2011). "Interview with Debbie Ferreday and Adi Kuntsman." *Borderlands* 10(2): 1–17.

Lazarus, Neil (2011). *The Postcolonial Unconscious*. Cambridge: University of Cambridge Press.

Marshall, T. H. (1950). *Citizenship and Social Class*. Cambridge: University of Cambridge Press.

McLeod, John (2004). *Postcolonial London: Rewriting the Metropolis*. New York: Routledge.

Natarajan, Radhika (2013). *Organizing Community: Commonwealth Citizens and Social Activism in Britain: 1945–1982* (10075515). PhD diss., University of California, Berkeley. ProQuest Dissertations Publishing.

Premnath, Gautam (2002). "Lonely Londoner: V.S. Naipaul and the God of the City." In Pamela Gilbert, ed., *Imagined Londons*. Albany: State University of New York Press.

Procter, James (2003). *Dwelling Places: Postwar Black British Writing*. Manchester: Manchester University Press.

Rose, Nikolas (1999). *Governing the Soul: The Shaping of the Private Self*. New York: Free Association Books.

Webster, Wendy (1998). *Imagining Home: Gender, Race and National Identity, 1945–1964*. New York: Routledge.

19 Navigating the *regime of illegality*

Experiences of migration and racialization among 1.5-generation Mexican migrant women

Heidy Sarabia, Laura Zaragoza, and Alejandra Aguilar

Introduction

Children have always been part of the migrating flows to the United States—sometimes they come with parents and sometimes they come unaccompanied (Sarabia and Rodriguez 2016). Although mainstream accounts of unauthorized migration into the United States tend to focus on the largest share of this type of migration—Mexican adult migrants who come to seek work—growing attention has turned to unaccompanied minors and families. In this chapter, we look at these immigrant children who come to the United States before the age of 16—members of what is known as the 1.5 generation—in order to highlight how the circumstances of migration, the exposure to racialization in the country, and the experiences living undocumented are consequential in the everyday lives of immigrants.

Mainstream media has focused on the waves of Central American migrants that have come recently, but it is also important to highlight that by size, Mexican migration has been a significant wave in the United States during the twenty-first century, and remains so. For example, in the current fiscal year (through August 2020), close to a third of all family unit apprehensions along the US–Mexico border originated from Mexico (27%), third only to Guatemala (31%) and Honduras (30%); and during the same period, close to half of all unaccompanied minors encountered at the border were from Mexico (48%), followed by Guatemala (29%) and Honduras (15%) (CBP 2020).

Karen Silva[1] and Natalia Medina both left Mexico before the age of 16, and thus would be considered members of the 1.5 immigration generation. But their *circumstances of migration* were radically different. Karen left when she was 8 years old with her whole family, fleeing economic instability in their native Michoacán, Mexico. Thus she migrated as part of a family unit—as a dependent "minor"—and was enrolled in school immediately after her arrival in Sacramento, California, an immigrant-friendly destination. Natalia, in contrast, left Mexico at age 14 without her parents. Although she was accompanied by her godmother, Natalia was still considered an independent "adult." The objective of migrating was finding employment in order to help

DOI: 10.4324/9781003170686-23

her family. She began working in a laundromat when she arrived in Philadelphia, Pennsylvania, instead of going to school. While Philadelphia is a reemerging hub of immigration, it is less welcoming compared to California—a sanctuary state that has enacted immigrant-friendly policies state-wide. In this chapter, we focus on the stories of 25 women like Karen and Natalia from the 1.5 generation; we highlight how the *circumstances of migration* (as dependent "minors" or independent "adults") and the *context of reception* (arriving at California or Pennsylvania) shaped their migration and racialization experiences in the United States.

In conjunction with our focus on these women, we provide a brief review of the literature and our contribution. Then, we discuss the methods used to gather and analyze the data used. We discuss the three main findings that focus on how different *circumstances of migration, context of reception,* and *living undocumented* shape migration and racialization experiences in the United States. Finally, we conclude by highlighting our contributions to the literature on immigrant racialization.

Examining the literature

We begin in this section by providing context for our work through the lens of existing literature and theory in areas that concern how life cycle shapes migration, the diversity of migratory experiences, and how institutions of socialization affect a developing a sense of belonging.

How life cycle shapes migration experiences

Classical explanations of migration have theorized movement (that is, global migration) as motivated by economic pulls and pushes (Massey et al. 1993). Such explanations include neoclassical economics, the new economics of migration, segmented labor market theory, and world systems theory. These theories have been criticized for overemphasizing economic forces (Boswell 2008). Other theories have focused on social capital, cumulative causation, and transnational networks to explain continuous migration in the absence of clear economic forces, and even the role other actors such as politicians, bureaucrats, and pundits play in shaping migration (Massey 2015). Mexican migration has typically been described as economic migration, structurally driven by the imbalance of labor supply and demand, and economic inequalities as well as the historic relationship of imperialism between Mexico and the United States (Barajas 2009), situations that also typically focused on adults migrating to the United States in search of work. But scholars have acknowledged the diversity within Mexican migration flows, pointing to the unique experiences of women (Hondagneu-Sotelo 1994), those from indigenous backgrounds (Fox and Rivera-Salgado 2004), and those based on generational differences (Barajas 2009; Abrego 2011).

By building on this body of literature and focusing in this chapter on the experiences of women who migrated from Mexico as children, younger than

16, what is called the 1.5 generation in the United States (Rumbaut 2004), we narrow the range of experiences, which allows us to explore their experiences in greater depth. Migration by men in significant numbers has a longer history than the migration of women, and more scholarly work has focused on their experiences (Hondagneu-Sotelo 1994). We further challenge the homogenization of Mexican migration by focusing on the diverse experiences these women report from arriving in two very different places of reception—California and Pennsylvania. In addition, we have chosen to highlight the unique and diverse experiences of Mexican women, particularly in the context of the 1.5 generation, because scholars have acknowledged that "gender relations shape immigration patterns, and in turn, migration experiences reshape gender relations" (Hondagneu-Sotelo 1994: 2).

We also focus on the 1.5 generation because this subsection of immigrants has begun to garner more attention, in part because of their activism. Although most of the research on these young immigrants tends to focus on the context of education (Abrego 2006; Negrón-Gonzales 2014; Terriquez 2015; Gonzales 2016), we compare women who remain in higher education and women who did not enter the school system in order to contrast differences and similarities in their experiences.

We theorize the concept of "circumstances of migration" as one of the most immediate reasons why a migrant decides to leave Mexico and relocate to the United States, and one that significantly shapes experiences in the country once they arrive. While we acknowledge that the "immediate reason" why people leave might also depend on underlying structural forces that propel that migration, the immediate circumstance of migration, we argue, shapes both how migrants leave Mexico and how they are integrated when they get to the United States. Using data from in-depth interviews with 25 women, we explore their varied experiences based on whether they migrated with their families, as Karen did, or as de facto adults, as Natalia did, and also consider the context of reception after they migrated. Thus, we build on the work of Pierrette Hondagneu-Sotelo—who provided a typology of migration patterns based on family stage migration, family unit migration, and independent migration—by adding age and reception context. We also highlight the social construction of the life course (Holstein and Gubrium 2007) to show that circumstances under which migrants leave their homeland, in addition to age, shapes experiences in the United States.

Finally, we want to highlight how experiences among undocumented migrants in the United States are shaped by a *regime of illegality*—that is, a socio-legal system established and maintained by the US government through the organization and management of migrants according to their legal status in the United States, and operated through the law and policies that render unauthorized migrants vulnerable to detention and deportation (Sarabia 2019: 43). Through this comparison, we highlight the commonalities and differences in how *illegality* is experienced every day, based on different *circumstances of migration*.

Diverse migratory experiences

Significant research has been devoted to diverse migration experiences. For example, Hondagneu-Sotelo (1994) argued that family and community relations play a key role in shaping migration, and that "gender is one of the fundamental social relations anchoring and shaping immigration patterns" (Hondagneu-Sotelo 2003). In addition, scholars have shown that immigration laws have specific effects on women (Abrego and Menjívar 2011), and that gender deeply affects undocumented young adults' dating, marriage, and parenting experiences (Enriquez 2017).

Age of migration also shapes experiences. While research has focused on adults, scholars are increasingly acknowledging the distinct experiences of the 1.5 generation and how different circumstances result in unequal pathways within this group (Gonzales and Burciaga 2018). The 1.5 generation refers to those who migrated to the United States before or during their early teens (Rumbaut 1997). Hence, these immigrants spend their formative years and acquire most of their cultural and social development in the host country (Portes and Rumbaut 2001; Rumbaut 2004). At the same time, research has identified protections for the 1.5 generation that first-generation migrants lacked because many had no access to school (Abrego and Gonzalez 2010). Undocumented 1.5-generation migrants are typically able to learn the language, the customs, and culture of the host county (Fernandez-Kelly and Curran 2001; Abrego 2006). While legal status limits their everyday lives, many 1.5-generation migrants claim social membership in the United States (Gonzales 2016). The particular experiences of the 1.5 generation give rise to unique modes of agency and belonging (Ellis et al. 2019), and they face particular challenges navigating varied *local terrains of illegality* (Burciaga et al. 2019). Scholars have also shown the importance of local experiences, arguing that a patchwork of municipal, state, and federal laws defines *immigrant illegality* (Flores et al. 2019).

In the context of these studies, we further explore the specific experiences of 1.5-generation immigrant women in two US states—California and Pennsylvania—in order to highlight divergent and convergent experiences based on different *circumstances of migration, contexts of reception,* and their experiences *living undocumented* in the United States. The divergences reflect the diversity of experiences migrants face in the United States, while the convergences highlight the national trends that tend to hold throughout the country, as well as the context-specific socializing experiences that seem to matter most.

Institutions of socialization and developing a sense of belonging

Abrego (2011) has shown that institutions play an important role in socializing migrants in that first-generation migrants to the United States tend to be socialized in the workplace while 1.5-generation migrants tend to be socialized in school settings. We expand this work by not only showing how specific institutions play a role in the socialization process of migrants, but also by highlighting how migrants experience racialization specifically as Mexican in the United States. We show the ubiquitous experience Mexican migrants share as they are racialized in multiple institutions, and we show the contextual differences (and similarities) that those who migrate to California and those who migrate to Pennsylvania experience.

This chapter also intersects the literature on migrant incorporation in the United States with theories of racialization. The former literature has, for the most part, convincingly shown that immigrant acculturation in terms of learning English, adopting the US culture, and identifying with the United States happens quickly—and can even result in the loss of the first language (Fillmore 1991). It has also shown that a sense of belonging in the host country is a social process that legal exclusion does not prevent. For example, Renato Rosaldo (1994) showed how migrants without legal status developed cultural practices in San Jose, California, that allowed them to build community and develop a sense of cultural citizenship, which facilitated a sense of belonging in their community in the United States. But this literature has been less precise about how the racialization of migrants shapes such incorporation and acculturation.

In addition, research has shown that Mexican migration in the United States has changed in an important way in the last 30 years due to the militarization of the border. Mexican migration has become more permanent in the United States. While still relying on transnational networks developed over decades (Schiller et al. 1995), Mexican migration is no longer circular. Given the escalation of danger along the border, most migrants have stayed permanently in the United States rather than migrating for work and returning to Mexico on a regular basis (Durand et al. 1999). Given these trends, questions about how migrants develop a sense of belonging in the United States are becoming more significant, especially in the context of *perpetual illegality* (Sarabia 2012)—as the opportunities for undocumented migrants to adjust their status are limited. In this chapter, we contribute to this literature on migrants' sense of belonging by showing how *circumstances of migration* shape that sense of belonging, how political climate shapes how migrants think about and plan for their future, how the context of reception and experiences of racialization shapes those experiences, and finally, how legal status shapes their sense of belonging among 1.5-generation immigrant women in the United States.

Methods: 1.5-generation women in two contexts

The interviews in Pennsylvania took place in 2015 and in those in California took place in 2018. Thus, the Pennsylvania interviews occurred before the election of Donald Trump, and they do not reflect the full force of anti-immigrant sentiment that became more explicit during his administration. Yet both samples were similar in that all the migrants interviewed had arrived before turning 16. At the time of the research, the participants were between 18 and

29 years of age. The average age of migration in the California sample, however, was younger (by about 2.5 years), while the average years of education in the California sample was significantly greater (by about 5.7 years). The Pennsylvania sample migrated at an older age (by an average of 7 years), and its members were more likely to have migrated as "independent adults" (not in terms of age but whether they arrived as independents or dependents upon arrival), to be now cohabitating with a partner, and to have children.

Pennsylvania

The data from Pennsylvania come from a larger project focused on understanding the experiences of belonging among Mexican migrants in Philadelphia, a new area of migrant destination. Heidy Sarabia conducted a total of 69 interviews with Mexican migrants in South Philadelphia in 2015. This included 17 who arrived before the age of 16, but two interviewees that arrived with legal status are not included here, leaving 15 Pennsylvania interviews in the data used for this chapter.

The average age of the Pennsylvania interviewees for this chapter was 24.4 (with a range of 18–29), the average age of migration was 13 (with a range of 6–16), and the average time living in the United States was 11.4 years (with a range of 7–16). All the interviewees were from Mexico. Most of the interviewees (13) were married or cohabitating, and most also had children (12). All eight participants who migrated as "independent adults" had children at the time of the interview; while about half (4 of 7) who arrived as "dependent minors" did not have children. At the time of the interview, only one respondent had applied for and was protected under Deferred Action for Childhood Arrivals (DACA).

Given that the growth of the Mexican immigrant population in Philadelphia was a relatively recent phenomenon, the reception was mixed in the city. In 2006, a time that correlates with the arrival of many of the participants,[2] one of South Philadelphia's most famous and iconic business shops, "Geno's Steaks," began displaying a sign that read: "This is AMERICA: WHEN ORDERING 'SPEAK ENGLISH'" (AP 2006). The sign highlighted the tensions that were emerging as the famous Italian market in South Philadelphia, where most of the migrants lived, underwent a demographic shift. Nevertheless, the reception of immigrants in the city was mixed; for example, since 1992 the police department had a directive in place to limit cooperation with federal immigration authorities (Armenta and Sarabia 2020: 2).

California

The data from California come from two projects focused on understanding the experiences of 1.5-generation college students during the Trump administration and in a particular context, given California's reputation as an immigrant-friendly state. Laura Zaragoza conducted a total of eight in-depth interviews between October 2017 and March 2018. All the participants of this project were members of the 1.5 generation since they all arrived in the United States as children. For the purpose of this project, only the data from the five female interviews was used. A third set of participants were interviewed by Alejandra Aguilar, who conducted a total of seven in-depth interviews between April and May 2018, including five with women. We also draw from those interviews here.

At the time of the interview, six of the respondents in the California sample were protected under DACA, while two did not qualify for this program, and two had not applied because of fear. The average age of the ten female participants in the California sample was 21.9 years old (with a range of 18–25), the average age of migration was 6.3 years old (with a range of 1–10), and the average time living in the United States was 16 years (with a range of 10–21). All the respondents were from Mexico. The participants in this study did not have children and three were married or cohabitating. All the interviews took place in Northern California.

The context in California was different from the context in Pennsylvania, in that California is a state historically receptive to migration. Unlike Pennsylvania, where the majority of Mexican migrants had arrived in the past 15 years, Mexican migration to California dates back to when the territory belonged to Mexico before 1848. Reflecting this more welcoming environment, California has established immigrant-friendly policies such as AB540 (Abrego 2008) and the California Dream Act (Whaley 2012) that allow undocumented immigrants to attend universities as residents and provide some financial aid as well. In addition, California became a sanctuary state in 2017.

Data analysis

We analyzed the interview data using NVivo software to find patterns on the data. The data were first coded using general themes including "integration," "reasons of migration," "discrimination," and "returning to Mexico," among others. Each author coded her own interview data. Once all the data were coded, further analysis was conducted to see differences and similarities in the codes. Three interesting findings emerged: *circumstances of migration*, *context of reception*, and *living undocumented* in the United States.

Findings of the Pennsylvania and California interview data

We have organized the findings into three sections. First, focused on differences, we show how the different *circumstances of migration*—that is, how migrants left as "minors" and as part of family units, or as independent "adults"—shape their subsequent life in the United States. In the second part, focusing on the commonalities among all women, we further explore how, regardless of *circumstances of migration*, most migrants tend to share similar racialized experiences in the United States. Finally, in the last section, focused on the local context of reception, we show the challenges 1.5-generation undocumented immigrants from Mexico face as they try to create a sense of belonging under the *regime of illegality* in the United States, and we highlight the way context of reception shapes those experiences.

Circumstances of migration: diverse experiences leaving Mexico

Beyond the importance of age of migration (Barajas 2009; Abrego 2011), we found that the level of responsibility interviewees assumed in migrating shaped their experiences. In the Philadelphia sample, we interviewed migrants who came to the United States as "minors" and migrants who came to the United States as "adults," whereas all the California interviewees migrated as "minors" with their parents or to join their parents.

For example, Sol Martinez (PA),[3] who left Mexico at the age of 7, explained, "I came with my mom and my sister first. My dad was already here in Philadelphia and then after we came, then my brothers came 2 years after that." She entered school and learned English quickly. Sol explained, "Fortunately, at the beginning since we were so little, we were able to pick up the English really fast and after 3 months we were able to speak and communicate with other students."

Liliana Martinez (CA) also had a typical 1.5-generation experience. She explained, "I came with my mom and my younger sister—my dad was already here—and I didn't have to go through the desert or anything, it was like under somebody else's paperwork, somebody else's documentation." With her father already in the United States, it was easier to make such arrangements, and Liliana and her sister avoided some of the dangers that can be present when crossing the border in harsher conditions. Like Sol (PA), Liliana (CA) learned English quickly in elementary school, which she began when she arrived.

Miriam Gomez (CA) did not even have to learn English as a second language as she migrated when she was a year old. Like Liliana, she came on someone else's paperwork. "I don't remember anything because I was only one," she explained, "but my family tells me they brought me in with a little boy cousin's paperwork. They just dressed me up as a little boy." Of course, Miriam has no memories of Mexico, and she is not even fluent in Spanish.

Entering the school system meant that participants who came as minors were socialized within the schools and became bilingual. We found that the context, arriving to California or Pennsylvania, did not matter much, but in sharp contrast, it did matter for those migrating as "adults" who joined the labor market upon arrival and were not bilingual. Julia Flores (PA), for example, did not go to school even though she was 16 and the state of Pennsylvania would have provided an education to her.[4] But she immediately started working in order to send her family remittances.

Like Julia, Natalia Medina (PA) migrated at the age of 14 with her godmother to Pennsylvania and began to work right away. She explained,

> I began to help my godmother in a laundromat a few days after I arrived. Then I started working cleaning homes, which was difficult because I did

not speak English, and it was difficult for people to explain to me what to do. That was really hard.

Natalia became an adult fast, moved in with a partner at 15, and had her first child at age 17. Migrating as independent adults sped up her growing process. Similarly, Belen Velasquez (PA), who migrated at the age of 14 to Pennsylvania, came with an older brother. While she attended an adult school to learn some English, she immediately started working at a restaurant. She moved in with a partner at age 16 and became pregnant right away. She stopped working to take care of her baby, and while she says she can "defend" herself in English—meaning that she understood and spoke some English—she was not fluent. Belen's experience in the United States is more similar to the typical first-generation immigrant experience than to the 1.5-generation experience, even though she arrived at the age of 14. Those who migrated as adults tended to start their families at an earlier age (between 16 and 18); while those who migrated as minors delayed that process.

In sum, we found that age of migration was an important factor that shapes how migrants experience a sense of belonging in the United States. Whether they were socialized through the school system or the labor market played a key role as well (also shown by Abrego 2011). Thus the *circumstances of migration*—whether they took on adult responsibilities upon migration—was a significant factor—especially in the acquisition of English and in patterns of family formation.

Context of reception: Common experiences at the intersection of illegality and racialization

Gonzales (2016) argues that migrant children tend to be "protected" from illegality since they do not have to deal with the same issues that adults do, and that they "learn to be illegal" when they turn 18. But this ignores the fact that some in the 1.5 generation do not attend school; instead, they enter the labor market immediately after they arrive, as we found in the Pennsylvania sample. We also found that all participants in the United States described a process in which they were "othered" as racially different from the main-stream. This process of racialization was intensified by the *regime of illegality* that rendered them vulnerable as outsiders by law.

Racialization that took place in institutional settings also solidified the perceived "otherness" of these migrants, although the school setting was different from other settings. For example, Noemi Rodriguez (PA), who arrived at the age of 14, explained the explicit "racism" she faced in high school that made her want to drop out of high school. She recalled:

> In school because people are racist, they said things, they said that immigrant people could not enter school, that is, university or college, that they could not study. So, I … kept thinking about it and I said to myself, "OK, so why am I going to continue studying? It will be better if I go to work so that I help my parents."

Participants who entered school learned the ideas and expectations of a society that did not consider them as having as much potential as others. This racialization was experienced both in the Pennsylvania and California context, as participants who reported similar experiences explained. For example, Cindy Lopez (CA), who arrived at the age of 4, said:

> I think I still didn't understand fully what it was to be undocumented until my junior year of high school, when I wanted to join a program [about going to college] and the director straight out told me, "There is no room for you in my program unless the Dream Act passes." That was the first time that I understood what it was to be undocumented.

Cindy's potential was directly measured by her legal status. She was explicitly told she would not be considered for a program because it, and she felt that her legal status caused her teachers to treat her differently. This was not an isolated incident.

In another case, Estrella Perez (CA), who arrived as a child, discovered that once she applied for a job she had to confront her legal status as well as the othering and dehumanizing label of "alien." She explained:

> When I was getting hired it asked, on the application, if I was an alien. And I was like, Oh my God! I wanted to cry! How is that? Why? You know you accepted me, you already hired me. Why does that matter? They are referring to us as aliens! That is dehumanizing. It's sad.

Similarly, Sol (PA) described being conscious of her legal status while traveling. When asked about when she realized that she did not have status in the United States, she said:

> As a young person, when we were traveling, we already knew that we had to lie all the time. So we always had to lie about our names and our status, and after that we already had acknowledged that we had no status in the world.

As Sol highlights, and Joanna Dreby (2015) also showed, children are increasingly becoming socialized and aware of their own and their family members' legal status.

Alma Ramos (CA), who arrived at the age of 5, attributed her experience of the intersection of the *regime of illegality* and the racialization experiences to a generalized culture of exclusion she experienced at school:

> When people ask you, "Where you come from?" they're fast to judge you. Like if you say, "Oh I came from Mexico," they're like, "Oh, wetback!" or they start making jokes about you and saying like, "You're illegal, you don't belong here, go back." It does affect the way you interact with people because it does change the way in which you talk to everybody. You become scared because you feel like they're going to judge you.

Alma expressed her perception that in the United States, being from Mexico is often equated with being undocumented. Hence, the racialization process of being read as a Mexican migrant is immediately tied to the process of being assumed to be undocumented as well. For those who are undocumented, the racialization as Mexican exposes their legal vulnerability. For those who arrive as children, and understand English, being subject to all the stereotypes and insults can be painful and traumatic.

We found that this intersection between the *regime of illegality* and racialization was also experienced by those who arrived as adults. Similarly, those who migrated as "adults" experienced racialized socialization in other settings and by different institutions. For example, Natalia (PA) reported facing discrimination at the clinic where she received medical services:

> In the clinics, sometimes they don't give you the same attention they give Americans because you don't speak English, they don't give you the same attention. But if [you are] an American and you speak English, then they give you attention fast and they help you. Because we are Latino they don't [help us].

Natalia linked her experiences of discrimination to language barriers, and clearly expresses that she was treated differently because she was not able to fluently communicate in English. In a similar way, Valeria (PA) explains her negative experiences in clinics:

> The clinic wasn't very good. It was like the receptionists were very ... I don't know if they were racist but it was like, they pretended like they didn't understand us very well. I don't know, when I went to the clinic I felt discriminated against for not speaking English, even though the receptionist spoke Spanish.

Valeria articulated a feeling shared among many, which again, was tied to the association between being Mexican and being undocumented. Hence, even when receptionists spoke Spanish but were not Mexican, immigrant women like Valeria still felt discriminated against or mistreated. Likewise, Maite Campos (PA), who arrived at the age of 15, also complained:

> Sometimes, because you don't speak English ... like, maybe you want to apply for a health program but there's no one there at the clinic who speaks Spanish and it's impossible. They have you there all day filling out paperwork that says that they can't discriminate. But you're sitting there in the clinic and think, "Who is going to pay attention to me?" So I just stay quiet.

Women often complain about negative experiences in clinics (Armenta and Sarabia 2020), and they have interpreted these negative experiences as discriminatory acts against them because of their background (as Mexican immigrants) or language ability (lack of English skills).

In sum, the circumstances of migration shape the institutions that shape the socialization of new migrants. At the same time, we found that all of the study participants experienced, albeit differently, racialized socialization in the United States as Mexican migrants, which equated being Mexican immigrants with being undocumented.

Undocumented: belonging under the regime of illegality

Scholars have shown that migrants who arrive in the United States as children tend to be socialized and acculturated as US-Americans—learning English and US culture fast. But they must also confront their legal status once they age into adulthood (Gonzales and Chavez 2012). While we did see that pattern, we found that circumstances of migration shaped family ties, which in turn shaped how these women experienced the *regime of illegality* and legal vulnerability. We found that circumstances of migration complicated immigrant women's sense of belonging.

Those who migrated as "minors" and entered the school system were much more "Americanized"—as they did not know much about Mexico and were not very fluent in Spanish. Hence, they could not think about leaving the United States. For example, Alma who arrived in California at age 5, said,

I know I'm Mexican but I feel like if I was ever to go back to Mexico, I don't know anything there. I'm scared because I speak Spanish but I don't think I speak it well enough to live there.

Similarly, Claudia Vasquez (CA), who migrated with her family at age 6, explained, "I would like to visit [Mexico], because even though I was born there, I do not have a lot of knowledge about the history. I would like to visit the pyramids and things like that." But she would never consider moving there.

In addition, because the women migrated as minors with their parents or to join parents, their strongest family ties were in the United States. Noemi (PA) explained:

I have been asked if I think about going to Mexico, and to be honest, I am not sure. My parents are here, my siblings are here, basically all my family is here. Except my grandparents—maybe if I had the opportunity to go back and visit them, I would do it. But I guess that can't happen right now [because of the border control], so they have to wait [to see me].

While Noemi wanted to visit Mexico to see relatives, her strong ties to the United States made it easier to be "patient." But migrants with strong ties in Mexico, such as parents, felt more urgently the desire to travel to visit them. For example, Natalia (PA) who migrated as an independent adult, explained that she misses her parents most. She said, "You should have the right to visit or see your parents, but unfortunately that does not happen. Since they are both alive, that is what I missed the most [from Mexico]." Such limitations made participants with more connections in Mexico feel "trapped," unable to travel and visit family in their country of origin. Hence, family ties shape how migrants think about their life in the United States. For those with strong family ties, separated by the border, unable to visit or see their parents, and getting used to not being able to travel, was difficult. Yet, having family and responsibilities in Mexico also forced migrants to stay in the United States and think of their migration as long-term.

Migrants who assumed adult responsibilities upon arrival also wanted to stay. They had partners and children, and they wanted to stay to continue to build on the lives they had established in the United States. Even Marita Chavez (PA), whose husband, also an undocumented migrant who wanted to return to Mexico, wanted to stay. She had arrived when she was 14 and had already lived 11 years in the United States. She saw better opportunities for her and her four children in the United States. At the same time, she was conscious of discrimination and danger in the United States. She shared, for example, how her husband is sometimes robbed at night as he returns from work with his weekly earning in cash. She explained: "As Hispanics, if they see you with something, they take it away, they rob you. ... My husband has been assaulted two times; they have not done anything to him but have taken his bike away." Because Marita and her husband lack documentation, they would be afraid to go to the police. Hence, the regime of illegality mattered. Legal status made participants feel vulnerable.

In addition, the political context also played a role. Even though California is a sanctuary state and has many progressive policies, the election of Trump made the threat of deportation more salient and the idea that undocumented migrants might have to return to Mexico became more pronounced as well. Most of the California respondents (6 of 10) had received DACA protections, meaning that the federal government knows exactly where to locate them. After the Trump administration rescinded the protections of DACA in 2017, while claiming to investigate the program as a whole, DACAmented[5] immigrants felt intensely vulnerable to the threat of deportation.

Liliana (CA), for example, who was very close to graduating from college, explained the uncertainty she is experiencing because she does not know if she will be able to get a job in her field if her DACA protections end. She said, "Well, I will have DACA for one year after I graduate. I don't know if I will be able to work after that, so that's a question mark." Since the future of DACA was unclear at the time of her interview, Liliana was experiencing a lot of anxiety, and she was forced to explore ways to pursue professional development opportunities without work authorization.

Even Claudia (CA), who was not protected under the DACA program, was still experiencing some of the effects of the anti-immigrant rhetoric at the national level. In planning for her future, she noted:

> We are waiting, maybe until after grad school to see what happens, like maybe a Dream Act or something. But if not, then the solution would

be to get married, but I don't want to get married yet. If I had my social security [number], I would not be getting married.

At the time of the interview, Claudia was waiting to hear back from two graduate programs she had applied for, and she was also aware that she would not be able to get a job in her field without legal status. Her partner had recently naturalized.

The stress over the end of DACA was such that Estrella (CA), described it as "feeling hunted." She said,

I just want to live in a world where I don't have to worry about who's hunting us down or what's going to happen. I don't want to be scared. I don't want to feel that way. I just want to be happy.

Clearly, the political context has radically changed for undocumented migrants in the United States since Trump took office, although the future looks brighter: in December 2020, a federal judge ordered the Trump administration to restore DACA as it existed under President Obama, and in January 2021 Joe Biden will become president.

In sum, while different institutions play a significant role in how migrants are socialized in the United States, the experience of being racialized was ubiquitous and predated entering the workforce for those who enrolled in school. In addition, specific sociopolitical changes, such as the election of Donald Trump and the elimination of the DACA program, and various legal rulings of DACA, highlighted the toll on migrants who have had to deal with the intensification of the anti-immigrant rhetoric and practices.

Conclusion

This chapter shows that the 1.5 generation is not a uniform group. Whether migrants arrived as adults or minors—and whether they took on adult responsibilities in the workplace or entered the school system—those experiences shaped their feeling of belonging in the United States. As other scholars have highlighted, these institutions—through the socialization process—further shape the experiences of migrants in the United States (Abrego 2011). Yet, as we have shown, regardless of the circumstances of migration, a common experience among all 1.5-generation migrants, and across all institutions, was the ubiquitous experience of being racialized in the United States, with mostly negative consequences. In addition, while there are strong pulls to develop a sense of belonging in the United States, specific anti-immigrant moments (such as the election of Trump and rulings on DACA) intensified the sense of vulnerability among undocumented immigrants.

We also found that family ties, length of time in the United States, and even language skills shape how migrants experience a sense of belonging in three important ways. First, family ties shaped migrants' compass of belonging.

Those with strong family ties in the United States were less likely to desire to visit or return to Mexico, while those with strong family ties in Mexico were more likely to want to visit Mexico to see those family members. Second, we found that all migrants desired to remain in the United States and none had plans to voluntarily return to Mexico. Third, living undocumented in the United States shaped their sense of vulnerability and hence belonging in their communities. On the one hand, they had cemented ties in their communities in the United States; on the other hand, their legal vulnerability made them feel in limbo about their futures.

In sum, the 1.5 generation is more diverse than past research has suggested in that the *circumstance of migration* played a significant role in their experience. Yet they all shared the experience of being *racialized*, and their *legal status* was a constant and influential factor that shaped their experiences in the United States.

Notes

1 All names used in the chapter are pseudonyms to protect the identities of the participants.
2 The participant arrived between 1999 and 2008, and about half (8) of the participants arrived between 2004 and 2008.
3 After each name, we signal (CA) for those interviewed in California, and (PA) for those interviewed in Pennsylvania.
4 In the *Plyer v Doe* (1982) US Supreme Court case, the court established that undocumented children are entitled to a free public education in the United States.
5 DACAmented refers to the migrants who applied for and received protections and benefits from the Deferred Action for Childhood Arrivals (DACA) program.

References

Abrego, Leisy J. (2006). "I Can't Go to College Because I Don't Have Papers: Incorporation Patterns of Latino Undocumented Youth." *Latino Studies* 4(3): 212–31.

Abrego, Leisy J. (2008). "Legitimacy, Social Identity, and the Mobilization of Law: The Effects of Assembly Bill 540 on Undocumented Students in California." *Law & Social Inquiry* 33(3): 709–34.

Abrego, Leisy J. (2011). "Legal Consciousness of Undocumented Latino: Fear and Stigma as Barriers to Claims-Making for the First- and 1.5 Generation Immigrants." *Law & Society Review* 45(2): 337–70.

Abrego, Leisy J. and Roberto G. Gonzalez (2010). "Blocked Paths, Uncertain Futures: The Postsecondary Education and Labor Market Prospects of Undocumented Latino Youth." *Journal of Education for Students Placed at Risk* 15(1): 14

Abrego, Leisy J., and Cecilia Menjívar (2011). "Immigrant Latina Mothers as Targets of Legal Violence." *International Journal of Sociology of the Family* 37: 9–26.

AP (2006). "Philadelphia's Geno's Steaks Adopts English-Only Ordering Policy." Associated Press. Accessed on April 26, 2020, https://www.foxnews.com/story/philadelphias-genos-steaks-adopts-english-only-ordering-policy

Armenta, Amada, and Heidy Sarabia (2020). "Receptionists, Doctors, and Social Workers: Examining Undocumented Immigrant Women's Perceptions of Health Services." *Social Science & Medicine* 246: 112788.

Barajas, Manuel (2009). *The Xaripu Community across Borders: Labor Migration, Community, and Family.* Notre Dame, IN: University of Notre Dame Press.

Boswell, Christina (2008). "Combining Economics and Sociology in Migration Theory." *Journal of Ethnic and Migration Studies* 34(4): 549–66.

Burciaga, Edelina M., Lisa M. Martinez, Kevin Escudero, Andrea Flores, Joanna Perez, and Carolina Valdivia (2019). "Migrant Illegality Across Uneven Legal Geographies: Introduction to the Special Issue of *Law & Policy*." *Law & Policy* 41(1): 5–11.

CBP (2020). "U.S. Border Patrol Southwest Border Apprehensions by Sector Fiscal Year 2020: Southwest Border Unaccompanied Alien Children (0–17 yr. old) Encounters." U.S. Customs and Border Protection. Accessed October 8, 2020, https://www.cbp.gov/newsroom/stats/sw-border-migration/usbp-sw-border-apprehensions?_ga=2.22807679.550117026.1602178745-1594647682.1602178745

Dreby, Joanna (2015). *Everyday Illegal: When Policies Undermine Immigrant Families.* Berkeley: University of California Press.

Durand, Jorge, Douglas S. Massey, and Emilio A. Parrado (1999). "The New Era of Mexican Migration to the United States." *The Journal of American History* 86(2): 518–36.

Ellis, Basia D., Roberto G. Gonzales, and Sarah A. Rendón García (2019). "The Power of Inclusion: Theorizing 'Abjectivity' and Agency Under DACA." *Cultural Studies↔ Critical Methodologies* 19(3): 161–72.

Enriquez, Laura (2017). "Gendering Illegality: Undocumented Young Adults' Negotiation of the Family Formation Process." *American Behavioral Scientist* 61(10): 1153–71.

Fernandez-Kelly, Patricia, and Curran, Sara (2001). "Nicaraguans: Voices Lost, Voices Found." In Ruben Rumbaut and Alejandro Portes, eds., *Ethnicities: Children of Immigrants in America.* Berkeley: University of California Press

Fillmore, Lily Wong (1991). "When Learning a Second Language Means Losing the First." *Early Childhood Research Quarterly* 6(3): 323–46.

Flores, Andrea, Kevin Escudero, and Edelina Burciaga (2019). "Legal–Spatial Consciousness: A Legal Geography Framework for Examining Migrant Illegality." *Law & Policy* 41(1): 12–33.

Fox, Jonathan, and Gaspar Rivera-Salgado, eds. (2004). *Indigenous Mexican Migrants in the United States.* La Jolla: University of California, San Diego, Center for Comparative Immigration Studies & Center for US-Mexican Studies.

Gonzales, Roberto G. (2016). *Lives in Limbo: Undocumented and Coming of Age in America.* Berkeley: University of California Press.

Gonzales, Roberto G., and Edelina M. Burciaga (2018). "Segmented Pathways of Illegality: Reconciling the Coexistence of Master and Auxiliary Statuses in the Experiences of 1.5-Generation Undocumented Young Adults." *Ethnicities* 18(2): 178–91.

Gonzales, Roberto G. and Leo R. Chavez (2012). "'Awakening to a Nightmare' Abjectivity and Illegality in the Lives of Undocumented 1.5-Generation Latino Immigrants in the United States." *Current Anthropology* 53(3): 255–81.

Holstein, James A., and Jaber F. Gubrium (2007). "Constructionist Perspectives on the Life Course." *Sociology Compass* 1(1): 335–52.

Hondagneu-Sotelo, Pierrette (1994). *Gendered Transitions: Mexican Experiences of Immigration.* Berkeley: University of California Press.

Hondagneu-Sotelo, Pierrette (2003). "Gender and Immigration: A Retrospective and Introduction," 3–19. In Hondagneu-Sotelo, Pierrette, ed., *Gender and US Immigration: Contemporary Trends*. Berkeley: University of California Press.

Massey, Douglas S. (2015). "A Missing Element in Migration Theories." *Migration Letters* 12(3): 79–299.

Massey, Douglas S., Joaquin Arango, Graeme Hugo, Ali Kouaouci, Adela Pellegrino, and J. Edward Taylor (1993). "Theories of International Migration: A Review and Appraisal." *Population and Development Review* 19(3): 431–66.

Negrón-Gonzales, Genevieve (2014). "Undocumented, Unafraid and Unapologetic: Re-Articulatory Practices and Migrant Youth 'Illegality'." *Latino Studies*, 12(2), 259–78.

Portes, Alejandro, and Ruben G. Rumbaut, (2001). *Legacies: The Story of the Immigrant Second Generation*. Berkeley: University of California Press

Rosaldo, Renato (1994). "Cultural Citizenship in San Jose, California." *PoLAR* 17: 57.

Rumbaut, Rubén G. (1997). "Assimilation and Its Discontents: Between Rhetoric and Reality." *International Migration Review* 31: 923–60.

Rumbaut, Rubén G. (2004). "Ages, Life Stages, and Generational Cohorts: Decomposing the Immigrant First and Second Generations in the United States 1." *International Migration Review* 38(3): 1160–205.

Sarabia, Heidy (2012). "Perpetual Illegality: Results of Border Enforcement and Policies for Mexican Undocumented Migrants in the U.S." *Analyses of Social Issues and Public Policy* 12(1): 49–67.

Sarabia, Heidy (2019). "Citizenship in the Global South: Policing Irregular Migrants and Eroding Citizenship Rights in Mexico." *Latin American Perspectives* 46(6): 42–55.

Sarabia, Heidy, and Aida Rodriguez (2016). "Unaccompanied Undocumented Minors." In Alvaro Huerta, Norma Iglesias-Prieto, and Donathan L. Brown, eds., *Contemporary Issues for People of Color: Surviving and Thriving in the U.S. Today* (5 of 5 volumes—Immigration/Migration). Santa Barbara: ABC-CLIO/ Greenwood.

Schiller, Nina Glick, Linda Basch, and Cristina Szanton Blanc (1995). "From Immigrant to Transmigrant: Theorizing Transnational Migration." *Anthropological Quarterly* 68(1): 48–63.

Terriquez, Veronica (2015). "Dreams Delayed: Barriers to Degree Completion Among Undocumented Community College Students." *Journal of Ethnic and Migration Studies* 41(8): 1302–23.

Whaley, William (2012). "California Dream Act: A Dream (not DREAM) Come True." *McGeorge Law Review* 43: 625.

Part V

Challenges to migration research

20 Refusal and migration research

New possibilities for feminist social science

Emily Mitchell-Eaton and Kate Coddington

Introduction

I (Kate) met Ali while he was in an Australian immigration detention center. His journey from his country of origin took him years, as he traveled through multiple countries and finally boarded a boat from Indonesia to Australia. By the time I met him in Australia in 2011, he'd been detained, transferred, and detained again, spending more than a year awaiting his asylum claim processing. I visited the detention center regularly as part of wider fieldwork on spaces of containment in Australia, and while I couldn't answer his legal questions, I was a novelty: we talked about Australian culture, driving habits, and favorite foods. One visit, however, he was late, then withdrawn. There had been a violent incident inside the cafeteria. Things had gotten out of hand. There were Afghan and Sri Lankan asylum seekers, guards, and other employees involved. I never heard the details. Ali refused to explain, saying only that it would "hurt my heart" to talk further. Years later, reflecting on Ali's pivot away from the incident that day, I began to consider more fully what his refusal to speak might have meant, and what it meant for my work in Australia.

In our respective research endeavors, work on migration within the context of empire has forced us (Emily and Kate) to grapple with different forms of trauma, and Ali's silence represented one of many. In Emily's research on Marshall Islanders living in diaspora in the US South, she, too, encountered migrants' experiences of trauma, particularly trauma based in the US nuclear testing of the Marshall Islands in the mid-twentieth century. Emily, like Kate, has struggled with whether to focus on this trauma in her writing; here, she turns her analysis instead to the *process* of producing research on trauma. In this chapter, we explore how research about the relationships between migration, trauma, and empire can be rethought through consideration of refusals like Ali's refusal to speak. We extend indigenous critiques of social science research to the field of feminist migration studies, asking how feminist scholars might differently approach such refusals and silences in the research process.

For Eve Tuck and K. Wayne Yang (2014a: 227), who synthesize a trajectory of indigenous anticolonial critique focused on refusal, "academe's

DOI: 10.4324/9781003170686-25

demonstrated fascination with telling and retelling narratives of pain is troubling, both for its voyeurism and for its consumptive implacability." Refusing involves "an analytic practice that addresses forms of inquiry as invasion," reframing social science research as the relationships forged among "communities who refuse, the researched who refuse, and the researcher who refuses—or who do not" (Tuck and Yang 2014a: 244, 2014b: 811). Tuck and Yang's (2014b) formulation of refusal builds on the insights and methods of Native scholars like Paula Gunn Allen (1998) and Audra Simpson (2014). In this chapter, we explore how indigenous theories of refusal force feminist migration scholars to reconsider the work we do, particularly within a context like the contemporary academy, where knowledge can be commodified and exchanged in highly unequal ways. Theories of refusal pose challenging questions for researchers who center the voice and agency of migrants in their work, and who interrogate some of the implicit assumptions of feminist research. We take up these questions while recognizing the legitimate desire of scholars—situated both within and beyond the Global South—who wish to write about *their own* trauma and the trauma experienced by their communities. Our critique is aimed more specifically at trends in scholarship produced by white, settler, and/or Global North social scientists *on the trauma of others*. To ground that critique, we draw on indigenous theories of refusal, to which we turn now.

Refusal is part of a body of conceptual work that centers Indigenous nationalism. Recognizing the "deep impossibility" of Indigenous sovereignty and justice within colonial and settler–colonial occupied territories that are built upon dispossession of Indigenous land, denial of Indigenous sovereignty, and the disappearance of Indigenous people, Simpson (2014: 18) argues that living in opposition to this violence is a form of refusal. Refusal is a means of contesting, and ultimately exceeding, the framework of colonial nation-building to comprehend Indigenous sovereignty. Indigenous sovereignty is almost impossible to reconcile with the Western academy, as the Indigenous anthropologist Zoe Todd describes. Describing her resignation from an anthropology professorship, she writes that the social sciences do "not take seriously the sovereignty and politics of Indigenous peoples in the US or Canada" (Todd 2020: n.p.). Social sciences are "still too busy speaking for and about us" (Todd 2020). The consumption of non-Western voices is foundational for the academy; little space for alternative futures exists amid the overwhelming whiteness of Western academic thought. Moreton-Robinson's (2015) framework of "white possessive logics" that make Indigenous dispossession invisible, and the practices of colonization that dispossess, can be translated into the Western academic context, where whiteness both erases the absence of Indigenous sovereign claims *and* the colonial academic practices that enable their absence.

Taking seriously Indigenous theories of refusal within the context of social science research is, perhaps, impossible, given the white possessive logics that structure these disciplines (Faria et al. 2019). Todd (2020) stresses the importance of being "able to dream beyond tenure and peer review and the structures that have been imposed in these lands by white supremacist colonial capitalism." As we discuss below, whiteness is the foundation of our academic positionalities, and whiteness permeates the disciplinary conventions and methodological expectations of the social sciences. We want to suggest, however, that taking seriously *aspects* of refusal that inform methods and analysis in the study of migration forces migration scholars to ask difficult questions about studying migration, and we have a lot of learning to do so that we may do this work better.

For us, taking seriously the methodological implications of refusal involved rethinking the relationships between migration, trauma, and empire. We asked questions about how refusal works, and which questions it *allows* us to ask. In doing so we refused to center the stories of pain that commonly structure academic narratives of migrant trauma; we pushed for a different form, one in which the ethnographic encounter is decentered. The trauma of migrants within the context of empire, while real and true, is not the focus of this paper: instead, we spiral outward to encompass the context, history, and structures of domination in which the pain stories of the displaced are situated. Tuck and Yang (2014b: 811) describe working around the edges of trauma as *tracing the perimeter of the refusal*: "Because we cannot, will not, share certain accounts, we sometimes trace the perimeter of the refusal."

By moving away from the trauma we observed, we became increasingly attentive to the structures of domination that produced it. The perimeter of the refusal was not simply a means to make the refusal legible or "gain access" differently, but rather redirected attention to the centrality of *empire*—not just to migrants' stories, but to social science research more broadly, and feminist migration research in particular. This methodological critique of feminist migration research focuses on the question of why we share the stories we do, and how we might begin to refuse the extractive imperatives of social science. We begin by outlining refusal as an analytic, reflecting on the challenge posed by Indigenous and anticolonial scholars. Next, we situate our thinking both within work on empire and feminist social science. We then "trace the perimeter of the refusal," exploring how refusal operates as an analytic to deconstruct assumptions about migration research.

Refusal as a research intervention

What does it mean to take refusal seriously as an analytic for research, particularly as imperially implicated researchers? Refusal may be enacted by research participants or researchers alike as an ethical positioning: a politics of questioning and resisting "the proliferation of damage-centered studies, rescue research, and pain tourism" (Tuck and Yang 2014b). Refusal in all of these cases is a movement away from something, a denial of access, legibility, or translation. Refusal can also be the active production of an alternative, creative politics. Refusal incorporates a critique of Western scholars' gravitation toward stories of suffering and their demands for interlocutors' intelligibility. We highlight three important methodological aspects of refusal in our analysis: (1) the refusal of various people to share stories; (2) the refusal of the moments that were shared to become fully intelligible; and (3) the refusal of the expectations of research that demands intelligible stories for consumption.

Taking seriously theories of refusals requires that we balance "advocating for Indigenous perspectives, and the dangers of appropriation, co-option and further colonization of Indigenous knowledges" (Barker and Pickerill 2019: 4). Theories of refusal simultaneously highlight the "white possessive logics" at the heart of social science research *and* the possible alternative outcomes of research done differently—the project cannot become about "de-centering" whiteness, but instead making visible the overwhelming whiteness of the academy and how migrant voices become employed within academic knowledge production, and working toward better accountability within research decision-making (Moreton-Robinson 2015). Our position as white scholars enmeshed in relations of historical privilege highlights the dangers of taking up Indigenous conceptual frameworks to understand research among precarious migrant populations—perhaps another instance of white settler appropriation. Taking seriously Indigenous theories of refusal risks cooption of the history of critical Indigenous anticolonial scholarship. Yet there are also demands precisely *to* engage and travel with Indigenous conceptual frameworks. Within geography, for instance, "non-Indigenous academics have a role in centering Indigenous ontologies in geography because Indigenous peoples should not carry the burden of decolonization by themselves" (Barker and Pickerill 2019: 4). Part of the process of decolonizing the social sciences is to take very seriously the questions Indigenous theories push scholars to ask. Migration scholars' engagement with refusal, therefore, must move beyond a desire to be perceived as having "good intentions" toward an explicitly anticolonial research praxis (de Leeuw et al. 2013).

Refusal also demands to be taken seriously because of its conceptual specificity: Indigenous theories of refusal are grounded in practices of historical displacement and trauma. Although the formerly colonized people we met in our research had been displaced through specific practices of occupation, domination, and mobility, these displacements were also projects of imperial power. Refusal as a specifically Indigenous analytic recenters analysis on these imperial entanglements and extends their temporal and spatial limitations. Our use of refusal attends to the historical specificity of Indigenous theorizing, yet we see potential in refusal to understand the colonial underpinnings of feminist migration research as well.

Highlighting our accountability and ongoing engagement within extractive research practices also requires us to explicitly address our positionality as white settler US Americans. As researchers, we grapple with the status of critical scholarship on indigeneity, which increasingly provides "academic currency" (Daigle 2019: 3). Addressing our own place in an imperial white discipline is important, but more important still is to consider what that positionality *does* for the work we do: our mobility to and within our research sites, as well as access to contacts, was facilitated by our whiteness, our

US-born status, and our US passports. Inherited wealth facilitated access to higher education for both of us, and our experiences represent part of the wide racial wealth gap that creates intergenerational inequality for historically marginalized students of color (Herring and Henderson 2016). With our advanced degrees in geography, we join the more than 65% of members of the largest geography disciplinary association who identify as white (AAG 2018), a field that as de Leeuw and Hunt (2018: 7) write, "is largely enacted by White scholars living off the spoils of colonialism, including White settler scholars, and in which Indigenous presence is largely facilitated by, or filtered through, non-Indigenous 'experts.'" These details matter not to demonstrate reflexivity as "an apology" but to consider how US empire, and our positionings within it, *constitute* the perimeter that we trace. Empire makes possible our position within geography as an imperial discipline. Therefore, empire also makes possible our attempts to recenter attention on empire through analytics of refusal.

Feminist social science research, trauma, and imperial entanglements

Social science research has been central to the practices of imperialism and imperial knowledge production, and the impacts on Indigenous and occupied peoples have been documented by Indigenous scholars (Smith 1999; Alfred and Corntassel 2005; Simpson 2014). Research practices have prioritized the needs of researchers above those of communities—many times to the extent of harming Indigenous communities—to the point where research is a "dirty word" within Indigenous communities (Tuck and Wayne Yang 2014a: 223). The social sciences continue to maintain many of these legacies of imperial domination today through the whiteness of faculty, staff, and students; the centrality of Euro-American thinkers in publications and power centers; and the siting of universities on stolen Indigenous lands (Daigle 2019). While decolonization efforts have begun to contest colonial knowledge production and theoretical and methodological traditions within the social sciences, these efforts occur in contexts that "continue to center White settler scholars over Indigenous scholars" (de Leeuw and Hunt 2018: 10). Feminist research is also anchored in these imperial disciplinary dynamics. Women were deeply implicated in imperial projects. Mullings and Mukherjee (2018: 1410) situate contemporary feminist scholarship within an "academy that has become more competitive, increasingly precarious, and susceptible to anti-Black racism, xenophobia, Islamophobia and anti-Indigenous sentiment," noting that while feminist disciplines can be spaces of refuge, they are also susceptible to these trends.

In this chapter, we situate the implicit demands for migrants' traumatic stories within the context of the whiteness and imperial legacies of social science scholarship. It's clear that in the right context, direct testimony from survivors of trauma has powerful implications for scholarship: Million (2009: 54), for instance, writes that residential school survivors' stories forced Canadians to reckon with the injustices of settler colonialism, as the "narratives were political acts in themselves." Yet highlighting migrants' trauma raises questions as well. The geographic, racialized, and class inequalities that shape migration research mean that often, white women from the Global North seek research collaborators or participants among marginalized communities and people of the Global South. The inclusion of marginalized voices may counter dominant perspectives but often serve instead to "authenticate" research findings. Feminist research retains many of the implicit structures of extractive scholarship.

Feminist research on trauma builds on the often-implicit extractive demands of social science research. Feminist assumptions about the empowerment research participants gain by giving voice to silenced traumas (Coddington 2017) sit uneasily alongside the imperial underpinnings of Western social science; survivors risk being retraumatized when under pressure to repeatedly voice their experiences. Recent geographical attention to vulnerability (Mitchell-Eaton 2019) and trauma has opened spaces for new kinds of research that focuses instead on the specificities of place and the underlying geopolitical connections producing traumatic environments (Pain 2019). Looking for the structures of power that traumatize, rather than focusing on the experience of survivors, risks recentering whiteness and overstating the implications of empire. Obscuring the specificity of migrants' lived experiences results from such "integral technique[s] of western knowledge production [that] fragment, decontextualize, recontextualize, and 'spin'" (Million 2011: 320).

Yet we argue that the implicit demands of feminist research to prioritize qualitative, in-depth interviews and long-term relationships (Cuomo and Massaro 2016) push researchers too far in the direction of extractive knowledge production. The social sciences crave stories of struggle, displacement, and trauma, and survivors feel pressured to repeatedly relate stories of pain for academic consumption. While feminist researchers critique masculinist bravado within the field, implicit demands within feminist scholarship center more intimate forms of extraction. From our experience researching migration, for instance, we often face demands from reviewers and audiences to share the personal stories of migrants, perhaps to demonstrate the cultivation of relationships that appeals to feminist research ethics. But this too must be situated within extractive research traditions. Narrating the experiences of traumatized migrants within scholarship does not automatically decolonize feminist research; it may instead unproblematically assimilate snippets of migrant experiences into white disciplinary frameworks. Turning attention to the taken-for-granted demands for migrant trauma within scholarship is an active rethinking of the extractive structures of domination underscoring feminist migration research.

Refusal, trauma, and empire in research

Within our research, we have encountered a spectrum of refusals that inform the suggestions we develop for migration researchers below. Like Ali, many of our participants refused connection, participation, or relationships, and we did too. Sometimes, we interpret the larger experiences of fieldwork as refusal, as they refuse intelligibility and easy translation into research findings and conclusions. Refusal shifted our focus, from the desire to fully understand the "meaning" of the moment through the experiences of individual trauma to examining the framework *for that desire itself.* Decentering individual trauma from the ethnographic encounter and comparative reflection allows us to focus on the larger structure of empire and its centrality in producing trauma. In this section, we consider how engaging refusal shaped the forms and outcomes of our research.

Filtering our focus through refusal forced us to explore wider structures of trauma—not the individual stories of displacement, but the imperial structures that drove displacement, and the imperial research that continues to benefit from that displacement. Empire produces trauma in its subjects while presenting itself as their salvation, the provider of safe haven. In Ali's case, neo-imperial wars in Afghanistan layered new forms of violence upon older colonial divisions, forcing him to seek refuge abroad, but he confided that the ultimately two years he spent in Australian immigration detention proved to be more traumatic than the violence he fled. For Marshall Islanders living in Arkansas, the links to (US) empire were clear as well: the United States conducted nuclear testing in the islands from 1946 to 1958, while the Marshall Islanders were a nonsovereign US territory. The testing produced health-related, environmental, and economic devastation that prompted many to flee their homeland, primarily to the United States.

In each of our research contexts, empire complicates the notion of migrants' journeys as simple linear moves from danger to safety, or from trauma to healing. A focus on the critical role of empire in producing both trauma and mobility highlights the fractured spaces of migration as well as the geopolitical and imperial linkages between those spaces—but, of course, employing refusal in the research process does not restrict researchers to *either* centering migrant voices *or* exploring wider structures of trauma. For instance, Speed's (2019) *Incarcerated Stories* uses Indigenous women migrants' stories to highlight the structures of racialized, gendered, and neo-liberal capitalist power across the Americas that render women vulnerable to racialized and gendered forms of violence. In this example, the trauma of empire is here *and* "there," past *and* present, and Speed's analysis is deeply accountable to the *work* that these stories do (Loyd et al. 2018).

Shifting our analysis from the "traumatized individual" to the larger traumatizing system(s) forced us to ask new questions: If traditional social science research models frame the refusal to acquire certain data as "failure," what do such "failures" produce instead? Could we consider refusal generative, instead of restrictive or limiting? Perhaps more critically, what does "successful" research on trauma produce, and what structures does such research *re*-produce? Asking these questions forced us to reconsider many of the feminist research practices we had been taught: Were intimate relationships with migrants necessary to produce feminist research? What did migrant voices add to research projects, and what unacknowledged costs did this research bring to bear on migrants themselves? As we argue, refusal as a methodological intervention can and should be practiced in every stage of the research process. We now turn to discuss opportunities for practicing refusal in research and reflect upon what these practices looked like in our own projects.

Refusal in research design

During preliminary research design, particularly for ethnographic research, we can practice refusal as we map out a study's participant groups, the questions we will be asking, and the larger assumptions guiding our work. For scholars working in the US context, the completion of the Institutional Review Board (IRB) application offers a rich opportunity to practice refusal (Coombes et al. 2014). Indigenous scholars have called attention to the imperial academic imperatives within institutional IRBs (Smith 1999), and social scientists have documented how IRB protocols perpetuate imperial, extractive" and often violent research (Martin and Inwood 2012). Models of risk based on biomedical research practices individualize notions of risk, participation, and consent. The IRB's accountability to the university rather than to individual researchers or their research participants suggests the IRB may prevent access to knowledge rather than facilitate it (Sabati 2019). Indeed, Sabati (2019: 1057) argues that the IRB process produces the "active erasure of academia's complicity in producing ongoing contexts of racialized social injustice."

"Tracing the perimeter of the refusal" involves reflecting upon the extractive and imperial foundations of social science research. At this stage, researchers should reflect not only upon gaining consent from participants, or even how to do so "ethically," but whether such endeavors should be attempted at all. Instead of pushing past concerns about being well-received in certain communities, researchers can embrace moments of doubt as opportunities to change direction, seek feedback from communities, or discontinue a research project altogether. For instance, Emily conducted a preliminary visit to "gain access" to key contacts during one research project, but she failed to ask about previous research that had been conducted in the Marshallese community, the effects of those projects, and the kinds of research desired (or unwanted) by community members going forward. If she had done so, she might have learned earlier that a recent research project had put off many migrants in the community, many of whom were rightfully wary of new researchers.

Researchers must do the legwork in advance of their projects to determine what kinds of research their participants desire and what kinds of research have been conducted in the past. Such questions not only address the possibility that research can reproduce trauma but also seek to minimize that potential. This approach can also lead to stronger research projects: by anticipating "dead-ends" and community resistance to certain research methods or inquiries, researchers can develop projects that respect the refusals of their potential participants, leading to more nuanced scholarly work or to the abandonment of politically problematic research (Simpson, 2014). Such work helps to flag some of the taken-for-granted assumptions that posit research centering migrants' voices as uniformly positive, beneficial, and empowering—assumptions dismantled by long-standing critiques of social science work by Indigenous thinkers (Smith 1999; Simpson 2014; Daigle 2019)

Refusal in the field

It is never too late to practice refusal during the research process. We may discover at any time that the research we had planned is politically or ethically problematic, or that it reproduces the structural violence we aim to critique and dismantle. Participants may close down, turn away, or leave the project altogether: all these refusals should prompt new questions from the researcher. While social science norms place heavy emphasis on researchers' fidelity to original research plans, methods, and hypotheses, "tracing the perimeter" of refusals—in exposing the extractive context in which these demands are centered—can illuminate better ways of proceeding (or not) with research projects. These questions are particularly important for researchers whose work focuses on migrant suffering. If feminist social sciences continue to focus on stories of pain, the assumptions behind such work need constant, critical unpacking.

For us, refusal in the field happened in a number of different ways. "Research fatigue" among Marshallese migrants who Emily met resulted in a change of focus, and a shift to interviews with community leaders rather than residents. Emily also became wary of conventional narratives of Marshallese nuclear trauma, narratives that risk reproducing Marshall Islanders as (exclusively) traumatized subjects and pressuring Marshall Islanders to relive painful experiences for external consumption. By avoiding telling "damage-centered stories" (Tuck and Yang 2014b: 811), Emily, as a non-Marshallese researcher, hoped to interrupt this cycle, making room for other narratives to circulate. Meanwhile, for Kate, reflecting on moments such as Ali's silence "in the field" raised questions about what could have been refused during fieldwork, both from the traumatized migrants she met as well as refusals she could have enacted based on her own discomfort. Was it necessary to push for details Ali was unwilling to share? What did Ali's silence reveal about his capacity to consent to other interactions? How did this silence more fully articulate the trauma of Australia as a supposedly safe haven for refugees?

Refusal in writing and publishing

Finally, researchers can refuse as they write and publish their research. During writing processes, researchers can refuse to incorporate certain stories from the field, stories shared by field contacts during vulnerable exchanges or witnessed by the researcher, either with or without the participants' knowledge. We can leave these "data" out. We can also incorporate these stories selectively, with the input of participants. If our research projects are designed through the tenets of Participatory Action Research (PAR) or similar community-based models (Coombes et al. 2014), this process of seeking feedback on research dissemination may already be central to our research design. If not, researchers still retain an ability and a *responsibility* to practice refusal when publishing their findings. Accountability to Indigenous praxis and vulnerable research participants also requires us to consider where, how, and *if* to disseminate our work. Following an approach geared toward minimizing harm may mean seeking alternative venues in which to publish or coauthor research with participants; translating published work into Indigenous or marginalized languages, and sharing our work in venues outside of historically White colleges and universities and academic associations (Ybarra 2019).

During the writing and publication of our ongoing research, we have both contended with these sorts of refusals. Emily chose not to include well-rehearsed narratives of community trauma into her dissertation, as she had not yet developed an alternative framework for critiquing the institutional demands and desires to reproduce them. From one perspective, these stories became less central to the narrative she developed, but from a traditional social science perspective, those stories constituted an unmined source of data, a rich depository of narrative, performance, and embodiment that Emily tried, but failed, to work into a coherent narrative (Mitchell-Eaton 2019). For Kate, discomfort with stories of pain from migrants increased, and she became increasingly unwilling to mobilize these stories in academic circles. Access to "closed" spaces within critical migration studies becomes a form of career enhancement (Pascucci 2017), suggesting the value of such stories for individual career trajectories as well as disciplinary scholarship more broadly. Kate continued visiting with migrants, but these visits no longer formed the basis of her written work.

For these kinds of refusals to become more commonplace within social science research, researchers must practice them regularly, reflecting *openly*—both in publications and conversations with colleagues and students—about the decisions they make to refuse and to recognize the refusals of others. We might also reflect honestly on how refusal can cost us career and funding opportunities, to illuminate the imperial, extractive imperatives of the neoliberal university. As Harrowell et al. (2018: 231) argue, social science research needs to become more open about its negotiation with doubt and failure, because while "the academic labor of fieldwork remains a central component of how … knowledge is produced," the pressures of peer review and precarious neoliberal working environments silence discussion of these discomforting emotions.

Conclusion

Ali's silence could be interpreted multiple ways: necessarily, such analysis of refusal remains incomplete. Refusing to push for clarity or narrative singularity and leaving room for ambiguity and discomfort is another way we practice refusal and aim to honor the refusals of our participants. We are still grappling with these refusals, and with the ones we did not practice. Reflecting on the missed opportunities to refuse, we emphasize the need for white/settler/ Global North feminist scholars to practice refusal regularly, proactively when possible, and retroactively when necessary.

Does taking refusal seriously mean never centering the voice and agency of migrants in research projects? Clearly, no, but it does mean asking ongoing, critical questions about the assumptions that underlie demands for migrant voices, and connecting the push for "authentic," "vulnerable" stories of pain with Indigenous critiques of Western social science practices. Rather than an attempt at perfecting allyship, solidarity, or reflexivity, refusal, when honored and practiced by settlers or other imperially positioned researchers, is about practicing accountability to Indigenous and colonized peoples. As Million (2011: 317) writes, "intellectual decolonization is an active, positional practice, a verb, a doing." Accountability means paying attention to the wider political framework of Indigenous theorizing. As Simpson (2014) and Alfred and Corntassel (2005) write, Indigenous survival is an intensely political act, an oppositional act of sovereignty and nationhood that exceeds settler colonial frameworks for recognition or multiculturalism. Therefore, taking refusal seriously means more than just choosing not to tell certain stories, not to mine certain "data," or not to ask certain questions while in "the field." Refusal must also involve actively contesting white patriarchal supremacy within wider disciplinary structures (Mansfield et al. 2019), reconsidering "canonical" scholarship, reconstructing syllabi, attending to the politics of citation in our work (Mott and Cockayne 2017), and taking seriously the politics of refusal for Indigenous sovereignty (Daigle 2019). Such forms of refusal can mean backing out or jeopardizing other forms of institutional support. Refusal may slow research down, as it necessarily impedes social science research processes rooted in imperial knowledge production.

The radical political possibility of refusal offers even more: for the context of migration, refusal highlights how migration is fundamentally shaped by the dynamics of empire. Empires drive migration streams, often compelling people to migrate by force of displacement and dispossession; imperial forces also restrict migration by deterring, detaining, criminalizing, and killing migrants. Displacement, dispossession, and deterrence produce trauma for migrants and displaced people as well as for their extended networks of family and friends. Migration scholars must therefore take trauma seriously as an inevitable, and often intentional, product of empire (Loyd et al. 2018), just as we must acknowledge empire as a central driver and determinant of migration. Yet, contemporary academic fascination with the pain stories of migrants must be challenged. Feminist scholarship in the Global North has privileged embedded ethnographic research that makes the voices of vulnerable interlocutors legible to formal academic audiences (Coddington 2017). Particularly within this context, because of their devalorization across other parts of the social sciences, emotionally vulnerable, intimate field stories are valued, and the implications of the demand for migrant voices need to be critically examined. The challenge is for migration research—and particularly for migration researchers who center migrant voices and agency—is to account for trauma and empire without compounding their effects. Migration scholarship reinforces how projects of knowledge production never are, and have never been, politically neutral. Imperial knowledge production reproduces imperial thinking, and feminist scholars must refuse the language of objectivity and neutrality in deeply political modes of knowledge production such as ethnographic research. Refusal can become a rigorous and decolonial approach to the study of migration.

Most importantly, theories of refusal are grounded in Indigenous scholarship (Simpson 2014; Tuck and Yang 2014b) and their use among settlers and within an imperial discipline requires taking seriously the "hierarchies of knowledge production" (Mott and Cockayne 2017: 954) that have traditionally excluded Indigenous knowledge creation, as well as a praxis of accountability that engages our status as white settlers from the Global North. Refusal as politics is about more than just research practices or the confines of academic disciplinary structures, but about building alternative futures. As Tanganekald, Meintangk, and Boandik scholar Irene Watson (2007) writes, the "impossibility" of ongoing Indigenous life among settler colonialism is where thinking begins. Perhaps the radical reframing of knowledge production demanded by Indigenous theorizing may become a site of new feminist thinking about migration, displacement, and trauma.

References

AAG (2018). *AAG Membership Report*. Washington, DC. Available at http://www. aag.org/galleries/disciplinary-data/AAG_Membership_Data_Report_2018.pdf

Alfred, Taiaiake, and Corntassel, Jeff (2005). "Being Indigenous: Resurgences against Contemporary Colonialism". *Government and Opposition* 40(4): 597–614.

Allen, Paula Gunn (1998). "Problems in Teaching Leslie Marmon Silko's Ceremony," 55–64. In D. A. Mihesua, ed., *Natives and Academics: Researching and Writing About American Indians*. Lincoln: University of Nebraska Press.

Barker, Adam J., and Jenny Pickerill (2019). "Doings with the Land and Sea: Decolonising Geographies, Indigeneity, and Enacting Place-agency." *Progress in Human Geography* 44. Doi:10.1177/0309132519839863.

Coddington, Kate (2017). "Voice under Scrutiny: Feminist Methods, Anticolonial Responses, and New Methodological Tools." *The Professional Geographer* 69(2): 314–20.

Coombes, Brad, Jay T. Johnson, and Richard Howitt (2014). "Indigenous Geographies III: Methodological Innovation and the Unsettling of Participatory Research." *Progress in Human Geography* 38(6): 845–54.

Cuomo, Dana, and Vanessa A. Massaro (2016). "Boundary-making in Feminist Research: New Methodologies for 'Intimate Insiders.'" *Gender, Place & Culture* 23(1): 94–106.

Daigle, Michelle (2019). "The Spectacle of Reconciliation: On (the) Unsettling Responsibilities to Indigenous Peoples in the Academy." *Environment and Planning D: Society and Space* 37(4): 703–21.

de Leeuw, Sarah, Margo Greenwood, and Nicole Lindsay (2013). "Troubling Good Intentions." *Settler Colonial Studies* 3(3–04): 381–94.

de Leeuw, Sarah, and Sarah Hunt (2018). "Unsettling Decolonizing Geographies." Geography Compass 12(7): e12376.

Faria, Caroline, Bisola Falola, Jane Henderson, and Rebecca Maria Torres (2019). "A Long Way to Go: Collective Paths to Racial Justice in Geography." *The Professional Geographer*, 71(2): 364–76.

Harrowell, Elly, Thom Davies, and Tom Disney (2018). "Making Space for Failure in Geographic Research." *The Professional Geographer* 70(2): 230–38.

Herring, Cedric, and Lauren Henderson (2016). "Wealth Inequality in Black and White: Cultural and Structural Sources of the Racial Wealth Gap." *Race and Social Problems* 8(1): 4–17.

Loyd, Jenna M., Patricia Ehrkamp, and Anna J. Secor (2018). "A Geopolitics of Trauma: Refugee Administration and Protracted Uncertainty in Turkey." *Transactions of the Institute of British Geographers* 43(3): 377–89.

Mansfield, Rebecca, Kendra McSweeney, Rebecca Lave, Anne Bonds, Jaclyn Cockburn, Mona Domosh, Trina Hamilton, Roberta Hawkins, Amy Hessl, Darla Munroe, Diana Ojeda, and Claudia Radel (2019). "It's Time to Recognize How Men's Careers Benefit from Sexually Harassing Women in Academia." *Human Geography* 12(1): 82–87.

Martin, Deborah G., and Joshua Inwood (2012). "Subjectivity, Power, and the IRB." *The Professional Geographer* 64(1): 7–15.

Million, Dian (2009). "Felt Theory: An Indigenous Feminist Approach to Affect and History." *Wicazo Sa Review* 24(2): 53–76.

Million, Dian (2011). "Intense Dreaming: Theories, Narratives, and Our Search for Home." *American Indian Quarterly* 35(3): 313.

Mitchell-Eaton, Emily (2019). "Grief as Method: Topographies of Grief, Care, and Fieldwork from Northwest Arkansas to New York and the Marshall Islands." *Gender, Place & Culture* 26(10): 1438–58.

Moreton-Robinson, Aileen (2015). *The White Possessive: Property, Power, and Indigenous Sovereignty*. Minneapolis: University of Minnesota Press.

Mott, Carrie, and Daniel Cockayne (2017). "Citation Matters: Mobilizing the Politics of Citation Toward a Practice of 'Conscientious Engagement.'" *Gender, Place & Culture* 24(7): 954–73.

Mullings, Beverley, and Sanjukta Mukherjee (2018). "Reflections on Mentoring as Decolonial, Transnational, Feminist Praxis." *Gender, Place & Culture* 25(10): 1405–22.

Pain, Rachel (2019). "Chronic Urban Trauma: The Slow Violence of Housing Dispossession." *Urban Studies* 56(2): 385–400.

Pascucci, Elisa (2017). "The Humanitarian Infrastructure and the Question of Over-research: Reflections on Fieldwork in the Refugee Crises in the Middle East and North Africa." *Area* 49(2): 249–55.

Sabati, Sheeva (2019). "Upholding 'Colonial Unknowing' Through the IRB: Reframing Institutional Research Ethics." *Qualitative Inquiry* 25(9–10): 1056–64.

Simpson, Audra (2014). *Mohawk Interruptus: Political Life across the Borders of Settler States*. Durham, NC: Duke University Press.

Smith, Linda Tuhiwai (1999). *Decolonizing Methodologies: Research and Indigenous Peoples*. London: Zed Books.

Speed, Shannon (2019). *Incarcerated Stories: Indigenous Women Migrants and Violence in the Settler-capitalist State*. Chapel Hill: University of North Carolina Press.

Todd, Zoe (2020). (An Answer). *anthrodendum*. Accessed January 30, 2020, https://anthrodendum.org/2020/01/27/an-answer/

Tuck, Eve, and K. Wayne Yang (2014a). "R-Words: Refusing Research," 223–48. In Paris, D. and M. T. Winn, eds. *Humanizing Research: Decolonizing Qualitative Inquiry with Youth and Communities*. London: SAGE Publications.

Tuck, Eve and K. Wayne Yang (2014b). "Unbecoming Claims: Pedagogies of Refusal in Qualitative Research." *Qualitative Inquiry* 20(6): 811–18.

Watson, Irene (2007). "Aboriginal Sovereignties: Past, Present and Future (Im) Possibilities," 23–44. In Suvendrini Perera, ed. *Our Patch: Enacting Australian Sovereignty Post-2001*. Perth: API Network.

Ybarra, Megan (2019). "On Becoming a Latinx Geographies Killjoy." *Society & Space*. Accessed June 24, 2019, http://societyandspace.org/2019/01/23/on-becoming-a-latinx-geographies-killjoy/

Index

Pages in *italics* refer figures; **bold** refer tables and pages followed by n refer notes.

Printed in the United States
by Baker & Taylor Publisher Services